The Structure and Evolution of
Neutron Stars

The Structure and Evolution of Neutron Stars

CONFERENCE PROCEEDINGS

Edited by

D. Pines
University of Illinois

R. Tamagaki
Kyoto University

S. Tsuruta
Montana State University

Addison-Wesley Publishing Company
The Advanced Book Program
Redwood City, California • Menlo Park, California
Reading, Massachusetts • New York • Don Mills, Ontario
Wokingham, United Kingdom • Amsterdam • Bonn

Physics Editor: *Barbara Holland*
Editorial Assistant: *Diana Tejo*
Production Manager: *Pam Suwinsky*
Production Assistant: *Karl Matsumoto*
Cover Design: *Mike Fender*
Electronic Formatting: *SuperScript Typography*

Library of Congress Cataloging-in-Publication Data

The structure and evolution of neutron stars: conference proceedings
   edited by D. Pines, R. Tamagaki, S. Tsuruta.
      p. cm.
   Includes bibliographical references.
   1. Neutron stars—Congresses.   I. Pines, David, 1924–.
   II. Tamagaki, R., 1932–.   III. Tsuruta, Sachiko.
   QB843.N4S76 1991   523.8'874–dc20   91-33686
   ISBN 0-201-56293-6

Copyright ©1992 by Addison-Wesley Publishing Company

All rights reserved. No part of this publication may be reproduced, stored
in a retrieval system, or transmitted, in any form, or by any means, electronic,
mechanical, photocopying, recording or otherwise, without the prior written
permission of the publisher. Printed in the United States of America.
Published simultaneously in Canada.

1 2 3 4 5 6 7 8 9 10-MA-95 94 93 92

Contents

Preface — xi

Acknowledgements — xiii

Photograph of Participants — xv

Opening Address — 3
R. Tamagaki

Structure and Dynamical Behavior — 5

Inside Neutron Stars: A 1990 Perspective — 7
David Pines and M. Ali Alpar

Radio Pulsar Timing — 32
R. N. Manchester

The Equation of State in Neutron Star Matter — 50
James M. Lattimer

Study of Neutron Stars in X-rays — 63
H. Inoue

Properties of Neutron Stars from X-ray Pulsar Observations — 77
Fumiaki Nagase

Magnetic Field of Binary X-ray Pulsars — 86
K. Makishima

Contents

Mass-Radius Relations in X-ray Pulsars
  R. Hoshi ... 98

Gamma-ray Bursts as Neutron Starquakes
  R. D. Blandford ... 104

Topics in the Physics of High Magnetic Fields
  C. J. Pethick ... 115

Instabilities in Rotating Neutron Stars
  Lee Lindblom ... 122

Microscopic Calculations of Superfluid Gaps
  J. W. Clark, R. D. Davé, and J. M. C. Chen ... 134

Some Topics in Neutron Star Superfluid Dynamics
  M. Ali Alpar ... 148

Superfluid Dynamics in the Inner Crust of Neutron Stars
  Richard I. Epstein, Bennett Link, and Gordon Baym ... 156

Various Phases of Hadronic Matter in Neutron Stars and Their Relevance to Pulsar Glitches
  R. Tamagaki ... 167

Static and Dynamic Properties of Pion-Condensed Neutron Stars
  Toshitaka Tatsumi and Takumi Muto ... 178

Ultrarelativistic Heavy-Ion Collisions and Neutron Stars
  Gordon Baym ... 188

Cyclotron Lines and the Pulse Period Change of X-ray Pulsar 1E2259+586
  K. Iwasawa, K. Koyama, and J. P. Halpern ... 203

New X-ray Sources Near the Galactic Bulge Region
  Shigeo Yamauchi and Katsuji Koyama ... 206

Equation of State for Neutron Star Atmospheres: Finite Temperature and Gradient Corrections
  Andrew M. Abrahams and Stuart L. Shapiro ... 209

Thermonuclear Flash Model for Long X-ray Tails from
AQL X-1
 *Ikko Fushiki, Ronald E. Taam, S. E. Woosley,
 and D. Q. Lamb*   212

Gamma-ray Bursts and Dead Pulsars
 T. Murakami   215

Strangeness in the Proton and Kaon Condensation in
High-Density Nuclear Matter
 T. Kunihiro   218

High-Energy X-ray Production in Accreting Neutron Stars
 Tomoyuki Hanawa   221

The Current System in the Magnetosphere of Neutron Stars
 Shinpei Shibata   224

Superfluid Effects on the Stability of Rotating Neutron Stars
 Lee Lindblom and Gregory Mendell   227

Dissipation of Vibrational Energy of Neutron Stars in the
Pion-Condensed Phase
 Takumi Muto and Toshitaka Tatsumi   230

Formation   233

Theory of Neutron Star Formation
 S. E. Woosley and T. A. Weaver   235

X-ray Observations of Supernova Remnants: The Birth
of Neutron Stars and Their Evolution
 K. Koyama   250

Equation of State of Dense Supernova Matter and Newborn
Hot Neutron Stars
 T. Takatsuka   257

Supernova Explosions and the Soft Equation of State
 M. Takahara, T. Takatsuka, and K. Sato   268

The Cooling of Proto–Neutron Stars and Neutrino Bursts
 Hideyuki Suzuki and Katsuhiko Sato   276

Pulsar Cavity
  Humitaka Sato — 284

How Will the Pulsar in SN 1987A Emerge?
  T. Shigeyama, T. Kozasa, and K. Nomoto — 290

Neutron Star Masses from Supernova Explosions
  Friedrich-Karl Thielemann, Ken'ichi Nomoto, and Masa-aki Hashimoto — 298

Evolutionary Origin of Binary Pulsars
  K. Nomoto, H. Yamaoka, and T. Shigeyama — 310

Compact Stars on the Supercomputer
  Stuart L. Shapiro and Saul A. Teukolsky — 329

Coalescing Binary Neutron Stars
  Takashi Nakamura and Ken-ichi Oohara — 343

Proton Mixing in Hot and Dense Neutron Star Matter
  S. Nishizaki and T. Takatsuka — 349

Magnetic and Thermal Evolution — 351

Neutron Star Crust Breaking and Magnetic Field Evolution
  M. Ruderman — 353

Thermal Evolution of Neutron Stars—Cooling and Heating
  Sachiko Tsuruta — 371

Elementary Processes in Neutron-Star Cooling
  Naoki Iwamoto — 385

Neutron Star Evolution with Internal Friction
  N. Shibazaki and F. K. Lamb — 394

Neutron Star Cooling and Pion Condensation
  Hideyuki Umeda, Ken'ichi Nomoto, Sachiki Tsuruta, Takumi Muto, and Toshitaka Tatsumi — 406

Simulation of Pulsar Evolution
  Shigeto Wakatsuki, Naohiro Sato, and Naoki Itoh — 409

Perspectives 413

Banquet Lecture: A Cartoonist's View of the Neutron Star
Minoru Oda 415

NON-SENS 1990
M. Ruderman 418

Program and List of Participants 421

Program of SENS '90 423

Participants 427

Preface

The present volume contains the proceedings of the joint US-Japan Seminar on "The Structure and Evolution of Neutron Stars." The seminar, with the acronym SENS '90, provided an opportunity for United States, Japanese, and European physicists, astronomers, and astrophysicists to discuss in some depth our present understanding of neutron stars. As may be seen from the proceedings, and as benefits to a comparatively young subfield of physics and astrophysics, there was not always complete agreement among the theorists present on a number of key aspects of neutron star formation, structure, dynamical nehavior, and evolution, e.g., the role played by pion condensation or crustal neutron superfluidity in determining glitches and post-glitch behavior, the physical origin of gamma-ray bursts, the proper description of superfluid dynamics in the crust and core, the physical processes involved in the formation of neutron stars, etc. It is our hope that the different perspectives presented here on these and related matters might encourage interested readers, theorists, and observers alike, to carry out the necessary calculations or observations which will help decide many of these issues.

To assist the reader, we have divided this volume into three parts: Structure and Dynamical Behavior, Formation, and Magnetic and Thermal Evolution; within each of these sections we have included both the invited papers and the poster presentations.

We should like to take this opportunity to thank the US National Science Foundation, the Japan Society for the Promotion of Science, the Inoue Foundation of Science, and the Inamori Foundation for their generous financial support, the scientific secretary Dr. T. Tatsumi, for his outstanding efforts, and our colleagues, Professors S. Hayakawa, K. Nomoto, M. Ruderman, H. Sato, K. Sato, and Y. Tanaka, whose hard work and dedication contributed so much to the success of the seminar.

David Pines
Ryozo Tamagaki
Sachiko Tsuruta

Acknowledgments

The organizers would like to express their thanks to the National Science Foundation of the USA and the Japan Society for the Promotion of Science for their joint sponsorship of the US-Japan Joint Seminar on "The Structure and Evolution of Neutron Stars." They also acknowledge the support provided by the Inoue Foundation for Science and the Inamori Foundation.

ORGANIZING COMMITTEE

UNITED STATES   D. Pines, Chairman (University of Illinois)
                 M. Ruderman (Columbia University)
                 S. Tsuruta (Montana State University)

JAPAN   R. Tamagaki, Chairman (Kyoto University)
                 S. Hayakawa (Nagoya University)
                 K. Nomoto (University of Tokyo)
                 H. Sato (Kyoto University)
                 K. Sato (University of Tokyo)
                 Y. Tanaka (ISAS)

SCIENTIFIC SECRETARY
                 T. Tatsumi (Kyoto University)

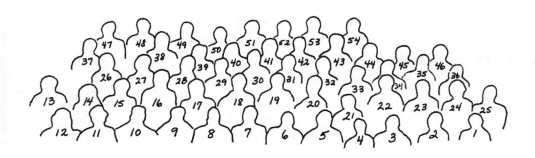

Key for Photograph (opposite)

1. H. Sato
2. Y. Tanaka
3. M. A. Ruderman
4. S. Tsuruta
5. R. Tamagaki
6. D. Pines
7. S. Hayakawa
8. G. Baym
9. K. Sato
10. N. Itoh
11. T. Tatsumi
12. S. Kuramoto
13. H. Suganuma
14. Y. Araki
15. F. Nagase
16. R. N. Manchester
17. R. Epstein
18. S. L. Shapiro
19. J. W. Clark
20. I. Fushiki
21. M. Takahara
22. L. M. Lattimer
23. D. G. Lamb
24. T. Nakamura
25. N. Iwamoto
26. Y. Tsue
27. Y. Suzuki
28. N. Shibazaki
29. H. Ögelman
30. C. J. Pethick
31. F.-K. Thielemann
32. L. S. Finn
33. T. Takatsuka
34. K. Nomoto
35. S. E. Woosley
36. H. Umeda
37. T. Muto
38. A. M. Abrahams
39. R. Hoshi
40. S. Shibata
41. T. Hanawa
42. H. Inoue
43. S. Teukolsky
44. K. Oohara
45. K. Iwasawa
46. S. Yamauchi
47. M. Ishii
48. S. Nishizaki
49. M. Fujimoto
50. T. Murakami
51. T. Shigeyama
52. H. Suzuki
53. L. Lindblom
54. R. Blandford

The Structure and Evolution of
Neutron Stars

R. Tamagaki
Department of Physics, Kyoto University, Kyoto 606, Japan

Opening Address

Dear Colleagues:

It is a great pleasure for me to address a few words, on behalf of the Organizing Committee, at the opening of the US–Japan Joint Seminar on "The Structure and Evolution of Neutron Stars".

First of all, I would like to extend my hearty welcome to all of you, in particular to those coming from far abroad.

This seminar, which we call SENS '90 in short, is sponsored by the National Science Foundation of the USA and the Japan Society for Promotion of Science. We are very grateful for the NSF and the JSPS for this joint sponsorship. Also we should like to thank Inoue Foundation for Science and the Inamori Foundation for the support to our seminar.

Here I wish to briefly mention the motivation of SENS '90. In recent years, several international meetings including neutron star problems as one of the major topics have been held in Japan. The US-Japan Joint Seminar on X-ray Sources, organized by Y. Tanaka and W. Lewin in 1985 in Tokyo, was one such meeting. In these meetings, the principal focus was on the observational aspects, on the basis of the successful results obtained by the Japanese X-ray satellites. Two years ago, I heard an idea from Prof. S. Tsuruta to have the US-Japan Joint Seminar in Kyoto, putting emphasis on theoretical studies of the neutron star interior. This idea was positively accepted both in the USA and Japan. In such background, as the co-chairmen, Prof. Pines and I proposed this seminar to the NSF and the JSPS, respectively, and the proposal was approved. Now I am very pleased to have the opening of SENS '90 today.

Neutron stars, as compact objects and one of the possible endpoints of stellar evolution, have a remarkable richness of internal structures, whose study encompasses a variety of research fields: astrophysics, nuclear physics, particle physics, and atomic physics. Because of such richness, neutron stars continue to furnish astonishingly fresh information in their phenomena, although most of them are old in the sense that the young Crab pulsar is already about 950 years old. Recently we have had the youngest member in SN1987A, which has given us valuable information on neutron star formation, and also is expected to be a source of new surprises in future. We, studying neutron stars, are now in a lucky and promising phase.

SENS '90 aims to review the recent advances of the studies on the following major topics: the internal structure and dynamical behavior, formation of neutron stars, and evolution of neutron stars—by focusing our attention on the internal properties of neutron stars and relevant phenomena. In addition to the astrophysical significance obtained from these studies, SENS '90 will contribute to a deeper understanding of some of the most fundamental problems in physics, through the investigations on various aspects under the extreme conditions realized only in neutron stars.

The most notable feature of this seminar is that here are gathered many excellent experts who are very active in the frontier of this field, although the size of the meeting is small. In spite of its compactness, SENS '90 is full of activities and variety, and so new interesting findings in facts as well as in logic are expected to emerge, in a manner similar to the behavior of our favorite objects, neutron stars. Therefore I believe that SENS '90 will certainly contribute much to the progress in this field, through the exchange of new results, the proposal of new ideas, and the discussions.

Finally I would like to thank all of you for your kind cooperation, and I hope that our foreign friends will enjoy their stay during one of the best seasons in Kyoto, brief as it may be.

Thank you for your attention.

Structure and Dynamical Behavior

David Pines† and M. Ali Alpar†‡
†Loomis Laboratory of Physics, University of Illinois, 1110 W. Green St., Urbana, IL 61801, USA; ‡Physics Department, Middle East Technical University, Ankara 06531, Turkey

Inside Neutron Stars: A 1990 Perspective

We present an overview of neutron star structure and superfluidity. Recent calculations of the equation of state and of the superfluid energy gaps inside the neutron star are discussed and related to observational constraints. Vortex motion in the neutron superfluid within the solid crust of the neutron star is proposed as the origin of glitches and a major source of energy dissipation in the star. A comprehensive model for glitches and postglitch relaxation is applied to the Vela and Crab pulsars. New approaches to the dynamics of the hadron superfluids in the neutron star core and observational constraints on crust-core coupling mechanisms are summarized.

1. INTRODUCTION

During the 20-plus years which have passed since the discovery of pulsars and their identification as rotating neutron stars, their structure, dynamic behavior, and evolution have been a focus of the theoretical and observational research by the astrophysics community. Since two decades is often regarded as a maturational period for humans, that leads us to examine the extent to which our understanding of neutron stars has matured over this time span, and to ask to what extent did the infant foreshadow the adult. There was no SENS '70 to mark the very early childhood of our field. However, one of us gave a talk entitled "Inside Neutron Stars" twenty years ago here in Kyoto (at the Twelfth International Conference on Low Temperature Physics) and we use that review (Pines, 1971) to provide perspective on what might have been discussed at a hypothetical SENS '70 and what we have learned since then.

We begin our review with a quite brief overview of some of the major observational and theoretical developments in the physics of neutron stars during the past twenty years. We next give a progress report on theoretical work on the crustal neutron superfluid and the information on its behavior which can be obtained from observations of pulsar glitches and postglitch behavior. In conclusion, we comment briefly on the core hadron superfluids and on some of the open questions concerning crustal and core behavior.

2. SENS '70 → SENS '90: AN OVERVIEW

As is evident from an inspection of Table 1, the past twenty years have seen remarkable progress in observations of neutron stars that are relevant to understanding their structure and evolution. Among the highlights noted are the following:

- The discovery of pulsating accreting neutron stars in binary systems, which led to the first determination of neutron star masses ($M \cong 1.5 \pm 0.5 M_\odot$), and, thanks to the ability to observe cyclotron features in their spectra, a direct determination of surface magnetic fields ($6 \times 10^{11} G \leq B \leq 4 \times 10^{12} G$).
- The discovery of millisecond pulsars orbiting neutron stars in binary systems, which has made possible quite accurate determinations of the masses of the neutron stars in those systems [$(1.32 \pm 0.03) M_\odot \leq M \leq (1.442 \pm 0.003) M_\odot$].
- Synoptic studies of the Vela pulsar following its superglitch [$(\Delta\Omega/\Omega) \sim 2 \times 10^{-6}$] in March 1969, which have led to the observation of seven subsequent "superglitches," of magnitude $1.1 \times 10^{-6} \leq (\Delta\Omega/\Omega) \leq 3.1 \times 10^{-6}$, with sufficiently detailed coverage of the postglitch behavior to enable one to identify four distinct postglitch recovery times in the data. Analysis of the most recent glitch (12/24/88), which was "caught in the act" by observers in both Tasmania and South Africa, makes possible the identification of a very rapidly relaxing region ($\tau \sim 0.4\ d$) and places an upper limit on the coupling of the superfluid neutron core to the crust; $\tau_{CC} \leq 2$ minutes.
- Observations of nearly identical postglitch behavior following two large glitches [$(\Delta\Omega/\Omega) = 3.7 \times 10^{-8}$; $(\Delta\Omega/\Omega) = 7.4 \times 10^{-8}$] of the Crab pulsar.
- Observations of glitches in six other pulsars that suggest that glitches are a way of life for all pulsars, with an interval between glitches determined, at least in part, by the pulsar spin-down rate.

TABLE 1 Some Observations Relevant to Understanding Neutron Star Structure and Evolution: 1970 and 1990 Compared

Property	SENS '70	SENS '90
Neutron star sample	• ~ 50 rotation-powered pulsars • ~ 2 glitching pulsars	• ~ 450 rotation-powered pulsars • ~ 30 accretion-powered pulsars • ~ 8 glitching pulsars
Rotation-powered pulsar periods	$33 \text{ ms} \leq P \leq 4s$	$1.6 \text{ ms} \leq P \leq 5s$
Accretion-powered pulsar periods	Unknown	$69 \text{ ms} \leq P \leq 835 \text{ s}$
Spin-down age $T_s \equiv P/2\dot{P}$	$1200 \text{ yr} \leq T_s \leq 10^6 \text{ yr}$	$1200 \text{ yr} \leq T_s \leq 10^9 \text{ yr}$
Surface magnetic fields B: • From pulsar spin-down • Direct observation	$\sim 10^{12} G$ None	$10^9 G \leq B \leq 1.7 \times 10^{13} G$ $6 \times 10^{11} G \leq B \leq 4 \times 10^{12} G$
Mass, $M/M_\odot$	Unknown	$1.32 \pm 0.03 \leq (M/M_\odot) \leq 1.442 \pm 0.003$ for ms pulsars in binary systems
Multiple glitches	None	Vela (9); PSR 1737–30 (5); Crab (4); PSR 0335+54 (2)
Postglitch relaxation times, τ	Vela: $\tau \sim 1$ yr Crab: $\tau \sim 10$ d	Vela: $\tau_1 \sim 0.4$ d; $\tau_2 \sim 3.2$ d; $\tau_3 \sim 33$ d; $\tau_4 \gtrsim$ yr Crab: $\tau_1 \sim 14$ d; $\tau_2 \gtrsim$ yr
Glitch observed	No	Vela (1988); Crab (1989)
Crust-core coupling time	$7 \text{ d} \leq \tau \leq \text{yr}$	$\tau \lesssim 300 \text{ s}$
Birth observed	By Chinese (7/4/1054)	SN 1987 A(?)

- The observation of SN 1987 A, which, it is hoped, signaled the birth of an observable pulsar, so that over time it may prove possible to study the behavior of a very young pulsar.

We summarize some of the corresponding changes in our basic theoretical perspective on neutron star structure and dynamics in Table 2. During this period there has been a continuing improvement in microscopic calculations of the equation of state of neutron matter, which, at densities exceeding the density ρ_0 of nuclear matter, determines the mass-radius relation, the crustal extent, the distribution of the stellar moment of inertia, and the central density of a neutron star (Pandharipande, Pines, and Smith, 1976). The more recent calculations (for a review, see Baym, 1991) have led to equations of state which are considerably stiffer than those circa 1970, which were based on the Reid potential for neutron-neutron interactions. As a result the radius of a $1.4 M_\odot$ star is now believed to be some 20%–70% greater than the 8 km radius calculated with the Reid potential, with a still more substantial increase in the extent of the inner crust.

TABLE 2 Aspects of Our Theoretical Understanding of Neutron Stars: 1970 and 1990 Compared.

Property	SENS '70	SENS '90
Neutron matter E.O.S.	Soft	Stiff
Radius of $1.4 M_\odot$ star	$R \sim 8$ km	$10 \text{ km} \leq R \leq 13$ km
Crustal extent ($1.4 M_\odot$)	800 m	$\sim 1.6 \pm 0.4$ km
Crustal superfluid properties	$\Delta \leq 4$ MeV $T_c \leq 2.3$ MeV	$\Delta \leq 1.3 \pm 0.3$ MeV $T_c \leq 0.7$ MeV
Leading candidates for physical origin of glitches	Oblateness-changing starquakes	• Catastrophic unpinning of pinned vortices in crustal superfluid • Starquakes induced by pinning of vortices in crustal superfluid • Oblateness-changing starquakes (Crab pulsar **only**)
Physical origin of postglitch behavior	Crust-superfluid core coupling	Crustal superfluid vortex creep

The cross section of a $1.4 M_\odot$ neutron star calculated using a moderately stiff Bethe-Johnson equation of state at $\rho > \rho_0$ (which is somewhat stiffer than that

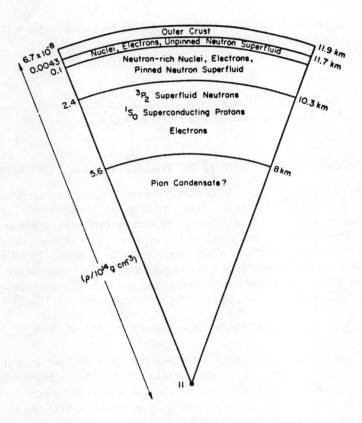

FIGURE 1 The cross section of a $1.4 M_\odot$ neutron star calculated using a moderately stiff (Bethe–Johnson) equation of state for neutron matter at densities $\geq \rho_0$

calculated recently by Wiringa, Fiks, and Fabrocini, 1989) is shown in Fig. 1. The composition of the quantum liquid core and the precise density at which the transition from a solid crust to a quantum liquid interior takes place continue to be somewhat uncertain. Thus, while a pion condensate at densities $\geq 2\rho_0$ has seemed possible for over a decade, there remain enough uncertainties in the microscopic calculations, despite many theoretical efforts, that one cannot be sure it is realized in practice (Baym, 1991). As we shall discuss later, the presence of proton superfluid in the quantum liquid core is also open to question, so that Fig. 1 should be regarded as a plausible scenario, to be confirmed by subsequent calculation or by observation.

The changes during the past two decades in our understanding of crustal superfluid properties, glitches, and postglitch behavior which are summarized in Table 2 are considerably more substantive than those described above and will form the subject of the remainder of this talk.

3. THE CRUSTAL NEUTRON SUPERFLUID

3.1 MICROSCOPIC CALCULATIONS OF THE SUPERFLUID ENERGY GAP

At distances ≥ 1 fm, the bare interaction between neutrons is attractive, while for distances ≤ 1 fm, it becomes strongly repulsive. At densities $\leq \rho_0$ in neutron matter, the short-range correlations brought about by that repulsion serve to keep the neutrons from sampling too much of the repulsive part of the interaction, so that the net effective interaction, V, between neutrons near the Fermi surface is attractive. As a result, the neutrons become superfluid, at a temperature T_c, with pairs in a $^1S_0$ state and a maximum energy gap in the single-particle spectrum, which, in the BCS weak-coupling approximation, is given by

$$\Delta = 1.76 k T_c \cong E_F \exp[-1/N(0)V] \qquad (3.1)$$

where E_F is the Fermi energy, and $N(0)$ the density of states per unit energy. At comparatively low densities ($\rho \leq \rho_0/500$, say) the effect of the medium in which the neutrons are immersed plays little or no role, and it is straightforward to calculate V and hence Δ, using, for example, a variational approach to take the short-range correlations into account. At these densities, V is comparatively large, but because $N(0)$ is low, the gap is small. As the density increases, $N(0)$ increases; however, because the neutrons begin to sample more of the short-range repulsion, V will decrease. As a result of this trade-off, the variational calculations reviewed by John Clark at this meeting (Clark, Davé, and Chen, 1992) show that the energy gap (and T_c) takes on its maximum value (~ 3.8 MeV) at densities $\sim \rho_0/10$. Beyond these densities it decreases gradually, until, near ρ_0, where the crust ends, it ceases to be energetically favorable for neutrons to pair in a $^1S_0$ state; rather, as a result of the attractive tensor force, they will tend to pair in a $^3P_2$ state, for which the influence of the strong short-range repulsion is less significant. (For a recent calculation see Baldo et al., 1991.)

Accurate calculations of the energy gap of the density region $\rho_0/500 \leq \rho \leq \rho_0$ are difficult because it has not proved easy to determine *polarization effects*, the way in which particle-hole correlations induced in the neutron liquid background by the neutron pair modify the long-range part of their interaction. As a result prior to 1989 one had estimates of the energy gap (Δ) in the vicinity of $\rho_0/10$ which differed by almost an order of magnitude, ranging from 0.5 MeV to 4.5 MeV. Fortunately, recent calculations of polarization effects are beginning to converge. One approach, which Ainsworth, Wambach, and Pines have followed (Ainsworth, Wambach, and

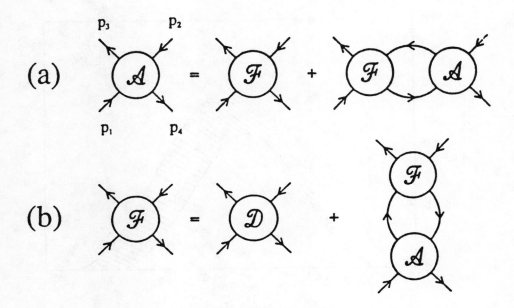

FIGURE 2 A diagrammatic representation of the nonlinear coupled equations for the scattering amplitude (A) and particle-hole interaction (F) driven by the direct interaction (D), which is particle-hole irreducible in all momentum channels

Pines, 1989; for a review see Wambach, Ainsworth, and Pines, 1991), is based on a Fermi liquid model. In the medium the particle-hole interaction (F) is decomposed into a driving term, the direct interaction (D), and an induced interaction which involves the exchange of the density and spin fluctuations which describe polarization effects. The set of nonlinear coupled equations for F and the scattering amplitude (A), which are represented diagrammatically in Fig. 2, is obtained and solved self-consistently.

In this way, Ainsworth, Wambach, and Pines are able to determine the quasiparticle effective mass and the pairing matrix element (V), from which they calculate, using weak coupling theory, the $^1S_0$ energy gap shown in Fig. 3. The result takes the same general form as that obtained using a variational approach with no polarization effects; however, the calculated gaps are about a factor of 3 lower.

A second approach is based on incorporating medium effects in variational calculations using correlated basis functions. Clark, Davé, and Chen (1992) conclude that this approach will yield results close to those of Wambach et al. for $\rho \leq \rho_0/5$, that is, up to densities a bit beyond those at which the gap is maximum. Thus, over this region, it appears likely that the gap is known quantitatively to some

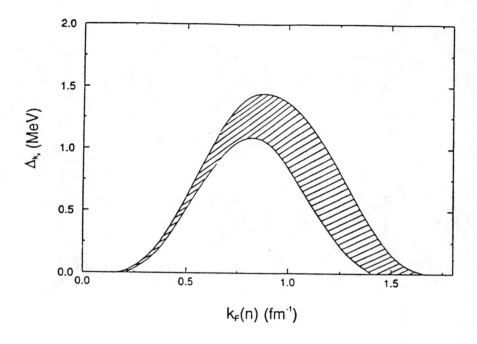

FIGURE 3 The calculated dependence of the $^1S_0$ energy gap, Δ (MeV), on the density, expressed as function of the neutron Fermi wave vector, $k_f \cdot k_{fo} = 1.36$ fm^{-1} corresponds to a neutron density equal to that of nuclear matter, $\rho_0 = 0.17$ fm$^{-3} \equiv 2.8 \times 10^{14}$ gcm^{-3}.

30%. For the important region, $\rho_0/5 \leq \rho \leq \rho_0$, it is likely that the gap is not yet known to within a factor of 2, so that the limits of error reflected in the hatched area of Fig. 3 do not appear overly conservative. The situation for densities near ρ_0 is particularly murky; here, near the transition from a solid crust to a quantum liquid core, it is quite possible that there is a narrow region of density near ρ_0 over which neutron matter might even be normal, since the recent calculations of Baldo et al. (1991) suggest that the ^1S_o energy gap becomes negligible at $\rho \sim \rho_0/2$, while $^3P_2$ superfluidity of neutron matter begins only for $\rho \geq \rho_0$.

3.2 VORTICES, PINNING, AND CREEP

A rotating superfluid mimics rigid body rotation by developing an array of quantized vortices, parallel to the axis of rotation, around which the circulation is quantized in units of $h/2m$. The density (n) of vortices per unit area is

$$n(h/2m) = 2\Omega \tag{3.2}$$

The superfluid can only change its angular velocity through motion of the vortices; thus vortices moving radially outward from the rotation axis at velocity v_r enable the superfluid to spin down at a rate

$$\dot{\Omega} = -n(h/2m)v_r/r \qquad (3.3)$$

The crustal neutron superfluid coexists with a lattice of neutron-rich nuclei; to the extent that the superfluid is aware of the presence of the nuclei, the latter present a highly inhomogenous medium for the motion of vortex lines and hence the spin down of the neutron superfluid.

Each vortex line in the crustal superfluid possesses a "normal" core within which the condensate wave function goes to zero over a distance

$$\xi = \frac{2E_F}{\pi k_f \Delta} \sim \frac{13 k_f (\text{fm}^{-1})}{\Delta(\text{MeV})} \text{fm} \qquad (3.4)$$

which is tiny compared with the distance between vortices,

$$d \sim 3 \times 10^{-2} \Omega^{1/2} \text{cm} \qquad (3.5)$$

Creating that normal core in a uniform rotating superfluid costs energy, the condensation energy, which is, per unit volume,

$$E_c \sim \frac{3}{8} \frac{\Delta^2(\rho)}{E_F} \qquad (3.6)$$

In the presence of the lattice of crustal nuclei, the question is then whether, for a configuration of vortices and nuclei, that energy cost can be reduced; where it can, the vortices may be expected to take on that energetically favored configuration and so be *pinned* to the crustal nuclei, in the sense that energy must then be supplied to the system for the vortices to move to another less advantageous position.

Quite generally we may expect that vortices will pay attention to the presence of crustal nuclei to the extent that ξ is comparable to the size of a nucleus ($2R_N \sim 14$fm) and to the extent that the neutrons inside the nucleus possess properties which are different from those of the crustal neutron matter. Neutrons lying outside closed shells in the neutron-rich nuclei of the crust possess a net attractive interaction, just as do free neutrons, and hence behave like a BCS superfluid, with strong pairing correlations and an energy gap for single-particle excitations. At crustal neutron superfluid densities such that the energy gap for neutrons outside nuclei is the same as that for those which are inside, nuclei become effectively invisible to the crustal neutron superfluid, and no pinning takes place. However, where that gap is greater and $\xi \leq 2R_N$, it is clearly energetically favorable for vortices to pass through, and hence *pin* to, nuclei in the crustal lattice (Pethick and Pines, as reported in Pines, 1971). The pinning energy per nuclear cluster depends on the overlap of the vortex core and the nuclei, on the imperfectly known difference in the energy gap (and hence condensation energy) of the neutrons outside and inside the nucleus, and on the way in which the condensate wave function for neutrons outside the nucleus match on to the condensate wave function for neutrons inside the nucleus. Various possible regions of pinning have been explored by Anderson et al. (1982). For the *weak pinning* regions, in which vortices pin to whatever nuclei they encounter (but do not act to pull nuclei out of their regular lattice array), and in which, moreover, the coherence length is sufficiently short that a vortex line pins to only one nuclear cluster at a time, Alpar et al. (1984b) estimate the pinning energy per cluster as

$$E_p^w = \gamma \frac{3}{8} \frac{\Delta^2}{E_F} \cdot \frac{4\pi R_N^3}{3} \frac{k_f^3}{3\pi^2} \sim \gamma k_f (\text{fm}^{-1}) \Delta^2 (\text{MeV}) \tag{3.7}$$

where $\gamma (\leq 1)$ is a factor which takes into account phenomenologically the difficult microscopic physics of pinning (the difference in energy gaps, possible proximity effects, matching condensate wave functions, etc.). The character of the pinning changes when the coherence length becomes comparable to, or larger than, the average spacing between nuclear clusters; one is then in a *superweak pinning* region, characterized by

$$\xi^{SW} \geq b_Z/2$$

where b_Z is the average spacing between nuclei. Under these circumstances since the normal core of a vortex is sufficiently large to encompass two or more nuclei, the pinning energy per nuclear cluster will be somewhat reduced, and it is easier for the vortex line to move, that is, to change its crustal nuclear configuration.

For a given pinning energy, the rate of vortex flow by thermal activation over the energy barriers opposing its motion can be calculated using an approach similar to that developed by Anderson and Kim (1962) for flux creep in Type II superconductors; it is

$$< v_r > = v_o \exp - \frac{E_p}{kT} \frac{\omega_{cr} - \omega}{\omega_{cr}} \tag{3.8}$$

where v_o is a characteristic microscopic approach velocity ($\sim 10^7$ cm s^{-1}) and ω_{cr} is the critical lag in angular velocity at which vortices will unpin, obtained by balancing the Magnus force produced by the lag against the force per unit length produced by the pinning potential,

$$\omega_{cr} = \frac{E_p}{\rho(h/2m) r b \xi} \tag{3.9}$$

where b is the average spacing between pinning centers along the vortex line. Anderson et al. (1982) have estimated the spacing between pinning centers for the weak-pinning regions as

$$b^w = \frac{b_z^3}{\pi \xi^2} \cong 230 (\frac{b_z}{50 \text{fm}})^3 \frac{\Delta^2 (\text{MeV})}{k_f^2 (\text{fm}^{-1})} \text{fm} \tag{3.10}$$

so that in the weak pinning regions the maximum lag is

$$\omega_{cr}^w \cong 0.4 \frac{\Delta(\text{MeV})}{r_6 k_f (\text{fm}^{-1})(b_z/50\text{fm})^3} \text{rad } s^{-1} \quad (3.11)$$

According to the energy gap calculations of Ainsworth et al. (1989), weak pinning should prevail over a density region $(\rho_0/10) \leq \rho \leq \rho_0/3$. At higher densities, where the pinning becomes superweak, there are no present microscopic calculations of pinning energies or of the effective spacing between pinning centers. Since the latter may be expected to be large compared with b^w, a substantial reduction in ω_{cr} is to be expected in this superweak region,

$$\omega_{cr}^{sw} \ll \omega_{cr}^w \quad (3.12)$$

A phenomenological description of vortex motion in an inhomogenous medium, the *vortex creep model*, was developed by Alpar et al. (1984a, 1984b) and applied to the postglitch behavior of pulsars by these authors and by Alpar, Cheng, and Pines (1989). As discussed elsewhere in this volume by Alpar (1992), the model is based on the idea that because there is a considerable variation in pinning energies throughout the crust, it is useful to consider a number of distinct pinning regions, each characterized by an average strength, I_i (the moment of inertia of the pinned superfluid in region i), and a drift velocity, $<v_r^i>$, which is related to $<E_p^i>$ and $<\omega_{cr}^i>$ by Eq. (3.8). The equation of motion of the observed outer crust, assumed to have an effective inertial moment $I_c(\sim I)$ which includes all components which couple to the crust on microscopic (i.e.. < 1 min) time scales, then reads

$$I_c \dot{\Omega}_c = N_{ext} + \sum_i I_i \dot{\Omega}_c \quad (3.13)$$

where N_{ext} is the external torque on the crust, and $\dot{\Omega}_i$ is related to $<v_r^i>$ by Eq. (3.3). From Eq. (3.8), it follows that the time scale which characterizes the response of a given pinning region depends on E_p^i; on how close the superfluid lag, ω, is to ω_{cr}^i; and on the temperature, T. For young, hot pulsars, in both weak- and superweak-pinning regions, a steady state of vortex creep can be established for values of ω which are small compared with ω_{cr}; vortex creep theory can then be linearized (Alpar, Cheng, and Pines, 1989), and the characteristic times in the observable range for the superfluid to respond to a glitch turn out to be

$$d \leq \tau_{lin} \leq \text{yr} \quad (3.14)$$

Still, given the expected substantial variation in pinning energies in the inner crust, such stars will also contain regions in which ω_{cr} is substantial and $\omega \sim \omega_{cr}$; the response of the crustal superfluid to changes in Ω_c is then inherently nonlinear. As discussed by Alpar, Cheng, and Pines (1989), a likely indicator of such regions is a "persistent shift" in Ω_c following a glitch. Quite generally, as a given pulsar evolves, the response of a given region will shift from being linear to nonlinear, so that primarily nonlinear response is to be expected for old and cold pulsars.

3.3 VORTEX CREEP HEATING

As a pulsar spins down, the pinned crustal neutron superfluid becomes an increasingly important reservoir of not only angular momentum but also rotational energy. As the vortex lines creep in response to a lag, ω, they transfer rotational energy to the crust. The energy dissipation rate in steady state is largest where the creep is nonlinear; it may be written as

$$E_{diss} \cong \sum_i I_{n\ell}^i \omega_{cr}^i \mid \dot{\Omega} \mid \tag{3.15}$$

Depending on the distribution of ω_{cr}^i and the extent of the nonlinear regions (as measured by $I_{n\ell}^i$), this source of energy dissipation may be significant. Indeed, once the initial heat of a pulsar has been radiated away, vortex creep may determine its temperature and surface luminosity (Alpar, Nandkumar, and Pines, 1985; Shibazaki and Lamb, 1989; Tsuruta, 1992) and so make possible the direct observation (in the soft X-ray region) of nearby old pulsars. A search for such emission from the nearby radio pulsar PSR 1929+10 was carried out by Alpar et al. (1987) using *EXOSAT*. From their failure to detect its thermal luminosity, these authors concluded that the star possessed no appreciable pinning regions with $< \omega_{cr} > \geq 0.7$ rad s^{-1}. For neutron matter energy gaps ≤ 1.75 MeV, as calculated for example by Ainsworth et al. (1989), this limit is consistent with the upper limit on ω_{cr}^w given in Eq. (3.11). Perhaps more important, *it rules out any appreciable regions of strong pinning*, such as those proposed by Epstein and Baym (1989).

4. GLITCHES AND POSTGLITCH BEHAVIOR

During the past decade, the accumulation of observational data on the Vela, Crab, and other glitching pulsars, which is discussed by Manchester (1992) in these proceedings, makes it possible to use observations of postglitch behavior to confirm and constrain theory to a greater extent than was possible during the decade following the initial observation of the Vela and Crab pulsars. Much of our work in Urbana and Aspen has been focused on understanding superfluidity in neutron stars and what can be learned about it from observations of glitches and postglitch behavior. Since there will be further discussion of vortex motion later in this meeting (Alpar, 1992; Epstein, Link, and Baym, 1992), and since some of our joint work has recently been reviewed elsewhere (Alpar and Pines, 1989; Alpar, 1991; Pines, 1991), we shall summarize here only the main arguments which have led us to conclude that as first proposed by Anderson and Itoh (1975), vortex pinning plays a significant role in causing glitches and that postglitch behavior is uniquely explained as the response of the pinned crustal neutron superfluid to a glitch (Pines and Alpar, 1985). The Vela pulsar will be treated in some detail, to give a first account of a comprehensive treatment of all its glitches and postglitch behavior and an explanation of the long term remnant spin-up from its glitches (Alpar et al. 1992).

4.1 THE VELA PULSAR

The Vela pulsar has been observed to glitch nine times during the past 22 years. As may be seen in Table 3, eight of these are "superglitches," with magnitudes $1.14 \times 10^{-6} \leq |\Delta\Omega_c/\Omega_c| \leq 3.06 \times 10^{-6}$. As befits the pulsar that has displayed the highest degree of glitch activity (glitch activity is the mean fractional change in period per year [McKenna and Lyne, 1990]), it has been the most carefully studied, so that even though its most recent glitch occurred on Christmas Eve, 1988, it was "caught in the act" by observers in both Tasmania (McCullogh et al., 1990) and South Africa (Flanagan, 1990). It is perhaps not unreasonable to regard the Vela pulsar as a kind of "Rosetta stone" for glitching pulsars in the hope and expectation that a scenario which enables one to understand the physical origin of its glitches and its remarkable dynamic response to a glitch might prove sufficiently robust to explain the "glitch phenomena" in other pulsars.

The observations of the March 1969 glitch of the Vela pulsar and of its postglitch behavior, in which part of the initial remarkably large jump in Ω and $\dot{\Omega}$ decayed away in a time of order months, led a group working together that summer at the Aspen Center for Physics to propose that the physical origin of the glitch was a starquake induced by the buildup of strain in the solid outer crust as the pulsar spins down, and that the postglitch behavior signaled the presence of a superfluid interior (Baym et al., 1969). They proposed a simple two-component theory to describe the response of the neutron superfluid core to the sudden change in the stellar crust rotation and attributed the observed long relaxation time to the comparatively weak, and hence long–time-scale, coupling of the core neutron superfluid to the stellar crust. The observation a few months later of a glitch in the Crab pulsar, with a decay time of the order of days, that could be fitted with our two-component model appeared to provide further support for this interpretation and led to detailed calculations of strain buildup, the expected time between oblateness-changing starquakes (Baym and Pines, 1971), and the leading candidate for crust-core coupling, the scattering of core electrons by the normal cores of the interior superfluid neutron vortices (Feibelman, 1971). However, the observation of a second superglitch in the Vela pulsar in August 1971 ruled out spin-down induced change in crustal oblateness as a cause of the Vela glitches. Based on the possibility of a solid neutron core in heavier neutron stars, corequakes were suggested as an alternative glitch mechanism (Pines, Ruderman, and Shaham, 1973), an explanation which was soon superceded by the proposal of Anderson and Itoh (1975) that glitches originate in the intrinsic pinning of the crustal neutron superfluid to the inner crustal nuclei.

The attention of theorists, therefore, shifted to the pinned crustal neutron superfluid, which can be responsible for glitches in one of two ways: by the sudden catastrophic release of pinned vortex lines (Anderson and Itoh, 1975; Pines et al., 1980) or by stressing the crustal lattice sufficiently to induce starquakes (Ruderman, 1976, 1992). In either scenario, the calculated glitch activity for the Vela pulsar is about right. Since angular momentum is conserved in a glitch, we may write

$$I_c \Delta\Omega_c(O) = I_p \delta\Omega \qquad (4.1)$$

where I_p is the net moment of inertia of the pinned neutron superfluid which has undergone an average net decrease in crustal neutron superfluid velocity, $\delta\Omega$, associated with the glitch. As proposed by Ruderman (1976) and Pines et al. (1980), the time required to restore the pinned crustal superfluid to its preglitch configuration, and hence set the pulsar for its next glitch, is then

$$t_g \cong \delta\Omega/|\dot{\Omega}| \qquad (4.2)$$

The glitch activity is, from Eqs. (4.1) and (4.2),

$$\frac{\Delta\Omega_c}{\Omega t_g} = \frac{I_p}{I_c}\frac{|\dot{\Omega}_c|}{\Omega} = \frac{I_p}{I_c}\frac{1}{2t_{sd}} \qquad (4.3)$$

where t_{sd} is the spin-down age of the pulsar. On taking $I_p \sim 2 \times 10^{-2} I_c$, a reasonable estimate for pinned crustal superfluid in the case of a moderately stiff equation of state, the expected glitch activity for the Vela pulsar is $\sim 9 \times 10^{-7} \text{yr}^{-1}$, which compares favorably with the observed value (McKenna and Lyne, 1990) for the Vela pulsar of $8 \times 10^{-7} \text{yr}^{-1}$. (For a derivation of Eq. [4.3] using a starquake scenario, see Ruderman, 1992.)

Still more observational evidence for the presence of pinned crustal superfluid in the Vela pulsar comes from an examination of the postglitch behavior observed following each of its eight glitches. From Eqs. (3.3) and (3.8), it is straightforward to show that for a region in the pulsar in which the response of the pinned crustal superfluid to a glitch is linear, the internal torque produced by the superfluid gives rise to a change in the crustal spin-down rate

$$\Delta\dot{\Omega}_c(t) = -\frac{I_i}{I}\frac{\delta\omega_i}{\tau_i}e^{-t/\tau_i} \qquad (4.4)$$

where $\delta\omega_i = \Delta\Omega_c + \delta\Omega_i$ is the decrease in the superfluid lag ω_i induced by the glitch, and τ_i is the associated relaxation time which, as noted in Eq. (3.14), may be expected to be of order days to years. For regions through which vortices have not moved, there can be no change in the superfluid velocity; $\delta\Omega_i = 0$, and $\delta\omega_i \equiv (\Delta\Omega_c)_0$. On the other hand, for regions through which vortices have moved, the reduction in the superfluid angular velocity, $\delta\Omega_i$, will typically be large compared with the increase in Ω_c, so that $\delta\omega_i = \delta\Omega_i$. One approach, then, to analyzing the postglitch data on $\dot{\Omega}$ is to consider times of roughly a few months and to determine the minimum number of distinct regions, with distinct times, τ, required to fit the observations, with the constraint that after every glitch one gets a distinct response $I_i\Delta\Omega_c/\tau_i$ from those regions through which vortices do not move. This is the approach which Alpar (1991) have followed. When account is taken of the fact that apart from the December 1988 glitch, observers did not catch the glitches as they occured, that is, observations begin a time Δ after each glitch (a further fitting parameter), they obtain the results shown in Table 3.

TABLE 3 Observed and Deduced Parameters for the Vela Pulsar (from Alpar et al., 1991), Assuming a Distinct Linear Response from Regions with Relaxation Times $\tau_1 = 0.417$ d; $\tau_2 = 3.2$ d; $\tau_3 = 32.7$ d.

Glitch date:	2/28/69	8/29/71	9/28/75	7/3/78	11/11/81	8/10/82	7/12/85	12/24/88
$[\Delta\Omega_c(o)/\Omega_c]_{-6}$	2.35	2.05	1.99	3.06	1.14	2.05	1.30	1.81
$[\Delta\dot\Omega_c(o)/\dot\Omega_c]_{-3}$	1.3	1.8	1.1	1.8	0.90	2.0	1.5	16
Δ(d)	4.9	4.8	5.0	2.8	0.44	0.91	0.22	0.00
$(I_1/I)_{-3}$	5.7	5.7	5.7	5.7	5.7	5.7	5.7	5.7
$(I_2/I)_{-3}$	1.5	1.5	1.5	1.5	1.5	1.5	1.5	1.5
$(I_3/I)_{-3}$	0.51	0.64	0.44	0.92	0.33	1.2	0.90	0.96
$(I_{nt}/I)_{-3}$	0.71	0.72	0.72	0.66	0.63	0.6	0.65	0.47
$(I_B/I)_{-2}$	1.2	1.2	1.6	1.9	1.5	1.2	1.1	1.0
$(\delta\Omega/\Omega_c)_{-4}$	1.9	1.6	1.2	1.6	0.73	1.7	1.2	1.7
$(I_p/I)_{-2}$	2.0	2.1	2.4	2.8	2.3	2.1	2.0	1.9
t_g (d)	1624	1375	1036	1371	616	1485	972	1422
t_{obs} (d)	912	1491	1009	1227	272	1067	1261	

When this approach is followed, there remain two additional contributions to postglitch behavior: a torque which increases linearly with time and a glitch-induced initial offset in $\dot{\Omega}_c$. Both can be explained by the phenomenological theory of vortex creep developed by Alpar et al. (1984a, 1984b), in which the long-term, nonlinear response to a glitch of regions through which vortices have moved gives rise to a torque,

$$\Delta N_{n\ell}(t) = I_c[\Delta\dot{\Omega}_c(t)]_{n\ell} = \frac{I_{n\ell}}{I_c}\left(1 - \frac{t}{\delta\Omega/\dot{\Omega}_c}\right)I_c\dot{\Omega}_c \qquad (4.5)$$

In this model, the reduction in the superfluid velocity ($\delta\Omega$) produced in the glitch can then be obtained directly from the coefficient of the torque, which increases linearly in time. (For an alternative analysis which the present one supercedes, see Pines, 1991.) A further constraint on the description of glitch phenomena comes from the condition that angular momentum be conserved in the glitch. In the above model, this takes the form

$$I_c\Delta\Omega_c(o) = (I_{n\ell}/2 + I_B)\delta\Omega \qquad (4.6)$$

where $I_B\delta\Omega$ is a glitch-induced change in Ω_c associated with regions through which vortices pass in a glitch, but in which there is *no glitch-associated net change in the number of vortices present*, or in the vortex creep current. The regions comprising the moment of inertia I_B are regions in the crust superfluid where the distribution of pinning centers and ω_{cr} is such that a vortex density and vortex creep current cannot be sustained (Cheng et al., 1988) so that the superfluid torque from regions I_B is zero both before and after the glitch. Although such regions cannot contribute to any change in the internal torque exerted by the superfluid on the crust, the passage of vortices through them gives rise to a change in the crustal superfluid angular momentum $I_B\delta\Omega$. The values of I_B determined in this way are likewise displayed in Table 3. Also shown there are t_g, the calculated time to restore the pinned crustal superfluid to its preglitch configuration, Eq. (4.2), and the observed time to the next glitch, t_{obs}.

Several features of the results of this analysis deserve further comment:

- Both the 3.2 d and 0.47 d response appear with the same strength after each glitch; the latter, which was only clearly seen in the December 1988 glitch, is responsible for the remarkably large initial jump in $\dot{\Omega}_c$ after that glitch. The fact that the previous three glitches were caught within less than a day of their occurrence serves to determine the nature of the 3.2 d response.
- I_3, the strength of the 32.7 d response, varies somewhat from glitch to glitch, as do $I_{n\ell}$ and I_B. However, the net pinned moment of inertia of the crustal neutron superfluid which is deduced from our fit to observation,

$$I_p = \sum_{i=1}^{3} I_i + I_{n\ell} + I_B = (0.023 \pm 0.005)I_c \qquad (4.7)$$

lies within 20% of the median value for all eight glitches.

- The lower bound on the net moment of inertia associated with crustal densities, $(\rho_0/10) \leq \rho \leq \rho_0$, $(I_p/I) \geq 0.027$, obtained from observation provides a strong constraint on the hadron matter equation of state at densities greater than nuclear density. For example, comparatively soft equations of state (such as that calculated using the Reid potential), which lead to a stellar crust of a $1.4 M_\odot$ neutron star so thin that the *total* crustal moment is less than $0.02 I_c$, scarcely represent viable candidates for the Vela pulsar.
- Although for a given glitch, the deduced values of $\delta\Omega$ are such that t_g (the time needed for subsequent pulsar spin-down to restore the pinned crustal superfluid to its preglitch configuration, Eq. [4.2]), is only a fairly good predictor of when the next glitch will occur, overall the glitch activity for the Vela pulsar as determined by Alpar et al. (1992) from their fit to observation is remarkably close to that observed ($\sum_{i=1}^{7} t_g \cong 8479$ d, while $\sum_{i=1}^{7} t_{obs}$ = 7239 d). Thus, on average and in accord with the proposal of Pines et al. (1980), the next glitch comes along when the healing from the previous glitch is essentially complete.
- The variation in the relative size of the superglitches, $\Delta\Omega_c/\Omega_c$, appears to result in part from a variation in $\delta\Omega$, the number of vortices which unpin in a glitch, and in part from a variation in I_B, which reflects the size of the regions through which vortices have passed without changing the net numbers of vortices present. For a representative value of $\delta\Omega/\Omega_c \sim 1.6 \times 10^{-4}$, the number of vortices which have unpinned in the glitch is $\sim 6 \times 10^{11}$. A significant part of the discrete spin-up in each glitch is simply part of the coupling of the crustal superfluid to the crust and, as such, does not heal, remaining as a permanent spin-up contribution of the internal torque against the overall spin-down of the pulsar.

It is natural to inquire about the physical location inside the neutron star of the pinning regions required by the fit of the phenomenological model to observation. An admittedly speculative response can be built on the toy model of a glitch shown in Fig. 4, which, when combined with the phenomenological theory of Alpar et al. (1984b), provides a framework for understanding the postglitch behavior of the Vela pulsar (Alpar et al., 1992). Thus, the regions I_1 and I_2, whose rapid response is the same from glitch to glitch, must be located in the passive region 1, through which no vortices move; these are plausibly regions in which the pinning is superweak. In this view, the ~ 33 d response is then to be associated with the transition from the passive to the active region, a transition which may well differ from glitch to glitch and be associated with a transition from superweak to weak pinning; hence the variation in I_3, which, it should be noted, is at times an order of magnitude smaller than I_1; I_B, and the nonlinear response described by $I_{n\ell}$ are then to be associated with the active region through which vortices move; this region of lower neutron density is plausibly one in which the pinning is weak.

Alpar, Cheng, and Pines (1989) (see also Alpar et al., 1992) have taken the further step of combining the results of the phenomenological analysis sketched above with the microscopic theory of crustal neutron superfluidity presented in

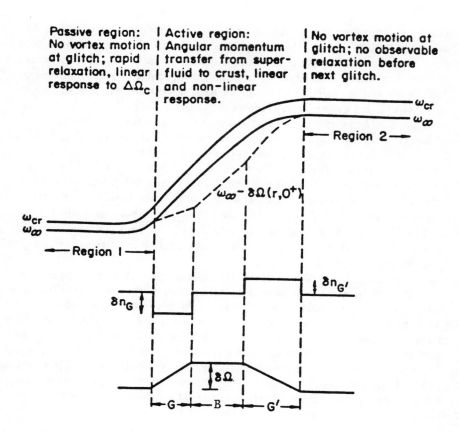

FIGURE 4 A toy model of a glitch, in which vortices unpin at the boundary of region 1 and repin at the boundary of region 2. The distance from the rotation axis increases to the right and the density to the left; regions 1 and 2 are "passive" regions, through which vortices do not move, and the response of the superfluid to a glitch is set by the change in $\Delta\Omega_c$. Based on the microscopic calculations of $\Delta(r)$, pinning energies increase to the right: In region 1, and possibly parts of the "active" region, GBG', the pinning of vortices is expected to be superweak; elsewhere it is expected to be weak. The corresponding relaxation times for vortex creep response are assumed to increase to the right, in such a way that the time scale for the response of region 2, if it contains superfluid neutrons, is longer than the time between glitches. The upper part of the figure gives the possible variation of ω_{cr} and the steady-state lag, ω_∞; the middle shows the change in vortex density at a glitch; the bottom part shows the change in superfluid velocity at the glitch.

section 3 in order to obtain limits on pinning energies and crustal angular velocities for the Vela and other pulsars. It is encouraging that in the weak-pinning region, the range of pinning energies they deduce (≤ 1 MeV) is consistent with those obtained by combining Eq. (3.7) with the microscopic energy gap calculations of Ainsworth, Wambach, and Pines (1989). Indeed a scenario in which the boundary between the passive and active regions of Fig. 4 corresponds to the boundary between superweak and weak pinning and is to be found near crustal densities $\sim 8 \times 10^{13}$ g cm^{-3} (where $\xi \sim 20$ fm, corresponding to $\Delta \sim 0.7$ MeV) is consistent with both existing microscopic calculations and observation. In this view, the active region of Fig. 4, through which vortices move, is a weak-pinning region which is found for crustal regions in which the density lies between 3×10^{13} g cm^{-3} and $\sim 8 \times 10^{13}$ g cm^{-3} (and might reasonably possess an inertial moment ratio $\geq 1.7 \times 10^{-2}$), while the passive region of Fig. 4 corresponds to crustal regions in which the density lies between $\sim 8 \times 10^{13}$ g cm^{-3} and $\sim 2.8 \times 10^{14}$ g cm^{-3} (and thus might reasonably possess a total inertial moment ratio in excess of 7×10^{-3}).

4.2 THE CRAB PULSAR AND PSR 0540−69

The model of pinned crustal superfluid response described above provides a simple physical explanation for the postglitch behavior of the Crab pulsar following its 1975 and 1989 glitches. Lyne and Pritchard (Lyne, private communication) have shown that an expression of the following form,

$$\frac{\Delta \dot{\Omega}_c(t)}{\dot{\Omega}_c} = \left[\frac{\Delta \dot{\Omega}_c(o)}{\dot{\Omega}_c}\right]_\ell \exp\left(-\frac{t}{\tau_\ell}\right) + \left(\frac{\Delta \dot{\Omega}_c}{\dot{\Omega}_c}\right)_{n\ell} \quad (4.8)$$

with $\tau_\ell = 14$ d, provides an excellent fit to the observational data (see Table 4) following both glitches; there is no observational evidence for a torque of the form of Eq. (4.5). In our model, the first term on the rhs of Eq. (4.7) describes the linear response of a pinning region of strength I_ℓ, and the second term on the rhs of Eq. (4.7) represents a long-term offset associated with the nonlinear response of a pinning region of strength $I_{n\ell}/I \equiv (\Delta\dot{\Omega}_c/\dot{\Omega}_c)_{n\ell}$. According to Eq. (4.4), the linear response must come from a region through which vortices have moved and is therefore

$$[\Delta\dot{\Omega}_c(0)]_\ell = \frac{I_\ell}{I}\frac{\delta\Omega}{\tau_\ell} \quad (4.9)$$

($\delta\omega_i \cong \Delta\Omega_c(0)$ is ruled out by the relative magnitudes of $\Delta\Omega_c(0)$ and $\Delta\dot{\Omega}_c(0)$.) Since I_ℓ corresponds to a region through which vortices have moved, the equation which expresses angular momentum conservation is

$$I_c\Delta\Omega_c(0) = (I_{n\ell} + I_\ell)\delta\Omega \quad (4.10)$$

The difference from the Vela pulsar is the absence in this angular momentum balance of any regions like I_B, which do not contribute to the change in torque,

Eq. (4.9). Eq. (4.10) may be combined with Eq. (4.8) to determine I_ℓ and $\delta\Omega$ for the two glitches. The results are displayed in Table 4, where one sees that while the glitch-induced change in $\delta\Omega$ is smaller for the 1989 glitch, a larger pinning region is involved. Thus, although fewer vortices are unpinned in this 1989 glitch, they travel farther. As is the case for the Vela pulsar, the change in the superfluid torque after a glitch is associated mainly with linear response.

TABLE 4 Crab Pulsar Glitch and Postglitch Parameters. The observational quantities $|\Delta\Omega_c(0)/\Omega_c|$, $(\Delta\dot{\Omega}_c/\dot{\Omega}_c)_\ell$ and $(\Delta\dot{\Omega}_c/\dot{\Omega}_c)_{n\ell} \equiv (I'_{n\ell}/I)$ are those determined by Lyne and Pritchard; the remaining quantities are derived using Eq. (4.4).

Glitch	$\left(\dfrac{\Delta\Omega_c(0)}{\Omega_c}\right)_{-8}$	$\left(\dfrac{\Delta\dot{\Omega}_\ell(0)}{\dot{\Omega}_c}\right)_{-3}$	$\left(\dfrac{I'_{n\ell}}{I}\right)_{-4}$	$\left(\dfrac{\delta\Omega}{\Omega}\right)_{-5}$	$\left(\dfrac{I_\ell}{I}\right)_{-4}$
2/3/75	3.7	1.9	2	3.3	9.2
8/29/89	7.4	4.1	4	2.1	31

If we assume that all glitching pulsars possess similar masses ($\sim 1.4 M_\odot$) it is reasonable to ask whether the glitch behavior of the Crab pulsar is what might be expected for a younger and hotter neutron star. Our conclusion is "perhaps." Consider first pinning regions and glitch sizes. Both $\delta\Omega$ (which reflects the number of vortices unpinned in a glitch) and the size of the regions through which vortices have moved are an order of magnitude smaller for the Crab pulsar than for Vela; hence the two orders of magnitude difference in glitch sizes. This is qualitatively what might be expected from the vortex pinning model; as a neutron star cools, a given set of pinning sites becomes more effective in holding vortices back. Next, one might ask whether the specific distinct pinning regions seen in one pulsar can also be seen in the other. Apparently not, since the expected difference in temperatures of the two pulsars is sufficiently large that, for example, the three regions with relaxation times that are clearly observable for the Vela pulsar would have, for the Crab pulsar, relaxation times that are much too short to be observable, whereas the 14-d region in the Crab pulsar may be expected to evolve into a region with a response time long compared with any seen for the Vela pulsar; that is, the I_ℓ region of Crab would become part of $I_{n\ell}$ or I_B in Vela.

A significant difference between the Crab and Vela pulsars is that in the Crab pulsar, since $\delta\Omega/|\dot{\Omega}| \sim 1$ month, the next glitch comes along well after the pulsar has spun down sufficiently for vortex motion in the superweak region to have restored the pulsar to its preglitch status. Put another way, the glitch activity of the Crab pulsar is some two orders of magnitude smaller than that observed for the Vela pulsar. These differences suggest that in a hot young pulsar, pinned vorticity plays little or no role in initiating a glitch; it simply responds to one. Indeed, for the Crab pulsar the spin-down rate is large enough, and the time between larger glitches sufficiently long, that the starquake scenario of Ruderman (1969), as refined by Baym and Pines (1971), might provide the physical origin of the glitches. It would seem, therefore, that as a pulsar evolves, somewhere between the ages of 10^3 yr and 10^4 yr pinned vorticity begins to play a significant role in the initiation of a glitch.

4.3 OTHER GLITCHING PULSARS

McKenna and Lyne (1990) have discussed the glitch activity of the ten youngest pulsars. As may be seen in Table 5, there is a clear dichotomy between the observed glitch activity of the three youngest pulsars and that observed for the next four. As we have seen, the overall agreement between the values predicted by Eq. (4.2) and observation makes vortex pinning a strong candidate for the physical process initiating a large glitch. As the table makes clear, to the extent that vortex pinning plays a role, glitches as large as those typically seen in the Vela pulsar may be expected only every decade or so for pulsars with spin-down ages comparable to those of PSR 1930−22, and at still greater intervals for pulsars with a larger spin-down age.

TABLE 5 Observed and Predicted Glitch Activity for the Ten Youngest Pulsars. The Observed Activity $[\Delta\Omega_c(0)/\Omega_c t_g]$ is taken from McKenna and Lyne (1990). The predicted values are those calculated from Eq. (4.3) with $I_p/I = 2 \times 10^{-2}$.

PSR	t_{sd} (10^3 yr)	Glitch activity Observed	(10^{-7} yr^{-1}) Predicted
Crab	1.2	0.06	80
1509−58	1.5	0	60
0540−69	1.7	0	60
Vela	11	8	9
1800−21	16	?	6
1737−30	20	4	5
1823−13	21	5	5
1930+22	40	0	2.5
2334+61	40	0	2.5
1727−47	80	0	1.3

As of this writing, despite the many glitches which have been observed in PSR 1737−30, observational coverage has not been sufficiently frequent to permit detailed fits to its postglitch behavior. If this pulsar possesses a mass similar to that of the Vela pulsar and was formed spinning at a similar initial angular velocity, observation of postglitch behavior will provide very useful information on the thermal evolution of vortex creep (increases in relaxation times, changes from linear to nonlinear response, etc.). The postglitch behavior of two considerably older pulsars, PSR 0355+54 and PSR 0525+21, appears to be consistent with the expected thermal evolution of vortex creep (Alpar, Cheng, and Pines, 1989) in that the former displays linear response with $\tau \sim 44$ d, while the latter pulsar appears to possess only nonlinear response with a relaxation time close to that predicted by vortex creep heating of the star (Alpar, Nandkumar, and Pines, 1985).

5. THE CORE HADRON SUPERFLUIDS

The calculation of the superfluid transition and superfluid properties of the interpenetrating neutron and proton quantum liquids in the core is difficult because many-body corrections, such as the induced interaction corrections discussed in Section 3 and the density-dependent effective mass of a quasi-particle near the Fermi surface, almost certainly play a significant role. Baldo et al. (1991) have very recently calculated the energy gap for both liquids, using realistic two nuclear potentials; they conclude, in agreement with pioneering variational calculations, that for densities $\geq \rho_0$ the neutron liquid will be superfluid, with pairing in the $^3P_2$ state, while the comparatively dilute proton liquid will be a superconductor, with pairing in the $^1S_0$ state. While their calculations take effective mass corrections into account, the extent to which many-body corrections will reduce their calculated energy gaps remains an open question. Such corrections are particularly important for the dilute proton liquid, since the induced proton-proton interaction resulting from the coupling between the proton and neutron liquids has been found capable of reducing the gap calculated using direct interactions only by a factor of 3-5 (Wambach, Ainsworth, and Pines, 1991). Thus while Baldo et al. (1991) find a density-dependent proton gap with a maximum ~ 1 MeV, Wambach et al. find a gap $\sim 0.3 \pm 0.1$ MeV which is nearly independent of density. Similarly, while agreeing on the sign of the effect, Wambach et al. find a reduction in the proton effective mass of $\delta m_p \sim -0.2 m_n$, which is comparatively weakly dependent on density, while Baldo et al. (1991) find a more density-dependent and somewhat larger reduction in the proton effective mass.

The matter is of more than academic interest. To cite but one example, as discussed by Alpar (1992) at this meeting, the coupling between the core neutron superfluid and the crust proceeds via a velocity-dependent coupling between the core neutron and proton superfluids and hence is proportional to $(\delta m_p)^2$ (Alpar, Langer, and Sauls, 1984); should δm_p be too small (and $\sim 0.2 m_n$ is on the borderline), the core neutron superfluid could not respond to a glitch within ~ 2 min, as the observations following the 1988 Vela glitch require. In this case, the latter observation would constrain the core proton superfluid (at least in the Vela pulsar) to be normal. If this is the case, then the question of whether neutron star spin-down is responsible for magnetic field decay by sweeping the proton flux tubes from the core (Srinivasan et al. 1990; see Ruderman, 1992, for a discussion) would be moot.

6. CONCLUDING REMARKS

While considerable progress has been made in understanding the crustal neutron superfluid, and in relating its behavior to observation at a phenomenological level, there remain a number of open questions concerning glitches, vortices, and crustal behavior. Among these are the following:

- What is the physical state which initiates a giant, Vela-size glitch? A crustquake (Ruderman, 1976, 1992)? Vortex unpinning at a transition from a region of vortex accumulation to one of vortex depletion (Cheng et al., 1989)?
- Can the superfluid rotational energy transferred suddenly to the crust in a Vela-size glitch ($\sim I_p \delta\Omega^2 \sim 5 \times 10^{39}$ ergs) be transferred sufficiently rapidly to the crust (via the mechanisms discussed here by Blandford [1992]) to give rise to an observable transient event at the stellar surface?
- How does an individual vortex line unpin (Epstein, Link, and Baym, 1992)?
- Once unpinned, does a vortex line repin readily at a nearby set of pinning sites, or does it move all the way to the boundary of the superfluid region?
- In the course of the combined spin, thermal, and magnetic evolution of a pulsar, which may be strongly influenced by the behavior of the core superfluids (Ruderman, 1992), do instabilities naturally arise (Shibazaki, 1992)?

In considering areas for future research, it is clear that much more work needs to be done on the microscopic description of vortex motion in the crust after a glitch or between glitches. The pioneering work of Shaham (1980) and the more recent calculations described by Epstein at this meeting (Epstein, Link, and Baym, 1992) represent useful beginnings, but at present there are few links between this body of work and observation. This is not surprising, since developing a consistent account of the dynamics of pinned vortices moving in a highly inhomogenous medium is a truly difficult problem, one which is almost certain to be with us for the indefinite future. Indeed, as one of us has noted elsewhere, given the many simplifications made in developing and applying vortex creep theory and the difficulty of carrying out microscopic calculations of energy gaps, pinning energies, and other properties of neutron matter, the extent to which present microscopic calculations of the consequences of hadron interaction are consistent with glitch observations is both pleasing and surprising (Pines, 1991).

We have learned a great deal about neutron stars from observation during the past two decades, enough indeed that we now can begin to use pulsars as cosmic laboratories for studying the behavior of hadron matter under conditions inaccessible in the laboratory. Given the planned program in X-ray astronomy (Tanaka, 1992) and the radio pulsar synoptic observations planned in conjunction with *GRO*, there is every reason to expect that this will continue to be the case. Indeed one might hope that during the next decade observations, as well as further theoretical calculations, will enable us to answer some of the many questions which have been raised here concerning the structure and evolution of the core of a neutron star (do the protons superconduct? is there a pion condensate? how does the magnetic field trapped in the core evolve and to what extent has this evolution influenced crustal behavior? etc.). Thus it is not unreasonable to expect that these and other questions will be on the scientific agenda of a meeting that would help mark the millennium, SENS 2000.

ACKNOWLEDGMENTS

This work was partially supported by a NATO Collaborative Research Grant and by NSF grants INT89-18665 and NSF PHY86-00377. We should like to thank R. Blandford, H. F. Chau, and K. S. Cheng for stimulating discussions and correspondence, and Andrew Lyne for communicating to us in advance of publication the results of his analysis with Pritchard of the timing behavior of the Crab pulsar.

REFERENCES

Ainsworth, T., Wambach, J., and Pines, D., *Phys. Lett. B* 228, 173 (1989).

Alpar, M. A., in *Neutron Stars: Theory and Observation*, eds. D. Pines and J. Ventura, Kluwer Acad. Pub. (1991).

Alpar, M. A., this volume (1992).

Alpar, M. A., Anderson, P. W., Pines, D., and Shaham, J., *Ap. J* 276, 325 (1984a).

Alpar, M. A., Anderson, P. W., Pines, D., and Shaham, J., *Ap. J* 278, 791 (1984b).

Alpar, M. A., Chau, H. F., Cheng, K. S., and Pines, D., in preparation (1992).

Alpar, M. A., Cheng, K. S., and Pines, D., *Ap. J.* 346, 823 (1989).

Alpar, M. A., Nandkumar, R., and Pines, D., *Ap. J.* 288, 291 (1985).

Alpar, M. A., Brinkmann, W., Kiziloglu, Ü., Ögelman, H., and Pines, D., *Astr. Ap.* 177, 101 (1987).

Alpar, M. A., Langer, S. A., and Sauls, J. A., *Ap. J.* 282, 533 (1984).

Alpar, M. A., and Pines, D., in *Timing Neutron Stars*, eds. H. Ögelman and E. P. J. van den Heuvel, Kluwer Acad. Pub. (1989).

Anderson, P. W., and Kim, Y., *Rev. Mod. Phys.* 36, 39 (1964).

Anderson, P. W., and Itoh, N., *Nature* 256, 25 (1975).

Anderson, P. W., Alpar, M. A., Pines, D., and Shaham, J. *Phil. Mag. A* **45**, 227 (1982).
Baldo, M., Cugnon, J., Lejeune, A., and Lombardo, U., preprint, (1991).
Baym, G., Pethick, C. J., Pines, D., and Ruderman, M., *Nature* **224**, 872 (1969).
Baym, G., and Pines, D., *Annals of Physics* **66**, 816 (1971).
Baym, G., in *Neutron Stars: Theory and Observation*, eds. D. Pines and J. Ventura, Kluwer Acad. Pub. (1991).
Blandford, R., this volume (1992).
Cheng, K. S., Alpar, M. A., Pines, D., and Shaham, J., *Ap. J.* **330**, 835 (1988).
Clark, J., Davé, R., and Chen, J., this volume (1992).
Epstein, R., Link, B., and Baym G., this volume (1992).
Epstein, R., and Baym, G., *Ap. J.* **328**, 680 (1989).
Feibelman, P., *Phys. Rev. D* **4**, 1589 (1971).
Flanagan, C., *Nature* **345**, 416 (1990).
Manchester, R. N., this volume (1992).
McCullough, P. M., Hamilton, P. A., McConnell, D., and King, E. A., *Nature* **346**, 822 (1990).
McKenna, J., and Lyne, A., *Nature* **343**, 349 (1990).
Pandharipande, V. R., Pines, D., and Smith, R. A., *Ap. J.* **208**, 550 (1976).
Pines, D., in *Proc. 12th International Conference on Low Temperature Physics*, ed. E. Kondo, Kaigeku, Tokyo, 7 (1971).
Pines, D., in *Neutron Stars: Theory and Observation*, eds. D. Pines and J. Ventura, Kluwer Acad. Pub. (1991).
Pines, D., and Alpar, M. A., *Nature* **316**, 27 (1985).
Pines, D., Shaham, J., and Ruderman, M., *Nature Phys. Sci.* **237**, 83.
Pines, D., Shaham, J., Alpar, M. A., and Anderson, P. W., *Prog. Th. Phys. Supp.* **69**, 376 (1980).
Ruderman, M., *Nature* **223**, 597 (1969).
Ruderman, M., this volume (1992).
Ruderman, M., *Ap. J.* **203**, 213 (1976).
Shaham, J., *J. de Phys.* **41**, C2-9 (1980).
Shibazaki, N., this volume, (1992).
Shibazaki, N., and Lamb, F. K., *Ap. J.* **346**, 808 (1989).
Srinivasan, G., Bhattacharya, D., Muslimov, A. G., and Tsygan, A. I., *Curr. Sci.* **59**, 31 (1990).
Tanaka, Y., this volume, (1992).
Tsuruta, S., this volume, (1992).
Wambach, J., Ainsworth, T., and Pines, D. in *Neutron Stars: Theory and Observation*, eds. D. Pines and J. Ventura, Kluwer Acad. Pub. (1991).
Wiringa, R., Fiks, V., and Fabrocini, A., *Phys. Rev. C* **38**, 1010 (1989).

R. N. Manchester
Australia Telescope National Facility, CSIRO, P.O. Box 76, Epping, NSW, Australia

Radio Pulsar Timing

Precise timing observations of radio pulsars have revealed a wealth of information concerning the electrodynamics of pulsar magnetospheres, the space motions of both single and binary pulsars, and the superfluid interiors of neutron stars. Following a review of the basic timing parameters for the 500 or so known pulsars, more detailed observations revealing higher order terms in the timing data for several pulsars are described. Notable among these are the relativistic effects detected in the binary pulsar PSR 1913+16. The final section describes the sudden glitches in period observed in many pulsars and the subsequent variations in period as the pulsar recovers from the glitch.

1. INTRODUCTION

There are now over 500 pulsars known, all but a few located within our Galaxy. The interpretation of pulsars as rotating neutron stars has considerable observational support and (almost) universal acceptance. Their most remarkable and distinctive property is the extremely high stability of the basic pulsar period. This great stability allows investigation of many important and interesting effects which explore physical conditions not otherwise accessible to observation. These include the extremely strong magnetic and electric fields within the magnetospheres of pulsars which result in the braking of pulsar rotation and, of course, the emission of the radiation beams which we observe as pulses as the star rotates. Motions of binary pulsars probe the strong gravitational fields in the vicinity of neutron stars. These observations have provided accurate determinations of neutron star masses, produced the first observational evidence for gravitational radiation, and allowed tests of theories of relativity. Pulsar periods are not always predictable however. Random irregularities in period occur on various time scales but most strikingly as sudden glitches, or discontinuous decreases in the pulse period. Such glitches have been observed in many, mostly relatively young, pulsars. While their overall character is the same in different pulsars, there are distinct differences in the postglitch recovery, which can take weeks, months, or even many years. These phenomena are attributed to activity in the interior of the neutron star, with the long time scales indicating involvement of superfluid components.

2. BASIC TIMING PARAMETERS

The known pulsars fall into two main groups. Most pulsars are relatively young by astrophysical standards with ages of up to about 10^7 years. The youngest are associated with supernova remnants confirming the suggestion, first made by Baade and Zwicky,[1] that neutron stars are formed in supernova explosions. There is another group of pulsars which appear to be much older, 10^8 or 10^9 years, which are characterized by short periods, often of only a few milliseconds. In sharp contrast to the young pulsars, a high proportion of these pulsars, nearly half, have a binary companion. Some are found within the general Galactic population, but most are members of globular clusters. These observations strongly suggest that these pulsars have been "recycled," that is, that they were old, possibly dead, pulsars which have been spun up and given a new lease on life as a result of accretion of material from a binary companion. The distribution of known pulsars among these various categories is summarized in Fig. 1.

Periods of known pulsars range between 1.55 milliseconds for PSR 1937+21, the first millisecond pulsar discovered, and 4.31 seconds. Most of the 500 known pulsars have periods of between 0.25 second and 1 second but the millisecond pulsars form a distinct tail to the distribution, as shown in Fig. 2. Recyling of pulsars is believed to take place in low-mass X-ray binary systems.[2] Motivated by the relative overabundance of such systems in globular clusters, searches for millisecond pulsars in globular clusters began in the mid-1980s and have been very successful in recent years. Fig. 2 shows that most of the pulsars found in such searches[3,4,5] have periods in the millisecond range. Most pulsar surveys have been conducted at relatively low radio frequencies, often about 400 MHz. These searches do not penetrate to the central regions of the Galaxy, largely because of the effects of scattering and

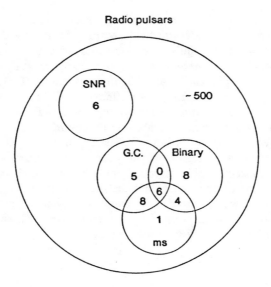

FIGURE 1 Venn diagram illustrating the main categories of currently known pulsars. The binary–millisecond–globular cluster group is believed to consist of recycled pulsars (old pulsars which have been reactivated by spin-up as members of accreting binary systems.)

dispersion by interstellar electrons. A recent survey of the southern Galactic plane at 1.5 GHz by Johnston et al.[6] overcame many of these problems and detected 46 pulsars, most of which were distant and of high luminosity. These pulsars have a shorter median period than that of previously known pulsars.

One of the most basic diagrams in pulsar astronomy is the plot of period derivative versus period shown in Fig. 3. At least in their own frame and on long time scales, all pulsars are slowing down. The basic energy source of pulsars, rotational kinetic energy, is steadily lost, mainly in the form of electromagnetic waves at the rotation frequency and/or energetic particles; the beamed high-frequency electromagnetic waves we observe as radio pulses form a tiny fraction of the total energy-loss budget. The ratio $P/\dot{P}$ is an indicator of pulsar age and is small for pulsars in the top-left corner. The named pulsars in this region are those which have convincing associations with supernova remnants; all are less than or about 20,000 years old. In contrast, all millisecond and binary pulsars are old with $P/\dot{P}$ ages of 10^9 years or more.

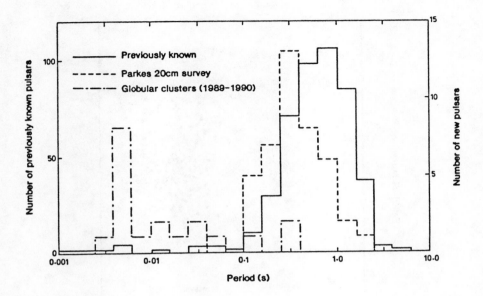

FIGURE 2 Period distribution for known pulsars. The solid line (left-hand axis) is for all known pulsars except those found in the recent high-frequency survey of the southern Galactic plane by Johnston et al.[6] (dashed line, right-hand axis) and those discovered within the last 2 years in globular clusters (dot-dashed line, right-hand axis).

3. HIGHER ORDER TIMING TERMS

For a dipolar magnetic field, the braking relation is

$$\dot{\nu} = -K\nu^n$$

where $\nu = 1/P$, K is a positive constant proportional to the square of the magnetic dipole moment, and n, the braking index, has the value 3. From this relation, the magnetic flux density at the surface of the neutron star, B_0, is proportional to $(P\dot{P})^{1/2}$; for typical pulsars $B_0 \sim 10^{12}$ gauss. Therefore, for constant magnetic field, young pulsars evolve down lines of constant $P\dot{P}$ into the region around $P = 1$ s, $\dot{P} = 10^{-15}$, where most observed pulsars lie. Clearly the millisecond pulsars cannot have been formed by simple aging of young pulsars. They have both much shorter periods and much lower magnetic field strengths than old normal pulsars. The spinning up almost certainly occurs in an accreting binary system; it is possible that the magnetic field decay is also a result of the accretion process.[7]

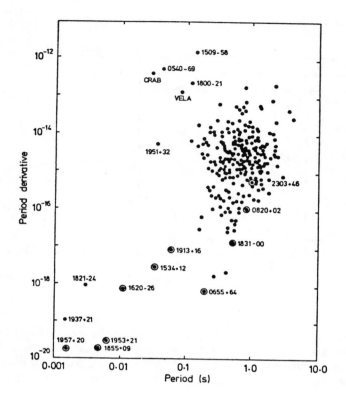

FIGURE 3 Plot of the first time derivative of pulsar period versus the pulsar period. Named pulsars in the upper-left corner are those associated with supernova remnants; and binary pulsars are indicated by ringed points.

The braking index is an important diagnostic of the energy-loss processes and can be computed from the second time derivative of the rotation frequency

$$n = \ddot{\nu}\nu/\dot{\nu}^2.$$

So far the braking index has been measured for three pulsars, the Crab,[8,9] PSR 1509–58[10] and PSR 0540–69.[11,12] The measured values are all less than 3, the value expected for braking by magnetic-dipole radiation or by acceleration of a stellar wind by a dipole field on the rotating star, and are 2.509 ± 0.001, 2.83 ± 0.03 and 2.02 ± 0.02 respectively. For PSR 0540–69, Ögelmann and Hasinger[13] have reported a different value, 2.74 ± 0.10, based on *EXOSAT* data, suggesting that the period of this pulsar may be somewhat irregular. The fact that all measured

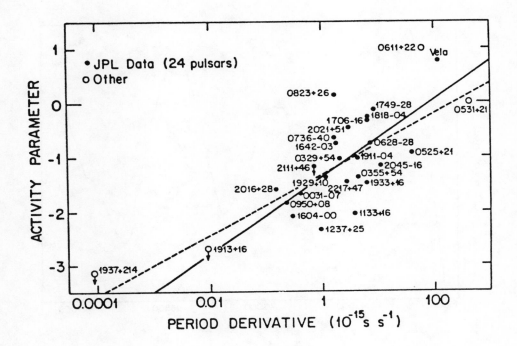

FIGURE 4 Plot of "activity" parameter (see text) versus period derivative for a sample of 28 pulsars.[14] The lines are fits to the filled points including (*solid*) and excluding (*dashed*) the Vela pulsar.

values are less than 3 indicates that magnetic-dipole radiation is not the dominant energy-loss process and that the pulsar magnetosphere is significantly affected by acceleration of a stellar wind that is most probably highly relativistic.

By ordinary standards, pulsar periods are extraordinarily stable. However, precise timing measurements made over long intervals show that most pulsars have small, quasi-continuous and unpredictable variations in period. The strength of these random irregularities relative to those of the Crab pulsar can be described by the activity parameter

$$A = \log(\sigma/\sigma_{Crab})$$

where σ is the rms timing residual from a second-order timing fit.[14] Fig. 4 is a plot of measured values of A showing that there is some correlation between activity parameter and period derivative. Young pulsars with high period derivatives such as the Crab and Vela pulsars tend to have strong period noise, whereas the recycled pulsars, which have very small period derivatives, have extraordinarily stable periods. Indeed, for 1937+21, the observed residuals can be attributed to uncertainties

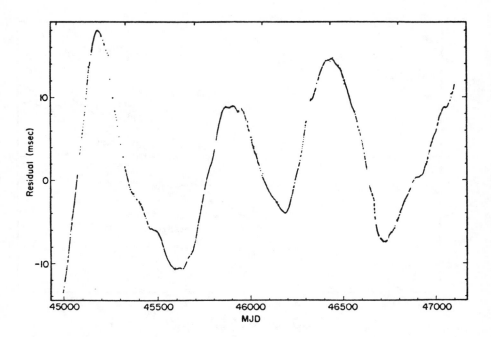

FIGURE 5 Phase residuals from a fit of a cubic timing model to data for the Crab pulsar showing quasi-periodic residuals[9]

in the standards of terrestrial time.[15] This opens up the possibility that millisecond pulsars may be used to *define* terrestrial time, at least for time intervals greater than a year or so. For the Crab pulsar, timing measurements made at Jodrell Bank over a 6-year interval[9] show that phase residuals from a cubic timing solution have a quasi-periodicity with period of about 20 months (Fig. 5). The origin of this quasi-periodicity is unclear but it may result from some long-period oscillation in the neutron superfluid in the interior of the neutron star.

The great period stability and very narrow pulses of millisecond pulsars allow many interesting effects to be measured. An example, shown in Fig. 6, is the measurement of the proper motion of PSR 1855+09.[16] For a fit which assumes a fixed position for the pulsar, the change in position introduces a sinusoidal oscillation of period 1 year and of linearly growing amplitude into the residuals. In this case the derived proper motion is 6.0 ± 0.2 milliarcseconds/year at a position angle of 210deg. Similar proper motion measurements have been made for a number of pulsars, but, in some cases, interferometric measurements[17] have shown that random period irregularities have dominated the proper-motion contribution to the residuals.

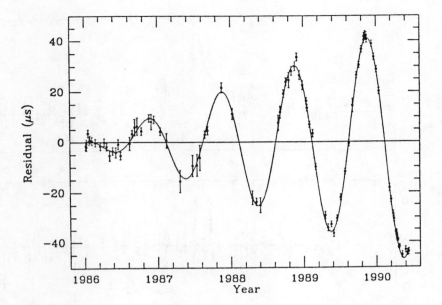

FIGURE 6 Timing residuals resulting from proper motion for PSR 1855+09.[16] The growing sine wave has a period of 1 year and results from errors in the correction for the Earth's motion about the Sun as the pulsar moves away from its assumed position.

4. BINARY PULSARS

As is now well known, precision timing measurements of the binary pulsar PSR 1913+16 by J. H. Taylor and his collaborators have been used to provide precise measurements of neutron star masses, the first observational evidence for gravitational radiation and the first confirmation that the theory of General Relativity describes the motions of masses in strong gravitational fields.[18,19] The relativistic effects with observable consequences are (to date) precession of the longitude of periastron, variations in the gravitational redshift and second-order Doppler effect as the pulsar moves around its elliptical orbit, decay of the orbit resulting from reaction to the emission of quadrupole gravitational radiation by the binary system and, finally, the Shapiro delay resulting from distortion of the ray path as the received pulsar signals pass near the pulsar companion. These effects all depend in different ways on the pulsar mass, the companion mass and the inclination of the orbit with respect to the observer's line of sight. Consequently, not only is it possible to determine these quantities, but the consistency of the model used to predict them, in this case the motion of two point masses under General Relativity, may

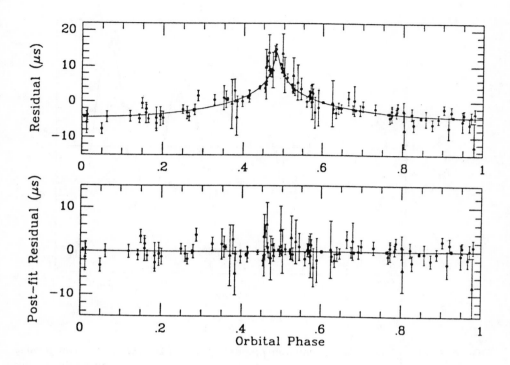

FIGURE 7 Phase residuals for PSR 1855+09 before (*upper*) and after (*lower*) correcting for the Shapiro delay of the pulsar signal as it passes close to the companion star[16]

also be checked. For PSR 1913+16, the derived value for the pulsar mass is 1.442 ± 0.003 solar masses, the companion mass is 1.386 ± 0.003 solar masses and the orbital inclination is about 45deg. The observed decay of the orbital motion is in accord with the prediction of General Relativity to better than 0.8% after allowance is made for the acceleration of the pulsar and the Earth toward the center of the Galaxy.[19]

Because of the orbital inclination for PSR 1913+16, the Shapiro delay is difficult to measure. However, for PSR 1855+09, which is in an orbit of period about 12 days, the inclination is close to 90 deg and so the Shapiro delay is much larger. From the observed delay, shown in Fig. 7, the pulsar companion, which is believed to be an old white dwarf,[20] can be shown to have a mass of 0.23 ± 0.02 solar masses.[16]

Two recently discovered binary pulsars have interesting implications for the evolution of pulsars, if not for tests of General Relativity. The orbit of PSR 1820−11 has the highest eccentricity known, 0.795, and the companion is almost certainly a neutron star[21]. Its long orbital period, just over 1 year, shows that the progenitor binary systems for such pulsars may also be of relatively long orbital period. The recently discovered eclipsing pulsar PSR 1744-24A in the globular cluster Terzan 5 has the very short orbital period of only 1.7 hours and provides a dramatic illustration of the variability of winds emanating from low-mass companions of neutron stars.[5]

5. PULSAR GLITCHES

Pulsar glitches—small, discontinuous and unpredictable decreases in the pulsar period—are believed to originate in the interior of neutron stars and hence are probably, within radio pulsar timing, the subject of most relevance to this symposium. Since the observation of the first such event in the Vela pulsar,[22,23] many glitches, both large and small, have been observed in many pulsars. Most, but not all, of these pulsars are young. Table 1 lists glitch parameters for the four pulsars which have been observed to glitch more than once.

To date, a total of eight major, or "giant" glitches and one minor glitch have been observed in the period of the Vela pulsar. Fig. 8 shows the observed variations in period for this pulsar over the past 21 years. The spacing of the giant glitches is variable, with the intervals ranging from 303 days to 4.1 years. The glitch size $\Delta\nu/\nu$ is also variable but in all cases is greater than a part in 10^6; the one observed minor glitch is two orders of magnitude smaller.

The University of Tasmania group has been monitoring the Vela pulsar for 18 hours a day for several years in order to catch the pulsar in the act of glitching. They were finally successful on December 24, 1988. Fig. 9a shows the residual phase, from a model fitted to the preglitch data, for a few hours around the glitch.[31] At the time of the glitch, the slope of the phase changed abruptly—the timescale of the glitch is less than 2 minutes. There were no prior irregularities and, within the precision of the observations and on time scales of up to an hour or so, no instability in the period after the glitch.

Although glitches are abrupt, significant variations in the period are observed following a glitch. The rotation frequency and its derivatives tend to recover toward their preglitch values, although as we shall see shortly, details of this recovery differ greatly from pulsar to pulsar. The long time scales of the recovery—days to years —imply that the neutron star contains significant inertia which is very weakly coupled to the crust. (The observed pulses are locked to the crust rotation by the very strong stellar magnetic fields.) Models for the postglitch recovery (e.g. Ref. 33) assume that the weakly coupled component consists of superfluid neutrons in the interior of the neutron sar.

Details of the postglitch recovery for the Vela Christmas 1988 glitch are shown in Fig. 9b. At the time of the glitch, the frequency derivative increased in magnitude by about 0.8%, but most of this increase quickly decayed with an exponential time constant of 4.64 days. This short exponential decay accounted for only about 0.5%

TABLE 1 Principal Observed Pulsar Glitches

PSR	Date	MJD	$\Delta\nu/\nu$ (10^{-9})	$\Delta\dot{\nu}/\dot{\nu}$ (10^{-3})	τ_d (days)	Ref.
0355+54	85Jan14	46079 ± 7	5.56 ± 0.03	1.8 ± 0.2		24
	86Feb07	46468 ± 35	4375 ± 10	105 ± 45	44	24
0531+21 (Crab)	69Sep29	40493.6 ± 0.9	9 ± 4	1.6 ± 0.9	4.8 ± 2.0	25
	75Feb03	42446.9 ± 0.1	37.2 ± 0.8	2.1 ± 0.2	15.5 ± 1.2	26
	86Aug22	46664.4	9.2 ± 0.1	2.5 ± 0.2	2.5 ± 0.5	27
	89Aug29	47767	67 ± 2	3.3 ± 0.1	15	28
0833−45 (Vela)	69Feb28	40280 ± 4	2340 ± 4	10.8 ± 0.1		29
	71Aug29	41192 ± 7	2059 ± 15	18 ± 7		29
	71Dec27	41312 ± 4	13 ± 1	2.2 ± 0.7		29
	75Sep28	42683 ± 3	1979 ± 4	11.0 ± 0.4		29
	78Jul03	43692 ± 11	3092 ± 32	23 ± 6		29
	81Oct11	44888.4 ± 0.4	1137.9 ± 0.4	8.52 ± 0.03	1.6 ± 0.2	29,30
	82Aug10	45191.6 ± 0.5	2051 ± 1	23.7 ± 0.5	3.2 ± 0.5	29,30
	85Jul12	46258 ± 1	1346 ± 1		6.5 ± 0.5	30
	88Dec24	47519.8036	1805	77	4.64	31
1737−30	87Jul27	47003 ± 50	420 ± 20	2.8 ± 0.8		32
	88Apr30	47281 ± 2	33 ± 5	1.7 ± 4.0		32
	88Jun20	47332 ± 16	7 ± 5	−1 ± 12		32
	88Oct24	47458 ± 2	30 ± 8	0 ± 4		32
	89May24	47670.2 ± 0.2	600.9 ± 0.6	2.0 ± 0.4		32

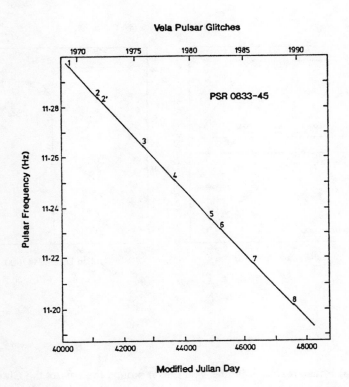

FIGURE 8 Variations in the period of the Vela pulsar over 21 years of observation. The average interval between giant glitches is about 2.9 years. Adapted from Cordes et al.[29]

of the glitch increment in frequency however. Flanagan[34] has found evidence for a more rapid component to the initial decay with a time constant of only 0.4 days. As shown in Fig. 9b, about 10% of the initial increment in derivative remained after these initial decays. McCulloch et al.[31] find that most of this decays away on a timescale of about 350 days, but even this decay accounts for only about 17% of the initial frequency glitch. On even longer time scales, Cordes et al.[29] find, for five of the previous Vela glitches, a linear recovery of $\dot{\nu}$ until the next glitch. Superimposed on the initial phase of the postglitch recovery, McCulloch et al. also found evidence for a damped quasi-sinusoidal oscillation in the rotation frequency with period of about 25 days, initial amplitude $\Delta\nu/\nu$ of about 6×10^{-10} and damping time of about 50 days. Clearly, the postglitch behavior of this pulsar is very complex.

The largest glitch observed to date, about 30% larger than the largest of the Vela glitches, was for the pulsar PSR 0355+54 (Table 1).[24] At the time of the glitch, the frequency derivative increased by 10%, much larger than the increments observed for the Vela glitches, but essentially all of it decayed away with a time

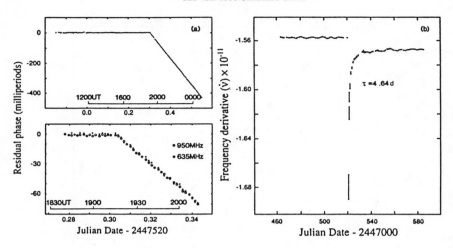

FIGURE 9 (a) Phase residuals for the Vela pulsar around the time of the Christmas 1988 glitch.[31] The residuals are with respect to a model fitted to the preglitch data. (b) Plot of the first time derivative of the pulsar rotation frequency from fits to running 8-day blocks of data.

constant of 44 days, leaving 99.9% of the initial frequency jump intact. Any further recovery of the frequency has a time constant in excess of 1000 years, implying an almost complete decoupling of the superfluid neutrons from the crust in this older pulsar.

The title of Champion Glitcher, however, must go to PSR 1737−30, a pulsar discovered recently at Jodrell Bank. This pulsar has large glitches with $\Delta\nu/\nu$ up to 6×10^{-7} and an average interval between glitches of only 220 days.[32] As shown in Fig. 10, both the amplitude of the glitches and the interval between them are variable. The pulsar has a period of 0.606 seconds but a very large period derivative of 4.6×10^{-13}, so it is relatively young. Over the interval shown in Fig. 10, the period derivative term added about 43 microseconds to the period, so, as for Vela, the jumps are small compared to the steady increase in period.

By the scale of the glitches described above, those in the Crab pulsar are small. Until recently, the largest was the 1975 glitch[26] which had an amplitude of 3.7×10^{-8} (Table 1). However, a glitch on 29 August, 1989, detected by Lyne et al.[28] was nearly twice as large as the 1975 glitch. The postglitch recovery from this 1989 event is

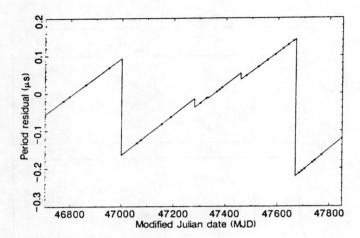

FIGURE 10 Residual period variations for the pulsar PSR 1737−30 *after* a fit for period and period derivative to the whole data set[32]

shown in Fig. 11. Most, but not all, of the increment in frequency derivative decayed exponentially with a time constant of about 15 days. The remaining $\Delta\dot{\nu}/\dot{\nu}$ of about 3×10^{-4} has persisted, resulting in the frequency actually dropping below the extrapolated prejump value 55 days after the glitch. This is not consistent with existing models for postglitch behavior.

Although glitches in all pulsars are the same to the extent that the rotation frequency and the magnitude of the frequency derivative increase at the time of the glitch, the postglitch behavior differs considerably from one pulsar to another. In a given pulsar, however, although successive glitches may differ greatly in magnitude, the character of the postglitch behavior remains the same. The different behaviors observed are summarized in Fig. 12. For the three pulsars shown, most of the increase in frequency derivative decays away on a timescale of weeks or months, but the fraction of the frequency jump accounted for during this period is large for the Crab, small for Vela, and tiny for PSR 0355+54. The Crab pulsar exhibits a persistent change in frequency derivative, the Vela pulsar has a second (or third) exponential term of longer time constant plus an apparently persistent change in frequency second derivative, and PSR 0355+54 has an apparently permanent change in rotation frequency.

The reasons for these differences are not clear. Age of the pulsar is apparently significant. Glitches in old pulsars have the same character as those in PSR 0355+54—a relatively old pulsar. McKenna and Lyne[32] have pointed out that glitch activity, defined to be the mean fractional change per year in period (or frequency),

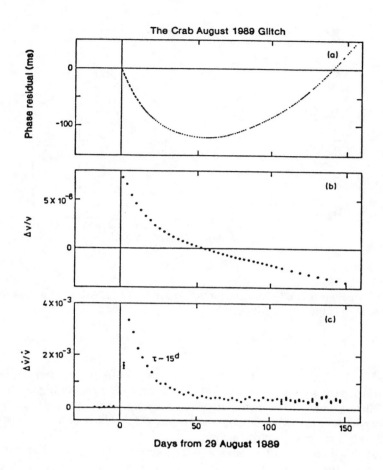

FIGURE 11 Variations of Crab pulsar in (a) phase, (b) frequency, and (c) frequency derivative, relative to a model of preglitch data, following the 29 August 1989 glitch[28]

summed over glitches, is greatest for pulsars of age between 10,000 and 20,000 years. Young pulsars such as the Crab have frequent glitches, but they are small, whereas old pulsars tend to have large but very infrequent glitches. It seems unlikely that a single mechanism can account for the diverse postglitch behavior observed. The range of behaviors illustrated in Fig. 12 may result from differences in the relative importance of different mechanisms in different pulsars. It remains a challenge for both observers and theoreticians to explore and explain the rich source of information provided to us by the glitching phenomenon.

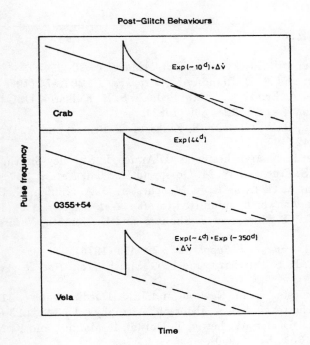

FIGURE 12 Schematic diagram illustrating the different postglitch behaviors of the rotation frequency for the Crab pulsar, PSR 0355+54, and the Vela pulsar. The relative scale of the frequency variations has been altered for clarity.

ACKNOWLEDGMENTS

I thank R. D. Blandford for suggesting the Venn diagram presentation of Fig. 1 to me. The Australia Telescope National Facility is operated in association with the Division of Radiophysics by CSIRO.

REFERENCES

1. W. Baade and F. Zwicky, *Proc. Nat. Acad. Sci.*, **20**, 254 (1934).
2. L. L. Smarr and R. D. Blandford, *Astrophys. J.*, **207**, 574 (1976).
3. A. G. Lyne, A. Brinklow, J. Middleditch, S. R. Kulkarni, D. C. Backer, and T. R. Clifton, Nature, **328**, 399 (1987).
4. S. Anderson, P. W. Gorham, S. Kulkarni, T. Prince, and A. Wolszczan, *Nature*, **346**, 42 (1990).
5. A. G. Lyne, R. N. Manchester, N. D'Amico, L. Staveley-Smith, S. Johnston, J. Lim, A. S. Fruchter, W. M. Goss, and D. Frail, *Nature*, **347**, 650 (1990).
6. S. Johnston, A. G. Lyne, R. N. Manchester, D. A. Kniffen, N. D'Amico, J. Lim and M. Ashworth, submitted to *Mon. Not. R. Astr. Soc.* (1991).
7. N. Shibazaki, T. Murakami, J. Shaham, and K. Nomoto, *Nature*, **342**, 656 (1989).
8. E. Groth, *Astrophys. J. Suppl. Ser.*, **29**, 453 (1975).
9. A. G. Lyne, R. S. Pritchard, and F. G. Smith, *Mon. Not. R. Astr. Soc.*, **233**, 667 (1988).
10. R. N. Manchester, L. M. Newton, and J. M. Durdin, *Nature*, **313**, 374 (1985).
11. R. N. Manchester and B. A. Peterson, *Astrophys. J.*, **342**, L23 (1989).
12. F. Nagase, J. Deeter, W. Lewis, T. Dotani, F. Makino, and K. Mitsuda, *Astrophys. J.*, **351**, L13 (1990).
13. H. Ögelmann and G. Hasinger, *Astrophys. J.*, **353**, L21 (1990).
14. J. M. Cordes and G. S. Downs, *Astrophys. J. Suppl. Ser.*, **59**, 343 (1985).
15. L. A. Rawley, J. H. Taylor, M. M. Davis and D. W. Allen, *Science*, **238**, 761 (1987).
16. M. Ryba and J. H. Taylor, *Astrophys. J.*, **371**, 739 (1990).
17. A. G. Lyne, B. Anderson, and M. J. Salter, *Mon. Not. R. Astr. Soc.*, **201**, 503 (1982).
18. J. H. Taylor and J. M. Weisberg, *Astrophys. J.*, **345**, 434 (1989).
19. T. Damour and J. H. Taylor, *Astrophys. J.*, **366**, 501, (1991).
20. S. R. Kulkarni, S. Djorgovski and A. R. Klemola, *Astrophys. J.*, **367**, 221 (1991).
21. A. G. Lyne and J. McKenna, *Nature*, **340**, 367 (1989).
22. V. Radhakrishnan and R. N. Manchester, *Nature*, **222**, 228 (1969).
23. P. Reichley and G. S. Downs, *Nature*, **222**, 229 (1969).
24. A. G. Lyne, *Nature*, **326**, 569 (1987).
25. P. E. Boynton, E. J. Groth, D. P. Hutchinson, G. P. Nanos, R. B. Partridge, and D. T. Wilkinson, *Astrophys. J.*, **173**, 217 (1972).
26. E. Lohsen, *Nature*, **258**, 688 (1975).
27. A. G. Lyne and R. S. Pritchard, *Mon. Not. R. Astr. Soc.*, **229**, 223 (1987).
28. A. G. Lyne, R. S. Pritchard, and F. G. Smith, in preparation (1991).
29. J. M. Cordes, G. S. Downs, and J. Krause-Polstorff, *Astrophys. J.*, **330**, 847 (1988).

30. P. M. McCulloch, A. R. Klekociuk, P. A. Hamilton, and G. W. R. Royle, *Aust. J. Phys.*, bf 40, 725 (1987).
31. P. M. McCulloch, P. A. Hamilton, D. McConnell and E. A. King, *Nature*, **346**, 822 (1990).
32. J. McKenna and A. G. Lyne, *Nature*, **343**, 349 (1990).
33. M. A. Alpar, P. W. Anderson, D. Pines, and J. Shaham, *Astrophys. J.*, **276**, 325 (1984).
34. C. S. Flanagan, *Nature*, **345**, 416 (1990).

James M. Lattimer
Department of Earth and Space Sciences, State University of New York, Stony Brook, NY 11794 USA

The Equation of State in Neutron Star Matter

Some effects of the equation of state of dense matter on the structure and early evolution of neutron stars are discussed. Focus is first placed on parameters of the nuclear equation of state and constraints on them that have been obtained from laboratory experiments. In turn, it is discussed how neutron stars and supernovae depend upon and could alternatively constrain these parameters.

1. INTRODUCTION

The equation of state of dense matter is crucial for the structure and early evolution of neutron stars. Neutron star observations, it is hoped, will eventually set limits on important nuclear parameters such as the bulk incompressibility parameter or the bulk symmetry energy. It may be possible to use observations of thermal emission from a neutron's star surface to constrain the equation of state above nuclear densities; this possibility is addressed in a note by Pethick in this volume. In the important regime just below nuclear density ($\sim 2 \times 10^{14}$ g cm^{-3}), nuclei become deformed and undergo possible phase transitions to other shapes such as an inside-out bubble phase or to uniform nuclear matter. The pressure and adiabatic index in this density regime depend on the incompressibility parameter of bulk nuclear matter. The supernova mechanism may be sensitive to these features. The so-called long-term supernova mechanism, in which a stalled supernova shock is revitalized by neutrino heating, may be sensitive to particulars of the equation of state. The effective neutrino-nucleus coherent scattering rate is a function of the nuclear size and of ion-ion correlations. Calculations of the coherent scattering rate are presented and compared with those that are currently used in other stellar collapse simulations.

2. NUCLEAR EQUATION OF STATE PARAMETERS

There are a number of nuclear parameters that the equation of state will include. All of them, in principle, can be experimentally determined, but in practice, some of them have relatively large uncertainties. They may, alternatively, be calculated from a given nuclear force by modeling heavy nuclei. Since matter in laboratory nuclei is cold ($T \simeq 0$), nearly symmetric ($x_i = Z/A \simeq 1/2$), and close to the saturation density ($n_s \simeq 0.16$ fm^{-3}), it is customary to write the following expansion for the free energy per baryon, f_{bulk}, of bulk nuclear matter:

$$f_{bulk}(n, x, T) \simeq -B + (K_s/18)[1 - (n/n_s)]^2 + S_v(1 - 2x)^2 - a_v T^2 + \cdots. \quad (1)$$

We can identify the following parameters:

1. n_s — The saturation density of symmetric nuclear matter. Values in the range 0.145–0.17 fm^{-3} have been derived.
2. $B = -f_{bulk}(n_s, 1/2, 0)$ — The binding energy of saturated, symmetric nuclear matter. Values for B, derived from nuclear mass formula fits,[1] lie in the range 15.85–16.2 MeV.
3. $K_s = (n_s^2/9)(\partial^2 f_{bulk}/\partial n^2)\big|_{n_s,1/2,0}$ — The incompressibility of bulk nuclear matter. Estimates in the literature are in the range $150 - 300$ MeV.[2]
4. $S_v = (1/8)(\partial^2 f_{bulk}/\partial x^2)\big|_{n_s,1/2,0}$ — The symmetry energy parameter of bulk nuclear matter. Estimates derived from mass formulae[1,3] and from the energy of the giant dipole resonance[4] range from 27.5 to 36.8 MeV.
5. $a_v = -(1/2)(\partial^2 f_{bulk}/\partial^2 T)\big|_{n_s,1/2,0}$ — The bulk level density parameter. Using the fact that nucleons are nearly degenerate, nonrelativistic fermions, it is

$$a_v = (2m^*/\hbar^2)(\pi/12n_s)^{2/3} \simeq (1/15)(m^*/m) \text{ MeV}^{-1}, \quad (2)$$

where m^* is the nucleon effective mass for saturated, symmetric matter. Determination of nuclear energy levels from Hartree-Fock calculations[5] and the position of the giant dipole resonance[4] imply that m^* is somewhat less than the bare nucleon mass m, $m^*/m \sim 0.7$–0.9. However, nuclear mass fits,[6] fission barrier fits,[7] and the reproduction of the density of single particle energies near the Fermi surface[8] argue instead that $m^*/m \sim 1$–1.2.

In its simplest form, the liquid drop model for an isolated nucleus is

$$f_N = f_{bulk} + f_s + f_C, \tag{3}$$

where the surface energy per baryon is $f_s(n_i, x_i, T) = (4\pi r_N^2/A)\sigma(x_i, T)$ and the Coulomb energy per baryon is $f_C(n_i, x_i) = 3Z^2 e^2/(5r_N A)$, where $A = (4\pi/3)n_i r_N^3$. The Coulomb energy is independent of T. The surface tension σ may be expanded in a fashion analogous to the bulk energy:

$$f_s A = 4\pi r_N^2 \sigma(x_i, T) \simeq 4\pi r_N^2 \sigma_s - A^{2/3}[S_s(1 - 2x_i)^2 + a_s T^2] + \cdots. \tag{4}$$

Note that there is no dependence of σ upon density (n_i), because the bulk nucleon fluid inside the nucleus is in equilibrium with that outside the nucleus (in this case, a vacuum). The equilibrium is only possible for one value of n_i for given values of x_i and T. Thus, the surface nuclear parameters are as follows:

6. $\sigma_s \equiv \sigma(1/2, 0)$ — The surface tension of symmetric nuclear matter. Values are again derived from nuclear mass formula fits,[1,3] and values in the range 1.06–1.34 MeV fm^{-3} have been obtained.

7. $S_s = -(A^{1/3}/8)(\partial^2 f_s/\partial x_i^2)|_{n_s,1/2,0}$ — The surface symmetry energy parameter. Estimates from nuclear mass formula fits[1,3] lie in the range 18–180 MeV. The reason for this large range is discussed below.

8. $a_s = -(A^{1/3}/2)(\partial^2 f_s/\partial T^2)|_{n_s,1/2,0}$ — The surface level density parameter. This is defined similarly to a_v, and typical values lie in the range 0.05–0.2 MeV^{-1}.

As is now discussed, these parameters are not all independent. In particular, theoretical or experimental relationships connect the pairs (K_s, σ_s), (K_s, a_s), and (S_v, S_s). Thus, in practice, only five of the above parameters should independently affect the equation of state.

An implicit connection between the bulk incompressibility parameter K_s and the surface tension can easily be derived by referring to a simple Thomas-Fermi model for the nuclear surface of a $T = 0, N = Z$ nucleus. We write the total free energy density of a nucleon fluid, taking into account the energy due to gradations in the fluid density between the interior and the nuclear surface, as

$$F = n f_{bulk} + Q(\nabla n)^2 = -Bn + (K_s n/18)[1 - (n/n_s)]^2 + Q(\nabla n)^2, \tag{5}$$

where Q is a constant parameter of the nuclear force and will be determined later. The density profile $n(r)$ is obtained by performing a Lagrange minimization of the total energy of a fixed number (A) of nucleons:

$$18Q(dn/dr)^2 = K_s n[1 - (n/n_s)]^2. \tag{6}$$

The Equation of State in Neutron Star Matter

The surface energy $4\pi r_N^2 \sigma_s$ is the difference between the total integrated energy and the equivalent bulk energy of A nucleons at the saturation density, which gives

$$\sigma_s = (8/15)n_s^{3/2}\sqrt{QK_s/18}. \tag{7}$$

Eliminate Q by defining the surface thickness $\tilde{a} \equiv (d\ln n/dr)^{-1}\big|_{n_s/2} = 6\sqrt{Qn_s/K_s}$:

$$K_s = (135\sqrt{2}/4)(\sigma_s/\tilde{a}n_s). \tag{8}$$

Experiments give $\tilde{a} \sim 1.1$ fm, which implies that $K_s \simeq 300$ MeV. This value is somewhat larger than those obtained by Blaizot from giant resonance data but is consistent with the results of the Groningen group for similar data. It is significantly larger than the theoretical surmise of Brown and Osnes that $K_s \sim 150$ or less, based upon Landau sum rule arguments.[2]

A relation also exists between K_s and the critical temperature of bulk matter. Two phases of nuclear matter, with different densities and, in general, different proton fractions, may coexist when their pressures, neutron chemical potentials, and proton chemical potentials are respectively equal. This so-called *bulk equilibrium*, which represents the equilibrium of nuclei with a vapor phase, is permitted for temperatures below the critical temperature T_c. T_c is determined by the solution of the equations

$$(\partial P_{bulk}/\partial n)\big|_T = (\partial^2 P_{bulk}/\partial n^2)\big|_T = 0, \tag{9}$$

for the case of symmetric matter. Using the definition $P = n^2(\partial f/\partial n)\big|_T$ and Eq. (1),

$$T_c = \sqrt{5K_s\hbar^2/96m^*}(5n_s/\pi)^{1/3} \simeq 14 \text{ MeV}$$

for $K_s \simeq 200$ MeV. A general result of Thomas-Fermi calculations of the nuclear surface at finite temperatures is[9] $\sigma(1/2, T) \simeq \sigma_s[1 - (T/T_c)^2]^3$, where the exponent 3 is approximate, so that

$$a_s \simeq 3\pi^{1/3}(6/n_s)^{2/3}/T_c^2 = (288m^*\sigma_s/K_s\hbar^2)\pi^{1/3}(6\pi/5n_s^2)^{2/3} \simeq 0.2 \text{ MeV}^{-1}.$$

Finally, the symmetry parameters S_v and S_s cannot be uniquely determined from nuclear mass fits. In the liquid drop model, the symmetry energy is

$$E_{sym} \equiv E_{total} + B - 4\pi r_N^2 \sigma_s - f_{Coul} = I^2(S_v - S_s A^{-1/3}), \tag{10}$$

where $I = (N - Z)/(N + Z) = 1 - 2x$ is the neutron excess. The problem is that the range of $A^{1/3}$ that encompasses known nuclei is too small; only a combination of the two parameters can be found. A small variation in the assumed value of S_v will lead to a large variation in the deduced value of S_s. Assuming two different parameter sets, both of which reasonably well fit the symmetry energy of known nuclei, one can show

$$S_s/S_s^o = 1 + (S_v/S_v^o - 1)q, \tag{11}$$

where (S_v^o, S_s^o) refer to a fiducial parameter set and (S_v, S_s) to another set, and $q = S_s^o A^{1/3}/S_v^o$. This relation may be written approximately as

$$S_s \approx S_s^o (S_v/S_v^o)^q. \tag{12}$$

Typical values of q are in the range 3–5. It should also be pointed out here that in liquid drop[1] formalisms, in which the symmetry energy takes the form

$$E_{sym} = I^2 S_v/[1 - (S_s/S_v)A^{-1/3}], \tag{13}$$

the exponent in Eq. (12) becomes $q+2$, so that an even stronger variation in S_s is predicted. This accounts for the large range in values for S_s noted earlier; nuclear mass systematics cannot resolve this problem. Fortunately, there is some hope that fits to giant resonance data in addition to mass data may eventually allow the independent determination of the symmetry parameters.[4] Until values for these symmetry energy parameters are agreed upon, however, calculations of supernovae, in particular, will be held hostage to them.

In summary, the relevant nuclear parameters are reduced to a smaller subset by using theoretical or experimental data. Since n_s and B are known to reasonable precision, this subset is essentially the triplet $K_s, m^*/m$, and S_v.

3. NEUTRON STAR STRUCTURE

Observations of neutron star masses, radii, moments of inertia, temperatures, rotation rates, and binding energies can be used to constrain the nuclear equation of state. However, due to the fact that most of a neutron star's mass lies above 2–$3n_s$, it is the supernuclear equation of state that is, in general, being tested.[10] Supernova calculations, on the other hand, are probably most sensitive to the equation of state at and below n_s.

One can limit the maximum neutron star mass by assuming that the maximum sound speed is given by c: This is the causal limit. Below some density, n_t, one assumes the equation of state is known. Using the TOV equation, one finds that with practically no dependence on the equation of state below n_t, the maximum neutron star mass is $M_{max} = 4.1\sqrt{n_s/n_t}M_\odot$. With the conservative choice $n_t \simeq 2n_s$, we see that the maximum mass must be less than $2.9 M_\odot$. The radius of the maximum mass configuration may also be found: $R_{max} = 18.5\sqrt{n_s/n_t}$ km, which leads to a maximum value of the parameter GM/Rc^2 of 0.328. In Figure 1, we show this

The Equation of State in Neutron Star Matter

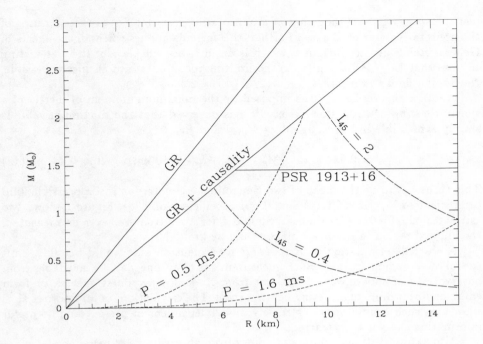

FIGURE 1 Valid equations of state must produce some neutron stars which lie to the right of curve "GR + causality". Observational constraints include $M > M_{1913+16}$, $I > 0.4 \times 10^{45}$ g cm² (Crab pulsar), and $P_{min} < 1.6$ ms (PSR 1957+20).

constraint together with that of the weaker assumption due to General Relativity alone, $GM/Rc^2 \leq 1/2$.

It is also possible to set general limits to the spin period of a uniformly rotating star. The minimum period must be at least equal to the Keplerian period (at which point mass would begin to be shed from the equator) and perhaps greater due to GR instabilities. The most rapid rotation rate for a given equation of state occurs at the maximum mass configuration, since then the radius is the smallest possible value. Haensel and Zdunik[11] first noted, using calculations performed by Friedman et al.[12] and by Lattimer et al.,[13] that a remarkably simple relation holds for the maximum angular rotation rate with a given causal equation of state:

$$\Omega_{max} = 0.77 \times 10^4 (M_{max}/M_\odot)^{1/2} (10 \text{ km}/R_{max})^{3/2} \text{ s}^{-1}, \qquad (14)$$

where M_{max} and R_{max} are the *nonrotating* maximum mass and the radius of this maximum mass star. It should be emphasized that the maximum mass of a rotating neutron star is greater than that of a nonrotating star, yet Eq. (14) is in terms of the nonrotating star properties. Further, this formula applies when all the effects of General Relativity are included, which is an intricate calculation. It is interesting to note that Eq. (14) is similar to the limiting rotation rate of an incompressible, Newtonian fluid,[14] which is $\Omega_{max} = (2/3)\sqrt{GM/R_0^3}$.

Another theoretical limit is imposed by the maximum moment of inertia of a slowly rotating neutron star. Haensel[15] has observed that the maximum value is simply expressible in terms, again, of M_{max} and R_{max}:

$$I_{max} = 0.984 \times 10^{45} (M_{max}/M_\odot)(R_{max}/10 \text{ km})^2 \text{ g cm}^2. \qquad (15)$$

This turns out to be the mean of two Newtonian estimates: an incompressible fluid (coefficient in Eq. [15] of 1.19) and $P \propto \rho$ (coefficient of 0.78). Figure 1 shows two estimates of I for the Crab pulsar. Should I for a neutron star ever be measured, the allowed region in the M-R plane would be limited.

A clear determination of R or M/R for a neutron star would possibly limit the allowed region further. Such measurements may one day be available from observations of supernova ν-bursts, or from X- and γ-ray burst spectra or from emission line profiles of accreting neutron stars. The calculations of Fujimoto et al.[16] show that such estimates are extremely model dependent at present and no useful information can yet be extracted.

A final observational constraint comes from observations of neutrinos from supernovae: the neutron star's binding energy. Essentially all the binding energy is released equally in neutrinos of all types ($\nu_e, \bar{\nu}_e, \nu_\mu, \bar{\nu}_\mu, \nu_\tau, \bar{\nu}_\tau$). To within 20%, the binding energy is given by the simple relation[17] $BE \simeq 1.5 \times 10^{53}(M/M_\odot)^2$ ergs. Although the data from SN 1987A could only determine the binding energy to within 50%, at best, and thereby limit the mass of the neutron star formed in this event to $1.3 \pm 0.3 M_\odot$, future supernovae within our Galaxy should be able to fix the neutron star mass to within $\pm 20\%$ (assuming a neutron star is indeed formed).

4. SUBNUCLEAR EQUATION OF STATE

At subnuclear densities, the pressure is dominated by relativistic, degenerate leptons (e^-, ν_e), for which the adiabatic index $\Gamma = d\ln P/d\ln \rho \simeq 4/3$. But three effects conspire to reduce Γ below this value. A slow conversion of electrons to neutrinos as the density is increased decreases the total pressure relative to that for electrons alone. Nuclear ionic (Coulomb) correlations lead to a softening of the equation of state. Finally, a succession of phase transitions around 10^{14} g cm^{-3} from nuclei to distorted nuclei (including bubbles) and then to uniform matter further reduces Γ. The difference $\Gamma - 4/3$ is extremely important in determining the size of the collapsing core, the mass of the newly formed neutron star, the location where the outgoing shock forms, and the shock's energy.[18]

The Equation of State in Neutron Star Matter

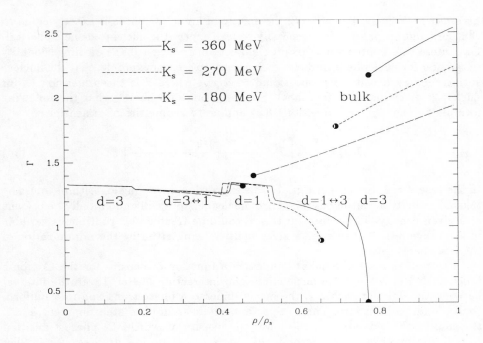

FIGURE 2 Adiabatic index as a function of density for three values of K_s. The nuclear energies of Ref. 20 were assumed. Dots show the location of the phase transition between nuclei and bulk matter. Values for the shape parameter d are shown.

In the subnuclear regime, large neutron-rich nuclei are immersed in a sea of interacting nucleons (mostly neutrons). The nuclear Coulomb energy is modified by neighboring nuclei. In the Wigner-Seitz approximation, in which matter is partitioned into spheres containing a single nucleus and enough electrons to neutralize it, the Coulomb energy is multiplied by a factor[19] $D(u) \equiv 1 - (3/2)u^{1/3} + (1/2)u$, where $u \approx n/n_s$ is the volume fraction occupied by nuclei. The nuclear virial theorem, which gives the size of the most energetically favored nucleus, is $f_{surf} = 2f_{Coul}$ and means both that $A \propto D(u)^{-1}$ and $f_{surf} + f_{Coul} \propto D(u)^{1/3}$. Therefore in the high-density limit, $u \to 1$ we find that nuclear masses get very large and that the energy of nuclei becomes more negative (more bound); that is, the pressure due to nuclei is negative and its magnitude increases.

When $u \approx 1/2$ a curious thing happens: The nuclei can turn inside out! Consider the energy of a rarified bubble of neutrons immersed in a dense sea of neutrons and protons. The bulk and surface energies are the same for bubble configurations as for nuclei (we neglect curvature effects), but the Coulomb energy is modified: u

becomes $1 - u$ in D. For $u = 1/2$ the energy of a nuclear phase is exactly equal to that for a bubble phase. More generally, we can expect the nuclei to be nonspherical. One interesting approach to this problem is to introduce a shape or dimensionality parameter for nuclei be introduced.[20] It is easy to see, for example, that the nuclear surface energy density is proportional to d/r_N, where d is the dimension (3 for nuclei, 2 for cylinders, 1 for slabs). Similarly, one may show[20] that the following formula for the Coulomb function $D(u)$ applies to geometries of dimension d:

$$D(u) = \frac{5}{9}d^2 \left[\frac{1 - (du^{1-2/d})/2}{d^2 - 4} + \frac{u}{2(d+2)} \right]. \qquad (16)$$

In the case $d = 2$, the proper limit of this expression, involving logarithms, must be taken. In reality a continuous mixture of nuclear shapes will undoubtedly be present at a given density. To account for this, d could be treated as a continuous variable in the range 3–1, its value for a given density being given by the minimization of the free energy.

Figure 2 shows the adiabatic index as a function of density for the Coulomb function of Ref. 20 and the nucleon-nucleon interaction of Ref. 10. In the interval between the dots for each K_s, the nuclei undergo a phase transition to a uniform (bulk) fluid, and Γ is zero. Three important results are immediately apparent. First, the smaller the value of K_s, the lower the density at which the phase transition to bulk matter takes place. Second, the changes in Γ are less dramatic for smaller values of K_s. Third, the region in which Γ is appreciably less than 4/3 is minimized for smaller values of K_s. Taken together, one sees that the pressure-density relation for small incompressibilities is much smoother. Cooperstein and Baron[21] suggest that such behavior is beneficial to shock formation and will result in more effective shocks.

5. NEUTRINO OPACITIES

The dominant source of opacity during stellar collapse and for at least a few tenths of a second following bounce is ν–nucleus coherent scattering. The cross section was originally derived by Freedman,[22] and the ratio of coherent scattering to ν–nucleon scattering is about $(N^2/6A)(X_H/X_n)$, where the X's are mass fractions. Note that the coherent scattering opacity is proportional to A^2. This ratio is generally of the order of 100 or more, until temperatures in the newly formed neutron star rise[23] to values sufficiently high to dissociate nuclei. A newly formed neutron star may be considered to consist of three zones: (1) an inner ($n > n_s$), unshocked, region in which neutrino mean free paths are large in spite of the high density because of degeneracy (Pauli blocking) and Fermi liquid effects;[24] (2) the unshocked region containing nuclei ($0.03n_s < n < n_s$). 3) The outer, shocked region where neutrino mean free paths are relatively large because the density is low. Heat diffuses from

The Equation of State in Neutron Star Matter

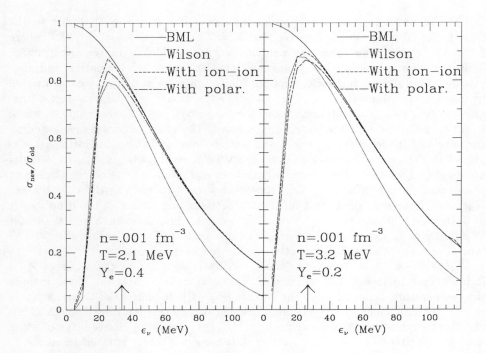

FIGURE 3 Ratio of coherent scattering cross sections σ_{new} to the uncorrected Freedman[22] value, σ_{old}. *BML* is with the form factor only, *Wilson* is the cross section from Ref. 31, *With Ion-Ion* includes the form factor and ion-ion correlations, and *With Polar* in addition includes polarization corrections. Fig. 3a (left) is along a typical collapse trajectory; Fig. 3b (right) is a typical post-bounce point at the boundary of the shock formation point.

region 3 to zone 2 after bounce,[23] and eventually the nuclei there will dissociate. Until this occurs, the neutrino fluxes from the proto-neutron star will be very sensitive to the opacity in zone 2. Accurate calculations of the extent of zone 2, the nuclear masses there, and the coherent scattering itself are essential for investigations of the delayed supernova mechanism discovered (but, as yet, not verified by others) by Wilson.[25]

There are three important corrections that must be applied to the Freedman cross section: (1) the nuclear form factor $\mathcal{F}$,[26,27] important when the ν wavelength $\lambda_\nu < r_N$; 2) liquid structure effects,[28 29] important when $\lambda_\nu > r_N/u^{1/3}$, the internuclear spacing; and (3) polarization of the medium by the neutrinos,[30] important when $\lambda_\nu > r_D$, the Debye radius.

The most important correction at high neutrino energy is that due to the form factor. Burrows et al.[27] derive that $\mathcal{F} = 3[L - 1 + (L + 1)e^{-2L}]/2L^3$, where $L \simeq 10^{-5}\epsilon_\nu^2 A^{2/3}$ and ϵ_ν is the neutrino energy in MeV. On the other hand, Wilson[31] does not perform any angular integrations in his evaluation and simply uses $\mathcal{F} \sim e^{-L}$. A further correction due to ion correlations may be computed by inserting tabulations of the liquid structure function[29] into the form factor integration. Itoh's early calculation[28] used the long-wavelength approximation, which is not very accurate for the conditions at hand. The effect of progressively adding the effects of the form factor, the ion-ion correlations, and the polarization are displayed in Figure 3. The cross section that Wilson employs is also shown. Results are displayed for two different combinations of n, T, and Y_e: one corresponding to a snapshot during infall (Fig. 3a), the other to a snapshot shortly after bounce (Fig. 3b). Note that significant differences between our results and Wilson's exist at both very high and low ϵ_ν. The former is due to differences in $\mathcal{F}$ and the latter to differences in ion correlations and Wilson's neglect of polarization effects. The typical neutrino energy for each case is indicated by an arrow.

Since the coherent scattering cross section is sensitive to the value of A, it is important to include all effects that influence nuclear masses. These include the temperature dependence of the surface [Eq. (4)], the effects of the translation energy of the nucleus,[32] and ion correlations (at zero temperature through $D(u)$; at finite temperatures, plasma corrections[33]). An interesting consequence of using the variable shape factor is that nuclear sizes are significantly reduced when $d < 3$. For example, for $d \simeq 1$, one finds that A may be reduced by a factor of 3 from the case in which spherical nuclei relations are used.

In summary, still uncertain nuclear and weak-interaction physics plagues current calculations of supernova explosions and restricts our understanding of the structure of neutron stars. This review has highlighted some of the most outstanding problems that are currently receiving attention, and for which resolutions may be forthcoming.

ACKNOWLEDGMENTS

This research has been supported in part by the U.S. Department of Energy under grant DE-AC02-ER-80317.

REFERENCES

1. W. D. Myers and W. J. Swiatecki, *Ann. Phys.*, **84**, 186 (1973); J. M. Pearson, Y. Aboussir, A. K. Dutta, R. C. Nayak, M. Farine, and F. Tondeur, *Nucl. Phys.* **A528**, 1 (1991).
2. G. E. Brown and E. Osnes, *Phys. Lett.*, **B154**, 223 (1985); J. P. Blaizot, Phys. Rep., **64**, 171 (1980); M. M. Sharma, W. T. A. Borghols, S. Brandenburg, S. Crona, A. van der Woude, and M. N. Harakeh, Phys. Rev., **C38**, 2562 (1988); P. Gleissl, M. Brack, J. Meyer, and P. Quentin, *Ann. Phys.*, **197**, 205 (1990).
3. P. Möller and J. R. Nix, *At. Data and Nucl. Data Tables*, **39**, 219 (1988); P. Möller, W. D. Meyers, W. J. Swiatecki, and J. Treiner, *At. Data and Nucl. Data Tables*, **39**, 225 (1988).
4. W. D. Meyers, W. J. Swiatecki, T. Kodama, L. J. El-Jaick, and E. R. Hilf, Phys. Rev., **C15**, 2032 (1977).
5. D. Vautherin and D. M. Brink, *Phys. Rev.* **C5**, 626 (1972).
6. F. Tondeur, M .Brack, M. Farine, and J. M. Pearson, *Nucl. Phys.*, **A420**, 297 (1984).
7. F. Tondeur, *Nucl. Phys.*, **A442**, 460 (1985).
8. M. Barranco and J. Treiner, *Nucl. Phys.*, **A351**, 269 (1981).
9. D. G. Ravenhall, C. J. Pethick and J. M. Lattimer, *Nucl. Phys.*, **A407**, 571 (1983).
10. M. Prakash, T. L. Ainsworth, and J. M. Lattimer, *Phys. Rev. Lett.*, **61**, 2518 (1988).
11. P. Haensel and J. L. Zdunik, *Nature,* **340**, 617 (1989).
12. J. L. Friedman, J. R. Ipser, and L. Parker, *Astrophys. J.*, **304**, 115 (1986).
13. J. L. Friedman, J. R. Ipser and L. Parker, *Astrophys. J.*, **304**, 115 (1986).
14. J. M. Lattimer, M. Prakash, D. Masak, and A. Yahil, *Astrophys. J.*, **355**, 241 (1990).
15. J. M. Lattimer, and A. Burrows, in *SN 1987A and Other Supernovae*, L. J. Danziger (ed.) (Reidel, Dordrecht, 1991), in press.
16. P. Haensel, Copernicus Astronomical Center preprint (1990).
17. M. Y. Fujimoto, M. Sztajno, W. H. G. Lewin, and J. van Paradijs, *Astrophys. J.*, **319**, 902 (1987).
18. J. M. Lattimer and A. Yahil, *Astrophys. J.*, **340**, 426 (1989).
19. J. M. Lattimer, A. Burrows, and A. Yahil, *Astrophys. J.*, **288**, 644 (1985).
20. G. Baym, H. A. Bethe, and C. J. Pethick, *Nucl. Phys.*, **A175**, 225 (1971).
21. D. G. Ravenhall, C. J. Pethick, and J. R. Wilson, *Phys. Rev. Lett.*, **150**, 2006 (1983).
22. J. Cooperstein and E. Baron, in *Supernovae*, A. Petschek (ed.) (Springer-Verlag, New York, 1989).
23. D. Z. Freedman, *Phys. Rev.*, **D9**, 1389 (1974).
24. A. Burrows and J. M. Lattimer, *Astrophys. J.*, **307**, 178 (1986).

25. N. Iwamoto and C. J. Pethick, *Phys. Rev.*, **D25**, 313 (1982); B. T. Goodwin and C. J. Pethick, *Astrophys. J.*, **253**, 816 (1982).
26. J. R. Wilson, in *Numerical Astrophysics*, J. Centrella, J. M. Le Blanc, and R. L. Bowers eds., Jones and Bartlett, Boston, (1985), p. 422.
27. D. Tubbs and D. N. Schramm, *Astrophys. J.*, **201**, 467 (1975).
28. A. Burrows, T. J. Mazurek, and J. M. Lattimer, *Astrophys. J.*, **251**, 325 (1981).
29. N. Itoh, *Prog. Theoret. Phys.*, **54**, 1580 (1975).
30. S. Ichimaru, H. Iyetomi, and S. Tanaka, *Phys. Reports*, **149**, 91 (1987).
31. L. B. Leinson, V. N. Oraevsky, and V. B. Semikoz, *Phys. Lett.*, **B209**, 80 (1988).
32. R. Bowers and J. R. Wilson, *Astrophys. J. Suppl.*, **50**, 15 (1982).
33. J. M. Lattimer, C. J. Pethick, D. G. Ravenhall, and D. Q. Lamb, *Nucl. Phys.*, **A432**, 646 (1985).
34. J. P. Hansen, *Phys. Rev.*, **A8**, 3096 (1973).

H. Inoue
Institute of Space and Astronautical Science, 3-1-1, Yoshinodai, Sagamihara, Kanagawa 229, Japan

Study of Neutron Stars in X-rays

The study of neutron star properties in X-rays is reviewed. Information about the mass is obtained from the orbital motion of X-ray pulsars, and the relation between the mass and radius is derived from analysis of X-ray bursts with photospheric expansion at the Eddington limit. The interior is investigated through fluctuations in the pulse period of X-ray pulsars and also through study of neutron star cooling. The strength of the magnetic field on the surface is measured through cyclotron features in the spectra of X-ray pulsars and gamma-ray bursts. The decay of the magnetic field is controversial with respect to the origin of low-mass X-ray binaries and millisecond pulsars.

1. INTRODUCTION

Most bright X-ray sources are close binary systems containing neutron stars. Supernova remnants are also bright X-ray sources and sometimes contain neutron stars at their centers. Gamma-ray bursts always emit X-rays and are probably phenomena related to neutron stars. Hence, a significant fraction of studies in X-ray astronomy has been devoted to phenomena around neutron stars and there is much diversity

2. MASS AND RADIUS OF NEUTRON STARS
2.1 MASS DETERMINATION THROUGH PULSE TIMING ANALYSIS OF X-RAY PULSARS

There are about 30 binary X-ray sources which have been found to be pulsating with a period from 69 ms to 835 s.[1] In those binary systems, neutron stars pulsate due to their rotation and the pulse period is modulated by the Doppler effect through the orbital motion. An analysis of the pulse arrival time gives us orbital parameters of a binary system of an X-ray pulsar. The semimajor axis of the neutron star motion around the gravity center, a_x, is derived from the amplitude of the Doppler modulation of the pulse period in terms of the inclination of the binary plane to us, i, and the mass function is estimated from the semimajor axis and the orbital period, P_{orb}:

$$(M_c \sin i)^3/(M_x + M_c)^2 = 4\pi^2(a_x \sin i)^3/GP_{orb}^2, \tag{1}$$

where M_x and M_c are the masses of the neutron star and the companion star respectively, and G is the gravitational constant. On the other hand, an optical spectroscopic observation can give us a semiamplitude of the Doppler velocity curve of the companion star, K_c, and the mass ratio of the neutron star to the companion star as

$$M_x/M_c = K_c P_{orb}(1 - e^2)^{1/2}/2\pi a_x \sin i, \tag{2}$$

where e is the eccentricity of the orbit. The inclination, i, can be obtained in terms of the duration of the eclipse of the neutron star by the companion star relative to the orbital period. Let us define an eclipse half angle, θ, by multiplying the ratio of the eclipse duration to the orbital period by 2π; then the inclination angle is derived from the following equation as

$$\sin i = [1 - (R_c/a)]^{1/2}/\cos\theta \tag{3}$$

where R_c and a are the radius of the companion star and the separation of the two stars respectively. If we assume that the companion star fills the Roche lobe, R_c/a can be represented as a function of the mass ratio of the two stars. From Eqs. 1 to 3 we can estimate the mass of the neutron star. The results are plotted in Fig. 1 for six X-ray pulsars[1] together with masses determined from the binary radio pulsar PSR 1913+16 by Taylor and Weisberg (1982)[2] for comparison.

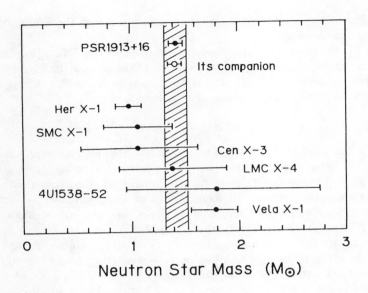

FIGURE 1 Neutron star masses derived from observations of six binary X-ray pulsars.[1] The mass of PSR 1913+16 and its companion are shown for comparison.[2] The hatched region represents the range 1.42 ± 0.10 M$_\odot$.

2.2 MASS/RADIUS DETERMINATION FROM X-RAY BURSTS

There are about 35 X-ray sources known to date which produce X-ray bursts. Most of the bursts exhibit significant cooling with time in the decay portion, and those properties are consistent with the thermonuclear flash model.[3]

2.2.1 BURSTS WITH PHOTOSPHERIC EXPANSION AT THE EDDINGTON LIMIT   In a given burst source, bursts which exhibit the largest peak flux commonly reveal a clear increase, followed by a decrease, of the apparent radius during the peak.[4] The behavior of the photospheric expansion can now be well understood through studies of neutron star envelopes radiating at the Eddington luminosity.[5-8] In the thermonuclear flash model for bursts, a burst usually takes place in a helium-rich envelope of a neutron star, which is covered by accreted matter with cosmic abundance of the elements. Since the Eddington limit has a dependency of $(1+X)^{-1}$ on the mass abundance of hydrogen, X, the limit of the helium-rich envelope is larger than that of the envelope with cosmic abundance by a factor of about 1.7. Hence, when the radiative flux near the Eddington limit in the burning helium-rich envelope flows outward into the hydrogen-rich envelope, the surface hydrogen-rich envelope will be ejected. Many observational facts are consistent with the understanding that the peak flux of bursts with photospheric expansion is just the Eddington limit in the helium-rich envelope.[9]

If the maximum peak flux indeed corresponds to the Eddington limit at the neutron star surface and the spectrum is indeed that of a blackbody, we can relate the effective temperature, $T_e(r)$, at the photospheric expansion during the maximum peak with the radius, r, of the photosphere as[10]

$$\sigma T_e(r)^4 = cGM[1 - (2GM/c^2r)]^{3/2}/\kappa_e r^2, \qquad (4)$$

where M is the mass of the neutron star, σ is the Stefan-Boltzmann constant, c is the velocity of light, and κ_e is the opacity for electron scattering. If we know $T_e(R)$ when the photosphere practically exists on the neutron star surface with radius of R, we have a relation between mass and radius of a neutron star.

However, a problem is that color temperatures observed from stellar surfaces are generally different from effective temperatures. In fact, it was pointed out that the color temperature should be higher than the effective temperature for the radiation from such an atmosphere in which the electron scattering opacity is dominant, as in the case of the bursting neutron star surface.[11]

A quantitative relationship between color temperature and effective temperature has been obtained by many authors.[12-17] By using the relation, we can determine the value of $T_e(R)$ by fitting the theoretical curve to the data points in the color temperature and the luminosity plane. Fujimoto and Taam (1986),[18] Ebisuzaki (1987),[17] Nakamura (1987),[19] and Ebisuzaki and Nakamura (1988)[20] performed such fits to the data during the decay part of bursts observed with Tenma. Once we have the value of $T_e(R)$ from the fits, we can obtain the mass-radius relation of the neutron star by using Eq. 4. However, since the result is still model dependent, we need further study on this important problem.

Recently, van Paradijs et al. (1990)[21] have derived information on the mass and radius of the neutron star in 4U2129+11 with a different method from that mentioned above. They have analyzed a very energetic X-ray burst from this source[22] and have found that the flux during the photospheric expansion slightly increased as the photosphere expanded. This flux change during the photospheric expansion can be understood to be due to the general-relativistic effect on the observed Eddington limit which depends on the radius of the photosphere as $[1 - (2GM/c^2r)]^{1/2}$. By comparing the flux change with the general-relativistic effect, they have obtained constraints on the mass and radius of the neutron star.

2.2.2 ABSORPTION LINES IN X-RAY BURST SPECTRA Tenma discovered significant features in X-ray burst spectra. Among 16 bursts from 1636-53, 4 bursts exhibited five cases of significant dips in the spectra, whose nature was consistent with an absorption line.[23] On the other hand, 3 bursts among 17 bursts from 1608-52 exhibited absorption lines in the spectra.[24] A significant spectral feature in a burst was also detected by *EXOSAT* from 1747-21.[25]

Study of Neutron Stars in X-rays

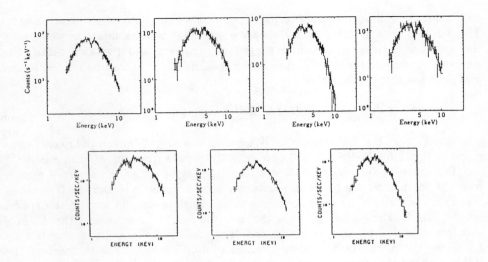

FIGURE 2 Absorption lines observed in four bursts from 1636–53[23] (*upper panels*) and those in three bursts from 1608–52[24] (*lower panels*). Histograms show the best-fit blackbody spectra including the absorption lines.

A line was observed at 5.7 keV only during the peak of a burst with a photospheric expansion from 1636–53. On the other hand, the line center in the other 8 cases from 3 sources are all consistent with 4.1 ± 0.1 keV, taking account of their errors (for 7 cases observed from Tenma, see Fig. 2). The equivalent widths of these absorption lines are about 100–500 eV. The observed line profile is well reproduced by the detector's response function for a single line without significant intrinsic line width, and the upper limit for the line width is about 500 eV (FWHM).

It is possible that the absorption lines are produced by atomic processes in the atmosphere surrounding the neutron star,[23] although serious difficulties in interpreting such large equivalent widths as observed have been discussed.[26,25] Then, for the plasma temperatures during a burst, the helium-like ions are considered to be most abundant for the heavy elements responsible for the absorption. Taking account of the possibility of the energy redshift due to the general relativistic effect on the neutron star surface, the candidate elements forming the absorption lines are those heavier than Ca for the 4.1-keV line.[23]

As discussed earlier, bursts which show the largest peak flux in a given burst source exhibit a clear photospheric expansion during the peak, and this can be understood to be due to the effect of radiation pressure when the luminosity is very close to the Eddington limit. This means that when the burst flux is well below the largest peak flux of the burst source, the effect of radiation pressure is negligible and hence the radius of the photosphere represents the radius of the neutron star. The 4.1-keV lines detected so far always appeared when the flux was well below the largest flux. Hence, if the above suggestion is correct, and if we can identify the element responsible for the absorption, the redshift factor, z, can be related to the gravitational redshift factor as

$$[1 - (2GM/c^2R)]^{1/2} = (1+z)^{-1}. \qquad (5)$$

The results from Eq. 5 are shown for five candidates for the line on the mass-radius plane in Fig. 3 for X1608-52, together with the results from Eq. 4 when $r = R$.[19]

The matter producing the absorption lines may be either the accreted matter from the companion star or the products of nuclear burning in the bursts. If accreted matter is responsible for absorption, the most plausible element for the absorption is iron, which is most abundant in the accreted matter. However, if the 4.1-keV absorption is due to iron, it means that the radius of the neutron star is only 1.6 times the Schwarzschild radius, which cannot be realized with most of the stable neutron star models available according to the current theory.[27]

Fujimoto (1985)[28] proposed a model to avoid the difficulty of the "stiff" neutron star. He considered the boundary layer between the neutron star and the accretion disk as the line formation region. Since the boundary layer is expected to rotate very rapidly with the accretion disk, the transverse Doppler effect due to this rotation is important in evaluating the redshift as well as the general-relativistic effect. He calculated the redshift factor expected from the boundary layer, and the mass-radius relation obtained gives good agreement with some neutron star models (see Fig. 3). However, this model has major difficulties.

The rotation of the line-producing matter causes not only the shift of the line center energy but also the broadening of the line profile. In order to accommodate the rotational broadening with the observed upper limit for the intrinsic line width, the line of sight should be within a small angle around the rotation axis. It is very unlikely that the rotation axes of the three sources from which the 4.1-keV lines were detected are all aligned toward us within a small solid angle.

On the other hand, Ti and Cr can be the most abundant element in the nuclear-burning products depending on the pressure of the burning shell,[29] and the corresponding values of mass and radius of the neutron star agree with some neutron star models (see Fig. 3). However, when the 4.1-keV lines were detected, the effect of the radiation pressure could be neglected, as discussed above, and no force seems to exist to prevent the freshly accreted matter from covering the neutron star surface. Hence, we need some unknown mechanism for mixing the nuclear burning products with the accreted matter and for exposing them on the surface of the neutron star.

Further study of the origin of the 4.1-keV line is needed.

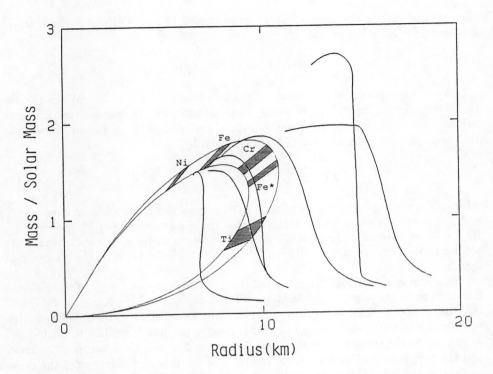

FIGURE 3 Error domains of mass and radius of the neutron star in 1608–52 for five candidates for the 4.1-keV absorption line.[19] The region with Fe* corresponds to the model including the transverse Doppler effect proposed by Fujimoto (1985).[28] The solid curves show the relation given by Eq. 4 obtained through fitting the observed relation between the color temperature and the energy flux to the theoretical relation by Ebisuzaki (1987).[17] Theoretical mass-radius relations of neutron stars[27] are also shown.

3. INTERIOR OF NEUTRON STARS
3.1 PULSE PERIOD CHANGE OF NEUTRON STARS

X-ray pulsars generally exhibit short-term fluctuations of pulse period.[1] Since these pulse-period changes represent changes in the rotation rate of the outer crust of a neutron star, we can diagnose the internal dynamical properties of the crust-superfluid system in the interior of neutron stars.

Let $P_{in}(f)$ represent the power density spectrum of an input noise torque disturbing the crust; then the observed power density spectrum of fluctuations in the rotation rate of the crust, $P_{obs}(f)$, is expressed in terms of the power transfer function, $F(f)$, by the relation[30]

$$P_{obs}(f) = F(f)P_{in}(f). \tag{6}$$

When the neutron star behaves like a rigid rotator, $F(f) = 1$. When the superfluid in the stellar core is, however, weakly coupled to the outer crust, the core will respond to changes in the rotation rate of the crust on a time scale longer than the coupling time, and hence $F(f)$ will be given by

$$F(f) = \begin{cases} 1 & \text{for } f > f_0 \\ (I/I_{\text{crust}})^2 & \text{for } f < f_0, \end{cases} \tag{7}$$

where f_0 corresponds to the coupling time between the crust and the core, and I and I_{crust} are the moment of inertia of the whole star and the crust respectively. Hence, if we observe a jump in the observed power density spectrum of the crust rotation rate, we can discuss the coupling time and the ratio of the moment of inertia.

Boynton et al. (1984)[31] investigated the power density spectrum of the residual fluctuations in the angular acceleration of the X-ray pulsar Vela X-1 after subtracting orbital effects. The pulse timing fluctuations were found to be consistent with white noise in a frequency range corresponding to periods from 0.25 to 2600 days. According to the results, it is suggested that if the coupling time between the crust and the core is in that period range, the ratio of the moment of inertia of the crust to that of the whole star cannot be less than 15%. Otherwise, the coupling time should be longer than 2600 days or shorter than 0.25 days.

3.2 NEUTRON STAR COOLING

The study of the thermal evolution of neutron stars after their birth is also very useful for understanding their interiors. Nomoto and Tsuruta (1986)[32] have calculated the "standard" cooling models of neutron stars incorporating the recent theoretical developments and compared them with the results of the X-ray observations. According to their results, the observed temperature upper limits for four supernova remnants (SNRs)—Cas A, Tycho, SN1006, and the Vela pulsar—are below the "standard" cooling curves. They argued that no neutron star is likely to be left in the Type I SNRs Tycho and SN1006 according to current theoretical predictions. On the other hand, in the case of Cas A, it is possible that a compact star left behind was too massive to avoid collapse into a black hole. However, the Vela pulsar, for which the temperature upper limit is below the "standard" cooling curves, necessitates some modification of the standard cooling models. Nomoto and Tsuruta[32] pointed out the possibility that greatly enhanced cooling may be due to the presence of "exotic" particles such as charged pion condensates and quarks. For details see Tsuruta (this symposium).

4. MAGNETIC FIELD OF NEUTRON STARS
4.1 X-RAY PULSARS

The strength of magnetic fields on the surface of neutron stars in X-ray pulsars can be directly measured through cyclotron features in the X-ray spectra. The electron-cyclotron energy, E_c, is proportional to the strength of the magnetic field, B, as

$$E_c = 12(B/10^{12} \text{ gauss})\text{keV}. \tag{8}$$

The presence of these features in X-ray pulsar spectra was first pointed out for Her X-1 by Trümper et al. (1978).[33] Wheaton et al. (1979)[34] reported evidence for a similar feature in the X-ray spectrum of 4U0115+63. Recent *Ginga* observations have confirmed these spectral features in the two X-ray pulsars.[35,36]

Ginga has further discovered cyclotron-line features in the spectra of several X-ray pulsars (for one example see Fig. 4[37]) and it has been found that the cyclotron-line features observed with *Ginga* are all consistent with a simple expression for cyclotron absorption lines (see Makishima, this symposium). Thus, the strengths of the magnetic fields on the surfaces of these X-ray pulsars have been estimated and lie in a fairly narrow range of 1–3 $\times 10^{12}$ gauss (see Makishima, this symposium).

4.2 LOW-MASS X-RAY BINARIES

Magnetic fields of neutron stars in low-mass X-ray binaries are believed to be much weaker than those of X-ray pulsars, although no direct measurement of the magnetic field strength has been performed for low-mass X-ray binaries. This belief is based on several arguments:

1. Low-mass X-ray binaries show no coherent pulsation, whereas X-ray pulsars do.
2. Low-mass X-ray binaries often exhibit X-ray bursts, whereas X-ray pulsars have never been observed to. Ebisuzaki and Nakamura (1988)[19,20] estimated the upper limit for the magnetic field of X-ray bursting neutron stars to be 3×10^{10} gauss.
3. X-ray spectra of low-mass X-ray binaries are generally softer than those of X-ray pulsars and are consistent with the picture of an accretion disk extending close to the neutron star.[38,39]

From these arguments, the magnetic fields of neutron stars in low-mass X-ray binaries can be considered to be no stronger than 10^9 gauss on their surfaces.

The difference in the magnetic field strength between X-ray pulsars and low-mass X-ray binaries has been considered to be due to a difference in age. X-ray pulsars generally have early type companions and are distributed in the Galactic arms.[40] Hence, X-ray pulsars are probably as young as about 10^6 years. On the other hand, low-mass X-ray binaries generally accompany late type companions and are distributed preferentially near the Galactic center and often in globular

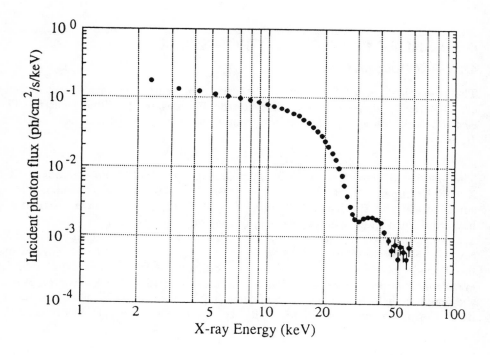

FIGURE 4 The inferred incident photon spectrum of 0331+53, displayed with the low-energy photoelectric absorption reset to zero for simplicity.[37]

clusters.[41] These facts indicate that neutron stars in low-mass X-ray binaries are as old as or older than 10^9 years. The inverse correlation of the magnetic field strength of neutron stars with age is consistent with the idea that neutron stars are born with magnetic fields of 10^{12} gauss which then decay with a time constant of $\sim 10^7$ years.[42] This idea arose from statistical analysis of radio pulsars[43,44] and has been strengthened by recent discoveries of millisecond pulsars which have magnetic fields as weak as about 10^9 gauss (see Manchester, this symposium). The canonical scenario for the millisecond pulsar is that old neutron stars with decayed, weak magnetic fields in low-mass X-ray binaries are spun up by mass accretion and become millisecond pulsars. In fact, some millisecond pulsars exist in close binaries. However, cyclotron line features in the spectra of gamma-ray bursts throw serious doubt on the idea of the magnetic field decay of neutron stars, as discussed below.

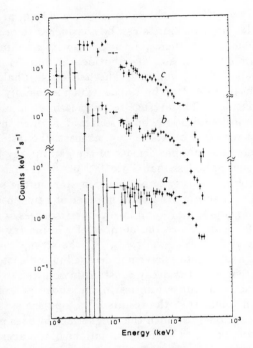

FIGURE 5 Raw counting-rate spectra for three consecutive time sections with 5-s intervals around the peak of GB880205 after background subtraction (not corrected for the counter response).[46] Spectral features centered at ∼ 20 and 40 keV can be seen for time section b.

4.3 GAMMA-RAY BURST SOURCES

Mazets et al. (1981)[45] first reported absorption features near 40 keV in the spectra of several gamma-ray burst sources. The gamma-ray burst detector on board *Ginga* has recently found clear evidence for absorption features consistent with first and second cyclotron harmonics (see Fig. 5[46]). The energies of the cyclotron absorption lines correspond to magnetic fields of a few times 10^{12} gauss and such a strong magnetic field suggests that gamma-ray bursts take place on neutron stars.

The number density of gamma-ray burst sources, n_{GB}, is estimated in terms of the average recurrence interval of a gamma-ray burst source, τ, the disk thickness, D, and the burst rate, $N(<D)$, in a sphere with a radius of D:

$$\begin{aligned} n_{GB} &= N(<D)\tau/(4\pi D^3/3) \\ &= 2.4 \times 10^{-4}[N(<D)/100\text{yr}^{-1}](\tau/10\text{yr})(D/100\text{pc})^{-3}\text{pc}^{-3}. \end{aligned} \tag{9}$$

$N(<D)$ can be estimated from the $\log N(>S) - \log S$ relation for gamma-ray bursts. The $\log N(>S) - \log S$ relation is consistent with the isotropic and uniform distribution, $N(>S) \propto S^{-3/2}$, in the range above a certain peak flux,[47–49] and at least a hundred of the brightest bursts can be considered to occur isotropically and uniformly around us in a year. Hence, for a sphere with a radius of the half-density half-thickness of the old neutron star distribution, the number of gamma-ray bursts in a year is not smaller than a hundred. The half-density half-thickness has been estimated by Paczynski (1990)[50] to be 190 pc, which is roughly consistent with that of the pulsar distribution.[43] The average recurrence interval is considered to be 10 years at least, since most gamma-ray burst sources for which the direction has been determined have not repeated bursts during, at least, the last 10 years.[51]

From these values, the number density of the gamma-ray burst sources in the solar neighborhood cannot be less than 3.5×10^{-5} pc^{-3}, whereas Paczynski (1990)[50] has calculated the volume density of old neutron stars in the solar neighbourhood to be 1.4×10^{-3} pc^{-3} assuming the total number of neutron stars in the Galaxy is 10^9. Hence, the average age of gamma-ray burst sources is estimated to be no younger than 3×10^8 years. Since the number of gamma-ray bursts which reveal cyclotron features is 10%–20% of the total,[45,52] the strong magnetic fields of the gamma-ray bursting neutron stars have not decayed in, at least, 3×10^7 years. This seems to be inconsistent with the decay constant introduced in the pulsar analysis, although there remain some uncertainties in the above parameter estimations.[53] However, it has been found that the results of the pulsar statistical analysis can also be explained by a model in which the magnetic field does not decay but aligns itself with the rotation axis.[54,55] Recent estimations of the ages of X-ray pulsars in the low-mass binaries 4U1626−67 and Her X-1 have also suggested that the neutron stars in these systems have not lost their magnetic fields in 10^8 years or longer.[56] Although observations do not require field decay of neutron stars, a problem is how to explain the weak fields of low-mass X-ray binaries and millisecond pulsars without the field decay. The field decay model can more easily explain the origin of weakly magnetized neutron stars. Further observational and theoretical study is necessary for this interesting problem.

ACKNOWLEDGMENTS

The author thanks Dr. T. Yaqoob for his careful reading of the manuscript.

REFERENCES

1. F. Nagase, *Pub. Astr. Soc. Japan*, **41**, 1 (1989), and references therein.
2. J. H. Taylor and J. M. Weisberg, *Ap. J.*, **253**, 908 (1982).

3. W. H. G. Lewin and P. C. Joss, *Space Sci. Rev.*, **28**, 3 (1981), and references therein.
4. See, e.g., H. Inoue et al., *Pub. Astr. Soc. Japan*, **36**, 831 (1984).
5. T. Ebisuzaki, T. Hanawa and D. Sugimoto, *Pub. Astr. Soc. Japan*, **35**, 17 (1983).
6. M. Kato, *Pub. Astr. Soc. Japan*, **35**, 33 (1983).
7. D. Sugimoto, T. Ebisuzaki and T. Hanawa, *Pub. Astr. Soc. Japan*, **36**, 839 (1984).
8. B. Paczynski and M. Proszynski, *Ap. J.*, **302**, 519 (1986).
9. H.Inoue, *Physics of Neutron Stars and Black Holes*, Y. Tanaka (ed.) (Universal Academy Press, Tokyo, 1988), p. 235.
10. E.g., I. Goldmann, *Astr. Ap.*, **78**, L15 (1979).
11. E.g., J. van Paradijs, *Astr. Ap.*, **107**, 51 (1982).
12. M. Czerny and M. Sztajno, *Acta Astr.*, **33**, 213 (1983).
13. R. A. London, R. E. Taam and W. M. Howard, *Ap. J. (Letters)*, **287**, L27 (1984).
14. T. Ebisuzaki and K. Nomoto, *Ap. J. (Letters)*, **305**, L67 (1986).
15. R. A. London, R. E. Taam and W. M. Howard, *Ap. J.*, **306**, 142 (1986).
16. A. J. Foster, R. R. Ross and A. C. Fabian, *M. N. R. A. S.*, **221**, 409 (1986).
17. T. Ebisuzaki, *Pub. Astr. Soc. Japan*, **39**, 287 (1987).
18. M. Y. Fujimoto and R. E. Taam, *Ap. J.*, **305**, 246 (1986).
19. N. Nakamura, Ph.D. thesis, Univ. of Tokyo (1987).
20. T. Ebisuzaki and N. Nakamura, *Ap. J.*, **328**, 251 (1988).
21. J. van Paradijs et al., *Pub. Astr. Soc. Japan*, **42**, 633 (1990).
22. T. Dotani et al., *Nature*, **347**, 534 (1990).
23. I. Waki et al., *Pub. Astr. Soc. Japan*, **36**, 819 (1984).
24. N. Nakamura, H. Inoue and Y. Tanaka, *Pub. Astr. Soc. Japan*, **40**, 209 (1988).
25. E. Magnier et al., *M.N.R.A.S.*, **237**, 729 (1989).
26. A. J. Foster, R. R. Ross and A. C. Fabian, *M.N.R.A.S.*, **228**, 259 (1987).
27. E.g. G. Baym and C. Pethick, *Ann. Rev. Astr. Ap.*, **17**, 415 (1979).
28. M. Y. Fujimoto, *Ap. J. (Letters)*, **293**, L19 (1985).
29. M. Hashimoto, T. Hanawa and D. Sugimoto, *Pub. Astr. Soc. Japan*, **35**, 1 (1983).
30. E.g. F. K. Lamb, *Accreting Neutron Stars*, W. Brinkmann and J. Trümper (eds.) (Max Planck Institute, Garching, 1982), p. 316; see also F. K. Lamb, D. Pines and J. Shaham, *Ap. J.*, **224**, 969 (1978).
31. P. E. Boynton et al., *Ap. J. (Letters)*, **283**, L53 (1984).
32. K. Nomoto and S. Tsuruta, *Ap. J. (Letters)*, **305**, L19 (1986).
33. J. Trümper et al., *Ap. J. (Letters)*, **219**, L105 (1978); see also W. Voges et al., *Ap. J.*, **263**, 803 (1982).
34. Wm. A. Wheaton et al., *Nature*, **282**, 240 (1979); see also N. E. White, J. H. Swank and S. S. Holt, *Ap. J.*, **270**, 711 (1983).
35. K. Makishima et al., *Nature*, **346**, 250 (1990).
36. F. Nagase et al., in preparation (1991).

37. K. Makishima et al., *Ap. J. (Letters)*, **365**, L59 (1990).
38. K. Mitsuda, et al., *Pub. Astr. Soc. Japan*, **36**, 741 (1984).
39. N. E. White et al., *M.N.R.A.S.*,, **218**, 129 (1986).
40. E.g. K. Koyama et al., *Nature*, **343**, 148 (1990)
41. E.g. T. Ebisuzaki, T. Hanawa and D. Sugimoto, *Pub. Astr. Soc. Japan*, **36**, 551 (1984); F. Verbunt, *Physics of Neutron Stars and Black Holes*, Y. Tanaka (ed.) (Universal Academy Press, Tokyo, 1988), p. 159.
42. R. E. Taam and E. P. J. van den Heuvel, *Ap. J.*, **305**, 235 (1986).
43. A. G. Lyne, R. N. Manchester and J. H. Taylor, *M.N.R.A.S.*,, **213**, 613 (1985).
44. G. M. Stollman, *Astr. Ap.*, **178**, 143 (1987).
45. E. P. Mazets et al., *Nature*, **290**, 378 (1981).
46. T. Murakami et al., *Nature*, **335**, 234 (1988); see also E. E. Fenimore et al., *Ap. J. (Letters)*, **335**, L71 (1988). T. Yamagami and J. Nishimura, *Ap. Spa. Sci.*, **121**, 241 (1986).
47. B. Paczynski and K. Long, *Ap. J.*, **333**, 694 (1988).
48. M. C. Jennings, *Ap. J.*, **333**, 700 (1988).
49. B. Paczynski, *Ap. J.*, **348**, 485 (1990).
50. O. N. Atteria et al., *Ap. J. Supple.*, **61**, 305 (1987).
51. A. Yoshida and T. Murakami, private communication (1991).
52. E.g., J. van Paradijs, *M.N.R.A.S.*, **238**, 45P (1989).
53. B. N. Candy and D. G. Blair, *Ap. J.*, **307**, 535 (1986).
54. A. F. Cheng, *Ap. J.*, **337**, 803 (1989).
55. F. Verbunt, R. A. M. J. Wijers, and H. M. G. Burm, *Astr. Ap.*, **234**, 195 (1990).

Fumiaki Nagase
Institute of Space and Astronautical Science, 3-1-1, Yoshinodai, Sagamihara, Kanagawa 229, Japan

Properties of Neutron Stars from X-Ray Pulsar Observations

Properties of neutron stars revealed from timing analyses of X-ray pulsars are summarized. Neutron star masses have been thus far determined for six X-ray binary pulsars. The masses determined are in the range of 1 $M_\odot$ to 2 $M_\odot$ and their average is about 1.4 $M_\odot$. Interior structures of the neutron star, such as the coupling between the outer crust and the superfluid inner core, have been studied from the changes of pulse periods based on data obtained over long periods of time for several pulsars. X-ray observations of the Crab-like pulsar PSR 0540-69 with *Ginga* provided a unique opportunity to derive an accurate value of the braking index of $n = 2.02 \pm 0.01$ for the neutron star.

1. INTRODUCTION

There are about 30 known X-ray pulsars powered by mass accretion in close binary systems and several X-ray pulsars with an isolated neutron star in a young supernova remnant like the Crab pulsar.[1] Many of the accretion powered pulsars contain a massive companion star and often exhibit an X-ray eclipse. Most X-ray pulsars, except for Her X-1, and LMC and SMC pulsars, are distributed almost uniformly in the Galactic plane.

Neutron stars in the X-ray pulsars contain strong magnetic fields of $(1-4) \times 10^{12}$ G at the surface, which has been confirmed for about 10 X-ray pulsars from recent *Ginga* observations.[2] Matter accreting onto the magnetic pole of the neutron star emits X-rays anisotropically, thus forming X-ray pulses. The pulse shapes in X-ray pulsars are usually broad with a duty cycle of about 50%. Their pulse periods distribute uniformly from fast pulsars to slow pulsars in the range of 68 ms to 835 s.

In addition to the intrinsic pulse period variations, we can measure, mainly for the sources with massive companions, the Doppler shift of pulse periods due to the binary motion of the neutron star. This enables us to determine orbital parameters of the binary systems, thus leading to the estimates of neutron star masses for several pulsars in combination with optical measurements of the amplitudes of radial velocity curves of companion stars. The long-term monitoring of the variations of intrinsic pulse period provides an estimate for acceleration (or deceleration) torques for the crust rotation. In the case of binary X-ray pulsars, an external accretion torque seems to play a dominant role compared with internal torques due to the coupling between the crust and superfluid inner core of the neutron star.

The 50 ms pulsar PSR 0540-69 in the supernova remnant in the LMC is considered to be a typical Crab-like pulsar. The pulses from this pulsar have been observed in the optical and X-ray bands, but not yet in the radio band. X-ray observations of the source with *Einstein* and *EXOSAT* and especially the long-term monitoring with *Ginga* provided the opportunity to investigate in detail the behavior of the pulse period change, thus leading to the determination of a finite value of the braking index.

2. DETERMINATIONS OF NEUTRON STAR MASS

Pulse periods observed from X-ray binary pulsars are usually affected both by the motion of the observer and by the motion of the pulsar. After removing the former effect, the barycentric pulse arrival times are used to obtained intrinsic pulse periods. Determination of the binary orbit is a necessary step for obtaining an intrinsic pulse period and provides in itself valuable information about the binary system and the neutron star.

Once a series of barycentric pulse arrival times is obtained for intervals of about an orbital period or longer, the intrinsic pulse period and its derivatives can be obtained together with the orbital parameters. From such pulse-timing analysis we obtain the orbital period P_{orb}, the projected semimajor axis $a_x \sin i$, the eccentricity e, and the periastron longitude ω of the neutron star. Precise orbital parameters have been obtained so far for 11 X-ray pulsars.[1] From the orbital parameters thus determined the mass function can be calculated as

$$f(M) = \frac{(M_c \sin i)^3}{(M_x + M_c)^2} = \frac{4\pi^2 (a_x \sin i)^3}{G P_{orb}^2}, \tag{1}$$

where G is the gravitational constant and M_x and M_c are the masses of the neutron star and the companion star, respectively.

Among the 11 pulsars, the semiamplitude of the Doppler velocity curve K_c of the companion star has been measured for 6 X-ray binary systems from optical spectroscopic observations. Hence, the mass ratio of the neutron star to the companion star is calculated as

$$q = \frac{M_x}{M_c} = \frac{K_c P_{orb} \sqrt{1-e^2}}{2\pi a_x \sin i}. \tag{2}$$

The inclination angle i can be estimated independently of a_x using the X-ray eclipse. The average radius of the companion star R_c is related to the X-ray eclipse half-angle θ_e as

$$R_c = a(\cos^2 i + \sin^2 i \sin^2 \theta_e)^{1/2}, \tag{3}$$

where $a = a_x + a_c$ is the separation of the centers of mass of the two stars.[3] From this the inclination angle can be estimated as

$$\sin i = \left[1 - \beta^2 \left(\frac{R_L}{a}\right)^2\right]^{1/2} \Big/ \cos\theta_e, \tag{4}$$

where β is the ratio of the companion star radius R_c to the critical potential lobe R_L. The critical potential lobe R_L can be computed as a function of the mass ratio q and the ratio Ω of the rotational frequency of the companion star to the orbital frequency.[5]

Using the orbital parameters derived from pulse timing analysis and the measured values of K_c and θ_e, the stellar parameters, M_c, R_c, and M_x have been estimated for six X-ray pulsars, SMC X-1, Her X-1, Cen X-3, LMC X-4, Vela X-1, and 4U1538-52. The neutron star masses thus derived for the six pulsars are shown in Figure 1. In the computations we adopted the Monte Carlo error propagation technique [1,5] to estimate the uncertainty of the calculated masses. In the computation we assumed that $\beta = 0.9 - 1.0$ and $\Omega = 0.0 - 1.5$; the justification of these assumptions is discussed by Rappaport and Joss.[5]

Present results suggest that the masses of neutron stars are likely to be in the range $1\,M_\odot \leq M_x \leq 2\,M_\odot$ and most of them are consistent with the canonical value of $1.4\,M_\odot$. These results are fundamentally consistent with theoretical models of neutron stars [6,7]. The lower limit of $1.56\,M_\odot$ (95% confidence) for the mass of Vela X-1, however, is somewhat higher than the theoretical maximum masses predicted with the softer equations of state.

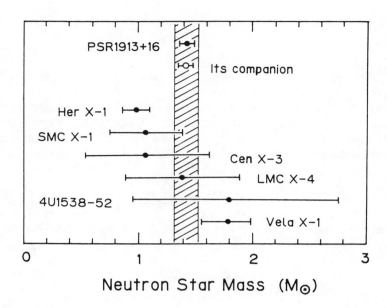

FIGURE 1 Empirically determined neutron star masses. Six of the masses are derived from the pulse timing analysis of X-ray pulsars.[1] The masses of the binary pulsar PSR 1913+16 and its companion are adopted from Taylor and Weisberg[4] for comparison. The hatched region represents the range 1.42 ± 0.10 $M_\odot$.

3. X-RAY PROBING OF NEUTRON STAR INTERIORS FROM PULSE PERIOD CHANGES

A trend of secular spin-up is expected for X-ray pulsars due to the acceleration torque of accreting matter. A simple binary model, involving a rotating neutron star with strong magnetic field and accreting matter forming a stationary accretion disk, predicts a rate of change of pulse period of

$$\dot{P}/P \simeq -3 \times 10^{-5} f P L_{37}^{6/7} \quad \text{yr}^{-1} , \qquad (5)$$

where L_{37} is an X-ray luminosity in units of 10^{37} ergs s^{-1} and the factor f is about unity for the case of a neutron star.[8-10] Ghosh and Lamb[11] and Wang[12] developed improved models which somewhat modify the relation between the spin-up rate and the X-ray luminosity.

From long-term monitoring of period changes for more than a dozen pulsars, however, it is revealed that only a few disk-fed pulsars such as 4U1626−67 exhibit a trend of secular spin-up as predicted by X-ray binary theory.[1] Wind-fed pulsars and Be-star binary pulsars show rather complex fluctuations of pulse period on a wide range of time scales, from a day to a few years, indicating that even a reversal in the angular acceleration torque occurs frequently on a short time scale of days. From recent observations with *EXOSAT* of an outburst of the Be-star binary pulsar EXO 2030+375, however, Parmar et al.[13] obtained a relation between the spin-up rate and X-ray luminosity which supports the X-ray binary theory,[11,12] suggesting that a disk is formed in the transient pulsar at least during the 3-month interval of the outburst.

Regardless of the origin of the exerted torque noise (i.e., external or internal), observations of pulse period fluctuations provide a unique opportunity to investigate the internal structure of a neutron star, as the changes of pulse period represent fluctuations in the angular acceleration of the crust in response to the torque noise exerted on the crust. The observed power density spectrum (PDS) $P_{obs}(f)$ of the fluctuations in the angular acceleration of the crust is expressed by

$$P_{obs}(f) = F(f)P_{in}(f), \tag{6}$$

where P_{in} is the PDS of an input noise torque distributed through the crust.[14] The power transfer function $F(f)$ is due to the dynamical properties of the neutron star. This function could be determined from observations related to the coupling time τ between the outer crust and inner superfluid core. When the neutron star behaves like a rigid body, $F(f) = 1$. When the superfluid core is, however, weakly coupled to the outer crust, the core will respond to changes in the angular acceleration rate of the crust on a time scale longer than the coupling time as

$$F(f) = (I/I_{crust})^2 \quad \text{for} f \leq 1/\tau, \tag{7}$$

where I and I_{crust} are the moments of inertia of the whole star and the crust respectively.

The pulse-timing data for Vela X-1 measured for the past 10 years from *SAS 3* to *Tenma* appear to show a trend of secular spin-up over a first 4-year period followed by a trend of secular spin-down.[1] However, the PDS of the pulse-frequency variations (i.e., the fluctuations in angular acceleration of the neutron star crust) derived from the data is relatively flat for periods ranging from 5 to 2600 days, as seen in Figure 2.[15,16] This indicates that the long-term trends of spin-up and spin-down are consistent with what would be expected as the result of a fluctuating torque. The result supports the hypothesis that the unresolved fluctuations in the angular acceleration of the neutron star crust (i.e., equivalent to a random walk in pulse period) are responsible for the entire history of the observed variation in pulse period of Vela X-1 on all time scales longer than a few days. This means that the neutron star is responding like a rigid body on the time scale explored.

Similar analyses have been performed to obtain power density spectra for Her X-1[17] and the Crab pulsar.[18] Both results so far obtained indicate rigid-body behavior of the neutron stars over the frequency ranges explored. Reduction in the measurement noise in pulse timing analysis will reveal more about the physical properties of neutron stars. Such as analysis is now being attempted on the data observed with *Ginga* from Her X-1 and SMC X-1.

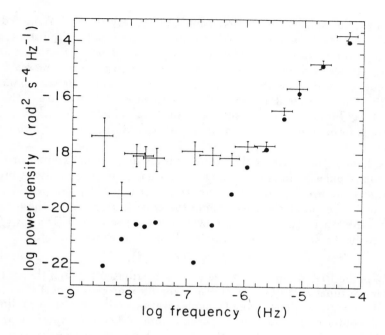

FIGURE 2 PDS of fluctuations in the angular acceleration for Vela X-1 as a function of analysis frequency (*crosses*). The dots indicate the noise contributed by measurement errors, which are mostly due to intrinsic variations in pulse shape.[16]

4. BRAKING INDEX OF PSR 0540–69

Radio pulsars are stationally spinning down due to the loss of rotational energy via electromagnetic or gravitational radiation. In most theoretical models the braking torque is expected to be proportional to some power n of the rotational frequency ν of the neutron star, $\dot{\nu} = -K\nu^n$. Thus the braking index, $n = \nu\ddot{\nu}/\dot{\nu}^2$, itself provides useful information for understanding the mechanism of energy loss from isolated rapidly rotating neutron stars. However, braking indexes have been measured from radio observations with reasonable accuracy for only two pulsars: $n = 2.509 \pm 0.001$ for the Crab pulsar[19,20] and $n = 2.83 \pm 0.03$ for PSR 1509–58.[21]

The 50 ms pulsar PSR 0540–69 was discovered by the *Einstein Observatory* in a supernova remnant in the LMC,[22] and pulsations are detected only in the X-ray and optical bands, not in the radio band. The determination of the braking index for the pulsar has been attempted using X-ray and optical observations,[23–26] as this pulsar is considered to be a Crab-like pulsar with a relatively large rate of

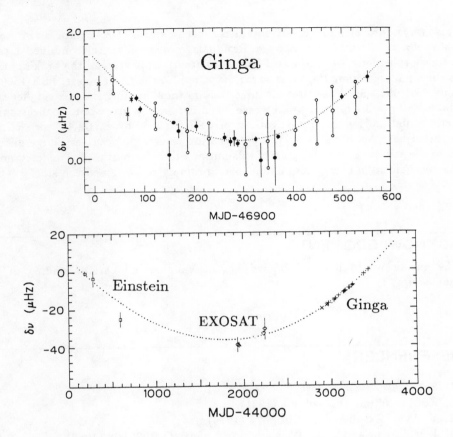

FIGURE 3 *(Top)* Residuals from a linear ephemeris for pulse frequencies determined from *Ginga* observations of PSR 0540–69. Filled circles indicate values obtained unambiguously from short data segments. Open circles, each three points connected by vertical lines, indicate values estimated for data with gaps longer than 20 days; the three data points are based on three alternative estimates of cycle counts between the gap. *(Bottom)* Residuals from a linear ephemeris for PSR 0540–69 for pulse frequencies estimated from a longer baseline with a combination of the *Ginga* results in top panel with the previously observed *Einstein* and *EXOSAT* results.[25]

period decrease. Among these observations, *Ginga* observations of PSR 0540–69 provide the most accurate determination of the braking index owing to continual observation since 1987 April.

Nagase et al.[25] derived a braking index of $n = 2.23 \pm 0.10$ from *Ginga* data alone taken between 1987 April and 1988 October. Residuals from the fit by a linear ephemeris for the pulse frequency are shown in Figure 3 *(top)*. Combining this *Ginga* data with previously obtained *Einstein* and *EXOSAT* data, an improved value of $n = 2.01 \pm 0.01$ was derived. Residuals from the fit for the combined data by a linear ephemeris are shown in Figure 3 (*bottom*). This value of the braking index is the smallest among the three so far measured for the Crab pulsar, PSR 1509-58 and PSR 0540-69. Note that all three pulsars show braking indexes smaller than the value $n = 3$ expected for the case when the energy loss is due to electromagnetic radiation. Further investigations of PSR 0540-69 by *EXOSAT* and *ROSAT* revealed complex behavior of the pulse period change[26] and a very large glitch[27] from the source. Further studies of this pulsar will be important for understanding the mechanism of energy loss in an isolated rapidly rotating neutron star.

ACKNOWLEDGMENT

The author would like to acknowledge B. Vaughan for careful reading of the manuscript.

REFERENCES

1. F. Nagase, *Publ. Astron. Soc. Japan*, **41**, 1 (1989).
2. K. Makishima, this volume (1991).
3. Y. Avni, *Astrophys. J.*, **209**, 574 (1976).
4. J. H. Taylor and J. M. Weisberg, *Astrophys. J.*, **253**, 908 (1982).
5. S. Rappaport and P. C. Joss, in *Accretion-driven Stellar X-ray Sources*, ed. W. H. G. Lewin and E. P. J. van den Heuvel (Cambridge University Press, Cambridge), p. 1 (1983).
6. W. D. Arnett and R. L. Bowers, *Astrophys. J. Suppl.*, **33**, 415 (1977).
7. P. A. Canuto and P. C. Bowers, in *Pulsars*, IAU Symp. No. 95, ed. W. Sieber and R. Wielebinski (D. Reidel, Dordrecht), p. 321 (1981).
8. J. E. Pringle and M. J. Rees, *Astron. Astrophys.*, **21**, 1 (1972).
9. F. K. Lamb, C. J. Pethick, and D. Pines, *Astrophys. J.*, **184**, 271 (1973).
10. S. Rappaport and P. C. Joss, *Nature*, **266**, 683 (1977).
11. P. Ghosh and F. K. Lamb, *Astrophys. J.*, **234**, 296 (1979).
12. Y.-M. Wang, *Astron. Astrophys.*, **183**, 257 (1987).
13. A. N. Parmar et al., *Astrophys. J.*, **338**, 359 (1989).
14. F. K. Lamb, in *Galactic and Extragalactic Compact X-Ray Sources*, ed. Y. Tanaka and W. H. G. Lewin (ISAS, Tokyo), p. 19 (1985).
15. P. E. Boynton et al., *Astrophys. J. Letters*, **283**, L53 (1984).
16. J. E. Deeter et al., *Astron. J.*, **93**, 877 (1987).
17. J. E. Deeter, Ph. D. thesis, University of Washington (1981).

18. P. E. Boynton, in *Pulsars,* IAU Symp. No. 95, ed. W. Sieber and R. Wielebinski (D. Reidel Publishing Company, Dordrecht), p. 279 (1981).
19. E. J. Groth, *Astrophys. J. Suppl.,* **29**, 453 (1975).
20. A. G. Lyne, R. S. Pritchard, and F. G. Smith, *Mon. Not. Roy. Astron. Soc.,* **233**, 667 (1988).
21. R. N. Manchester, J. M. Durdin and L. M. Neuton, *Nature,* **313**, 374 (1985).
22. F. D. Seward, F. R. Harnden, Jr., and D. J. Helfand, *Astrophys. J. Letters,* **287**, L19 (1987).
23. J. Middleditch, C. R. Pennypacker and M. S. Burns, *Astrophys. J.,* **315**, 142 (1987).
24. R. N. Manchester and B. A. Peterson, *Astrophys. J. Letters,* **342**, L23 (1989).
25. F. Nagase, et al., *Astrophys. J. Letters,* **351**, L13 (1990).
26. H. Ögelman and G. Hasinger, *Astrophys. J. Letters,* **353**, L21 (1990).
27. H. Ögelman, G. Hasinger and J. Trümper, *IAU. Circ.,* No. 5162 (1991).

K. Makishima
Department of Physics, University of Tokyo, 7-3-1 Hongo Bunkyo-ku, Tokyo 113, Japan

Magnetic Fields of Binary X-ray Pulsars

From *Ginga* observations of X-ray pulsars, spectral features due to electron cyclotron resonance scattering have been discovered from six (and possibly another) objects. Including two pre-*Ginga* detections, we now have field estimates for about eight X-ray pulsars. The cyclotron feature in these objects appears as a broad spectral dip. We have established a semiempirical analytic model describing the overall hard X-ray spectrum of X-ray pulsars, which suggests that the cyclotron resonance plays a fundamental role in the spectral formation in these objects. The surface magnetic field intensities implied by the resonance energy exhibit a rather narrow scatter, in the range $(1-4) \times 10^{12}$ Gauss. There is some indication that the cyclotron feature disappears at high source luminosity. There is no indication of magnetic field decay within typical lifetimes ($\sim 10^7$ years) of these massive binaries.

1. INTRODUCTION

Detection of effects due to electron cyclotron resonance (ECR) in the spectrum of mass-accreting X-ray pulsars provides the most direct and accurate measurements of the surface magnetic field strengths of neutron stars. We call such a spectral feature the *cyclotron resonance scattering feature* (CRSF).

Before launch of the *Ginga* satellite in February 1987, CRSFs had been observed in only two X-ray pulsars, Her X-1[1,2] and 4U0115+63,[3,4] as well as the possible detections from gamma-ray bursts.[5] Following the discovery of the harmonic cyclotron absorption lines by the *Ginga* gamma-ray burst detectors from several gamma-ray bursts,[6] we have newly discovered CRSFs with *Ginga* from at least six[7,8,19] (and possibly another [9]) X-ray pulsars (Table 1) and refined the previous two detections.[10,11] These results for the first time allow accurate estimates of magnetic field intensities of a fair number of neutron stars.

2. ELECTRON CYCLOTRON RESONANCE

2.1 LANDAU LEVELS

The ECR occurs when an electron in a magnetized plasma makes a transition between different Landau levels. The n-th Landau level energy U_n in a uniform magnetic field B is given as

$$U_n = \frac{mc^2}{1+z}\left[1 + \frac{p^2}{m^2c^2} + \frac{2n+1}{mc^2}E_0\right]^{1/2} \sim \frac{1}{1+z}\left[mc^2 + \frac{p^2}{2m} + \left(n+\frac{1}{2}\right)E_0\right], \tag{1}$$

where m is the electron mass, c is the light velocity, p is the electron momentum along the magnetic field, z is the gravitational redshift, and

$$E_0 = \frac{\hbar eB}{m} \quad \text{(in MKSA units)} = 11.6 \times \left(\frac{B}{10^{12}\,\text{G}}\right)\,\text{keV} \tag{2}$$

is the ECR energy with e the electron charge and h the Planck constant. Assuming $E_0 \ll mc^2$ and neglecting the effect of p, the Landau levels thus become equally spaced, with their separations given as integer multiples of $E_0/(1+z)$; in particular, the fundamental resonance will occur at $E_1 = E_0/(1+z) \sim E_0$.

2.2 FORMATION OF THE RESONANCE SPECTRAL FEATURE

In accreting pulsars, the infalling matter funneled onto magnetic poles becomes hot enough ($\sim 10^8$ K) to emit intense hard X-rays. Incidentally their surface field intensity, thought to be $B \sim 10^{12}$ G, is such that the ECR, Eq. (2), occurs right in the same hard X-ray range. Therefore we expect the ECR to profoundly affect the X-ray pulsar spectra. However what sort of spectral feature would result from the ECR is a complicated problem. In high-density plasmas of the accretion column, the resonance would work as a photon-scattering process rather than a photon emission/absorption process, because electron transitions are predominantly via radiation rather than via collision. In addition, the plasma must be treated as a self-emitting/self-scattering atmosphere. Therefore the resulting spectrum will strongly depend on the geometrical and physical conditions of the emission region.

2.3 SCATTERING CROSS SECTION

In spite of the above complication, the ECR is generally expected to produce a spectral *decrement* near $E \sim E_0$,[12-14] because resonant photons will be scattered into spectral wings while traveling through the accretion column. Thus the energy dependence of the cyclotron scattering optical depth $\tau(E)$ is a key concept in understanding the X-ray pulsar spectra.

Although many detailed calculations of $\tau(E)$ have been carried out, the simplest approximation may be provided by the classical cyclotron cross section in a cold plasma, namely,

$$\tau(E) = \frac{D(WE/E_1)^2}{(E-E_1)^2 + W^2} = \frac{D_T E^2}{(E-E_1)^2 + W^2}. \quad (3)$$

Here $E_1 = E_0/(1+z)$ and W denote the energy and the width of the ECR respectively, D is the optical depth at the resonance, and $D_T = D(W/E_1)^2$ is the optical depth at $E \gg E_1$, which is essentially identical to the ordinary Thomson optical depth. $\tau(E)$ takes the maximum value of $D[1+(E_1/W)^2]$ at $E = E_1[1+(W/E_1)^2]$.

3. OBSERVATIONS

The present observations have been performed with the *Ginga* LAC (large-area proportional counters),[15] which is composed of eight identical proportional counters. The LAC experiment covers an energy range of 1.2–36 keV when the detectors are operated at a normal high voltage (1830 V). The upper-bound energy can be expanded to 60 keV by reducing the high voltage to 1760 V. Although the detection efficiency rapidly decreases above 30 keV, the large effective area (4000 cm^2 in total) and the low intrinsic background make the LAC one of the most powerful cosmic X-ray experiments in the entire 2–60 keV energy range ever put into orbit.

Table 1 summarizes X-ray pulsars with certain or probable CRSFs. Using the *Ginga* LAC, we refined the CRSFs of Her X-1 [10] and 4U0115+63 [11] and discovered CRSFs from six pulsars. The suggested harmonic CRSFs of the peculiar pulsar X2259+586 [9] need further confirmation.

TABLE 1 X-ray Pulsars with Established or Suggested CRSFs

Source	Binary nature		Pulse	Spectrum[a]	
	Companion	Period (d)	Period (s)	E_{cut} (keV)	E_n (keV)
CRSF reconfirmed:					
Her X-1 [10]	A9-B (HZ Her)	1.7	1.24	17–21	35
4U0115+63 [11]	Be	24.3	3.6	7–9	12, 23, 36?
CRSF discovered with *Ginga*:					
X0331+53 [8]	Be (BQ Cam)	34.3	4.38	14–17	28.5, 56?
Cep X-4 [19]	Be?	?	66.3	15–17	32
4U1907+09	OB or Be	8.4	438	14–16	20
4U1538-52 [7]	B0I (QV Nor)	3.74	530	14–16	20
Vela X-1	B0.5Ib (GP Vel)	8.96	283	15–20	27, 53?
GX 301-2	B1.5Ia (BP Cru)	41.5	690	19–21	40
CRSF suggested:					
X2259+586 [9]	single?	?	6.9		~7?, ~13?

[a] The parameters E_{cut} and E_n refer to Eq. (5) and Eq. (7) in the text, respectively.

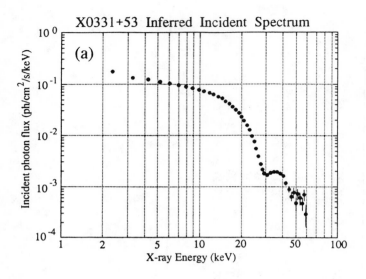

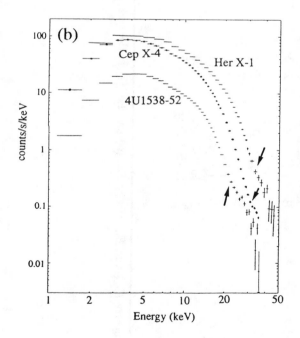

FIGURE 1 Examples of the CRSFs observed with the *Ginga* LAC. (a) Inferred incident X-ray spectrum of X0331+53[8], after removing the detector response. (b) Raw pulse-height spectra of Her X-1,[10] Cep X-4,[19] and 4U1538–52[7], displayed without removing the detector response.

4. DATA ANALYSIS AND RESULTS
4.1 DETECTION OF CYCLOTRON RESONANCE FEATURES

Fig. 1a shows the most prominent CRSF ever observed,[8] from the transient pulsar X0331+53 at 28.5 keV. Usually the CRSF is less prominent, as exemplified in Fig. 1b. The ECR in these cases produces a broad dip or a curvature change in the spectrum, rather than a sharp line feature as observed during gamma-ray bursts[5,6]. Sometimes the inferred CRSF is very shallow, or it appears near the xenon K-edge energy (35 keV), where the raw spectrum bears an instrumental feature. To confirm and quantify the CRSF, we therefore need to fit the spectrum with various mathematical and/or physical models, as described below.

4.2 FITTING WITH CONVENTIONAL SPECTRAL MODELS

We may model the hard X-ray photon spectrum of accreting pulsars, $f(E)$, as

$$f(E) = IE^{-\alpha}h(E), \qquad (4)$$

where E is the X-ray energy, I is the normalization, α is the photon index, and $h(E)$ is a function representing the high-energy spectral break. The form of $h(E)$ that has most commonly been used is the exponential cutoff model,[4] namely,

$$h(E) = \begin{cases} 0 & \text{for } E < E_{\text{cut}}, \\ \exp\left(-\dfrac{E - E_{\text{cut}}}{E_f}\right) & \text{for } E > E_{\text{cut}} \end{cases} \qquad (5)$$

where E_{cut} and E_f denote the start of cutoff and the cutoff steepness respectively. However, due to the abrupt kink at $E = E_{\text{cut}}$, this model usually fails to give acceptable fits to high-quality spectra of X-ray pulsars.[16]

An improved, smoother form of $h(E)$ may be provided with[17]

$$h(E) = \frac{1}{1 + \exp\left(\dfrac{E - E_{\text{cut}}}{E_f}\right)}, \qquad (6)$$

which is identical to the "Fermi-Dirac distribution function" although here we do not claim its physical meaning. This model can well reproduce X-ray pulsar spectra below the ECR energy. Therefore, significant negative deviations from Eq. (6) may be taken as evidence for a CRSF in the observed X-ray spectrum.

4.3 FITTING WITH THE CYCLOTRON SCATTERING MODEL

The simplest modeling of the CRSF would be a Gaussian or Lorentzian absorption line applied to a smooth continuum such as Eq. (6). However observations indicate that the spectral break and the CRSF are so closely interlinked (see section 5.1) that they should be treated in a unified way using a single model. So far the "cyclotron scattering" cutoff, defined as

$$h(E) = \exp[-\tau(E)], \qquad (7)$$

has served well for this purpose, where $\tau(E)$ is given by Eq. (3). Compared with Eqs.(5) and (6), this model has two advantages. One is that Eq. (7) has a clear physical basis, although it may be a fair oversimplification. The other is that Eq. (7) can simultaneously describe the cyclotron resonance trough at $E \sim E_1$ and the spectral break starting at $\sim E_1/2$. In most cases, the cyclotron scattering model (usually with the second harmonic term included; see section 4.4) has successfully reproduced the overall hard X-ray spectrum of pulsars that exhibit CRSFs[7,8,10] (see Fig. 2).

4.4 HIGHER HARMONICS

The spectra of 4U1538-52[7] and Her X-1[10] required the second harmonic term to be included in Eq. (3) through Eq. (7), with the second harmonic energy E_2 fixed at $2E_1$. In these cases the optical depth for the second harmonic (D_2) turned out to be larger than that (D_1) for the fundamental. This somewhat puzzling result may indicate that unlike the fundamental resonance (see section 2.2), the second harmonic resonance can be regarded as pure absorption. Presumably this is because an electron excited to the second Landau level, by absorbing a photon of energy $\sim 2E_1$, will return to the ground level by emitting two photons of energy $\sim E_1$ in cascade. These photons will in turn fill up the dip at the fundamental resonance and make it less prominent.

The reality of the second harmonic has been reinforced by observations of X0331+53 (Fig. 1a); Makishima et al.,[8] leaving the second harmonic parameters all free, obtained $E_2 = (1.85 \pm 0.07) \times E_1$. Still stronger evidence has been obtained with 4U0115+63[11]: a fit with Eq. (6) causes negative residuals at $\sim 12, \sim 23, \sim 36$, and even ~ 47 keV, well in harmonic ratios (Fig. 2a). Inclusion of up to the fourth harmonic in Eqs. (3) and (7), with E_n all left free, gives an acceptable fit (Fig.2b). We therefore believe in the physical reality of higher harmonics.

In some pulsars, the fundamental resonance may be quite shallow while the second harmonic is very prominent; an example is Vela X-1, exhibiting a subtle negative feature at ~ 28 keV (Fig. 2c) against Eq. (6). The same spectrum can be much better described with Eq. (7) incorporating a shallow ($D_1 \sim 0.2$) fundamental resonance at ~ 28 keV and a deep ($D_2 \sim 3.5$) second harmonic at 50-54 keV (Fig. 2d).

4.5 PULSE-PHASE DEPENDENCE

In many pulsars we have found that the resonance energy E_1 remains constant within 10–20% across the pulse phase. This indicates that E_1 is a well-defined quantity for each pulsar, and we regard it as representing the *surface* field strength at the magnetic poles, where the main body of the observed X-ray is thought to be produced. Our data are all consistent with X-ray pulsars having dipole magnetic configurations, without any indication of multipole configuration.

On the other hand, pulse-phase dependence of D and W is still somewhat confusing. In Her X-1 the CRSF is clearest (largest D and smallest W) around the pulse peak,[1,2,10] while in 4U1538−52 it is most prominent across the minor peak of the double-peaked profile.[7] In GX 301-2, Cep X-4, and 4U1907+09, which exhibit double-peaked pulse profiles, there is an indication that the CRSF is deepest at the declining slope of the major pulse. Such an asymetric phase dependence could not be explained easily. The pulse-phase dependence is rather complicated in 4U0115+63 and Vela X-1, while no phase-resolved information is available for X0331+53.

5. DISCUSSION

5.1 LUMINOSITY DEPENDENCE

It has been revealed that the accretion-powered pulsars exhibit CRSFs as a rule rather than as an exception. Actually, we have detected CRSFs from more than half the pulsars for which high-quality *Ginga* data have been available. However, our search for CRSFs has been negative in several pulsars, including Cen X-3, LMC X-4, SMC X-1, X-Per, and GX 1+4. The first three are the most luminous ($> 5 \times 10^{37}$ ergs/s) X-ray pulsars, whereas luminosities of the nine CRSF pulsars are all $< 3 \times 10^{37}$ ergs/s. Nevertheless, we can well fit the spectra of these luminous pulsars with Eq. (7), obtaining rather large (> 0.5) values of W/E_1. Note that Eq. (7) no longer gives sharp dips for $W > E_1$.

Presumably the CRSF does exist in the very luminous pulsars as well, but it becomes smeared out due to an enhanced thermal Doppler broadening and/or due to an increased accretion-column height leading to a larger field gradient over the emission region. Interestingly enough, the critical luminosity of the CRSF as inferred above, $\sim 4 \times 10^{37}$ ergs/s, is similar to the transition point from pencil-beam to fan-beam patterns.[18] Clark et al.,[7] in fact showed that the CRSF pulsar 4U1538−52 almost certainly has a pencil-beam pattern.

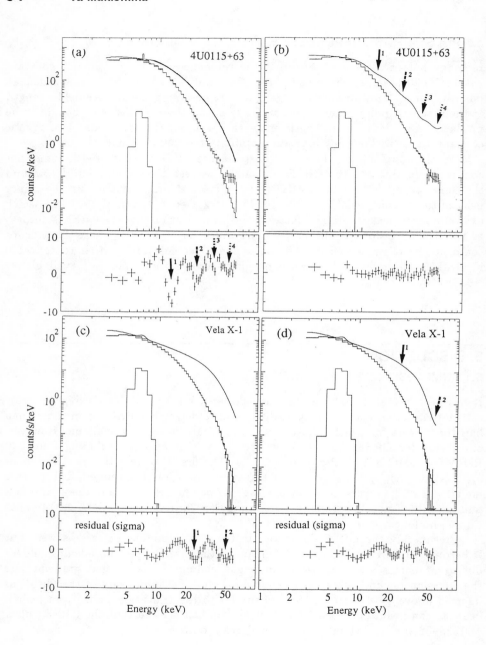

FIGURE 2 (a and b) Pulse-height spectra of 4U0115+63,[11] and (c and d) Vela X-1. Panels a and c are the fits with the conventional Fermi-Dirac cutoff, Eq. (6). Panels b and d are the fits with the cyclotron scattering model, Eq. (7), including up to the second and the fourth harmonics respectively. Smooth curves show the best-fit models, which are convolved with the detector response to give the histograms. Lower panels show fit residuals.

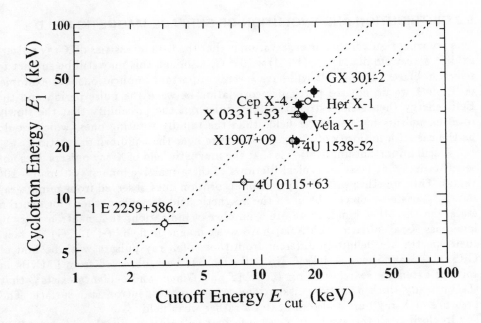

FIGURE 3 A scatter plot between the spectral cutoff energy E_{cut}, given by Eq. (5), and the fundamental cyclotron resonance energy E_1. There is a good correlation between these two quantities, indicating that the spectral cutoff is caused by the cyclotron resonance.

5.2 SPECTRAL FORMATION

We have established that, with an appropriate inclusion of higher harmonic terms, Eq. (7), which incorporates the classical cyclotron scattering cross section, can well reproduce the X-ray pulsar spectra. The parameter E_1 can be regarded as a good measure of the apparent (gravitationally redshifted) ECR energy, hence of the field strength, at the magnetic poles. The CRSF without doubt plays a fundamental role in the formation of X-ray pulsar spectra. This is revealed in Fig. 3, where the values of E_1 of the nine CRSF pulsars are plotted against their values of E_{cut} as obtained through fitting the conventional Eq. (5). The good correlation seen in Fig. 3 strongly suggests that the spectral cutoff is caused by the presence of harmonic CRSF series.

5.3 DISTRIBUTION AND EVOLUTION OF SURFACE MAGNETIC FIELD

One amazing result from our observation is that the field intensities of X-ray pulsars exhibit a very small scatter, $(1-4) \times 10^{12}$ G, although this may still be subject to selection biases. This result will have several important implications. For example, in Table 1 we do not see particular correlation between the pulse period and the ECR energy (hence field strength). This rules out the possibility that the slowly rotating pulsars have stronger fields than the rapidly rotating ones, which would be the case if a majority of X-ray pulsars were near the equilibrium rotation.

A still important implication is that the magnetic field of X-ray pulsars does not seem to decay, at least on typical lifetimes of these massive binary systems ($\sim 10^7$ years). This, together with the cyclotron absorption lines observed from gamma-ray bursts,[5,6] casts serious doubt upon the magnetic field decay hypothesis for neutron stars. On the other hand, there are a number of indications that neutron stars in low-mass X-ray binaries (LMXBs) have weak magnetic fields ($< 10^9$ G). In fact, their spectra are definitely different from those of X-ray pulsars, with no hint of CRSFs at least above ~ 1 keV. No pulsation has been observed from LMXBs in spite of extensive searches using *EXOSAT* and *Ginga*, and evidence exists[20] that the optically thick accretion disk extends close to the neutron star surface. The presence of X-ray bursts also indicates a rather weak field.

In closing, all the available X-ray information points to a rather bimodal field distribution among neutron stars. This suggests the presence of some sort of magnetic phase transition inside the neutron star.

REFERENCES

1. Voges, W., Pietsch, W., Reppin, C., Trümper, J., Kendziorra, E. and Staubert, R. 1982, *Astrophys. J.*, **263**, 803.
2. Soong, Y., Gruber, D. and Rothschild, R. 1990, *Astrophys. J.*, **348**, 641.
3. Wheaton, Wm. et al. 1979, *Nature*, **282**, 240.
4. White, N. E., Swank, J.H. and Holt, S.S. 1983, *Astrophys. J.*, **270**, 711.
5. Mazets, E. P, Golenskii, S. V., Aptekar, R.L., Guryan, Y. A. and Ilynskii, V. N. 1981, *Nature*, **290**, 378.
6. Murakami, T. et al. 1988, *Nature*, **335**, 234.
7. Clark, G., Woo, J., Nagase, F., Makishima, K. and Sakao, T. 1990, *Astrophys. J.*, **353**, 274.
8. Makishima, K. et al. 1990, *Astrophys. J. (Letters)*, **365**, L59.
9. Koyama, K. et al. 1989, *Publs. astr. Soc. Japan*, **41**, 461.
10. Mihara, T. et al. 1990, *Nature*, **346**, 250.
11. Nagase, F. et al. 1990, *Astrophys. J. (Letters)*, submitted.
12. Bonazzola, S., Heyvaerts, J. and Puget, J. L. 1979, *Astr. Astrophys.*, **78**, 53.

13. Mészáros, P. and Nagle, W. 1985, *Astrophys. J.*, **298**, 147.
14. Wang, J., Wasserman, I. and Salpeter, E. 1989, *Astrophys. J.*, **338**, 343.
15. Turner, M. J. L. et al., 1989. *Publs. astr. Soc. Japan*, **41**, 345.
16. Makishima, K. et al. 1990, *Publs. astr. Soc. Japan*, **42**, 295.
17. Tanaka, Y. 1985, in *Radiation Hydrodynamics in Stars and Compact Objects*, ed. D. Mihalas and K.-H. Winkler (Berlin: Springer), p. 198.
18. Parmar, A.N., White, N. E., Stella, L., Izzo, C. and Ferri, P. 1989, *Astrophys. J.***338**, 359.
19. Koyama, K. et al. 1991, *Astrophys. J. (Letters)*, **366**, L19.
20. Mitsuda, K. et al. 1984, *Publs. astr. Soc. Japan*, **36**, 741.

R. Hoshi
Department of Physics, Rikkyo University, Nishi-Ikebukuro 3, Toshimaku, Tokyo 171, Japan

Mass-Radius Relations in X-ray Pulsars

1. INTRODUCTION

Cyclotron line features have so far been detected in the spectra of seven X-ray pulsars. The *Ginga* satellite has played an important role in determining the cyclotron line energies in these pulsars (see Table 1). In some sources evidence of a cyclotron line feature of second harmonics has been found in spectra taken with the *Ginga* satellite. Two of the seven sources are low-mass binaries, three are X-ray transients with Be star companions and the remaining sources are massive binaries with early-type massive companions and spin periods longer than 100 s.

For an X-ray emitting neutron star, if the spin period P, the spin-up rate $\dot{P}$, the X-ray luminosity L_X and the cyclotron line energy ΔE_∞ at a distant observer are known, it can be possible to determine a semiempirical mass-radius relation, as has been done by Wasserman and Shapiro[1] for Her X-1 and 4U0115+63.

If we obtain mass-radius relations in some X-ray pulsars that compare tolerably well with theoretical mass-radius relations in model neutron stars, observations may constrain theoretical models and in particular equations of state for very high density material.

2. METHOD

Semiempirical mass-radius relations are calculated under several assumptions. X-ray pulsars are assumed to be neutron stars with dipolar magnetic fields and to accrete matter via accretion disks. Another assumption is that cyclotron lines arise in plasmas at rest at the magnetic poles of the pulsars. With these assumptions, the field strength B_s at the magnetic pole can be evaluated in terms of the cyclotron line energy as

$$B_s = \frac{2\pi m_e c}{eh} \frac{\Delta E_\infty}{(1-x^{-1})^{1/2}} , \qquad (1)$$

where x is the neutron star radius in units of the Schwarzschild radius, $x = R/(2GMc^{-2})$, and M and R are the mass and radius of the neutron star, respectively.

On the other hand, at the magnetic pole the field strength B_s is written as[1]

$$B_s = \frac{3\mu c^6}{4(GM)^3} \left[-\ln(1-x^{-1}) - \frac{(1+x^{-1}/2)}{x} \right] , \qquad (2)$$

where μ is the magnetic moment. If μ is evaluated in terms of a magnetospheric model combined with observed quantities such as P, $\dot{P}$, and L_X, we determine a semiempirical mass-radius relation.

3. DIPOLE MOMENT

We adopt the magnetospheric model presented by Wang.[2] According to this author the spin-up rate of an X-ray pulsar is given by

$$-\frac{\dot{P}}{P} = \frac{5}{2} \frac{L_X}{G^{1/3} M^{4/3} R} \left(\frac{P}{4\pi}\right)^{4/3} f(y_0) ,$$

$$f(y_0) = y_0^{1/2} + \frac{2}{9} y_0^{31/80} \left[1 - y_0^{3/2} - \frac{y_0^{9/4}}{(1-y_0^{3/2})^{1/2}} \right] , \qquad (3)$$

where $y_0 = r_0/r_c$, r_c is the corotation radius, and r_0 is some radius inside which the rotation velocity of the accretion disk plasma starts to drop below the Keplerian rate (the inner boundary of the accretion disk). Equation (3) determines y_0 as a function of M and R, if P, $\dot{P}$, and L_X are known. If a steady state is reached, y_0 approaches 0.971. For those X-ray pulsars with small $\dot{P}$ we adopt this equilibrium value of 0.971.

The magnetic dipole moment is written in terms of y_0 as

$$\mu \propto \left(\frac{\xi}{\gamma}\right)^{1/3} \frac{\alpha^{3/20}}{\eta} \frac{L_X^{8/15} M^{7/9} P^{211/180} [y_0^{211/80}(1-y_0^{3/2})^{-1/2}]^{2/3}}{[1-(1-x^{-1})^{1/2}]^{8/15}} . \quad (4)$$

Here, the parameters ξ and γ relate to the time scales of amplification and reconnection of magnetic fields in the disk and are taken as unity. The parameter α appears in the viscous stress used in the standard model of the accretion disk and is insensitive to the magnetic dipole moment as seen from equation (4). The most important parameter is the screening coefficient η. Ghosh and Lamb[3] found it to be of the order of 0.2 over a wide range of radial distance in the plane of the accretion disk. Since η is inversely proportional to μ, the value of η sensitively depends on the semiempirical mass-radius relation.

4. MASS-RADIUS RELATION

Using the above equations the semiempirical mass-radius relation can be written as

$$\frac{M}{M_\odot} = A\{F(x)\}^{9/20} , \quad (5)$$

$$A = 12.5 \left(\frac{\xi}{\gamma}\right)^{3/20} \frac{\alpha^{27/400} L_{x,37}^{6/25} P^{211/400} [y_0^{211/80}(1-y_0^{3/2})^{-1/2}]^{1/3}}{\eta_{0.2}^{9/20} (\Delta E_{\infty,\mathrm{keV}})^{9/20}} ,$$

$$F(x) = \frac{(1-x^{-1})^{1/2} [-\ln(1-x^{-1}) - (1+x^{-1}/2)/x]}{[1-(1-x^{-1})^{1/2}]^{8/15}} ,$$

and

$$R = \frac{2GM}{c^2} x , \quad (6)$$

where $L_{X,37}$ is the X-ray luminosity in units of 10^{37} ergs/s, $\Delta E_{\infty,\mathrm{keV}}$ is the cyclotron line energy in keV, and $\eta_{0.2} = \eta/0.2$.

Once the value of A is evaluated, a mass-radius relation can be determined as shown in Fig. 1, in which mass-radius relations are shown for several different values of A. Dotted curves are the theoretical mass-radius relations for the mean field (MF), three-nucleon interaction (TNI), and Reid + pion-condensate ($R+\pi$) equations of state. The values of A ranging from 4 to 7 seem to be consistent with the theoretical models.

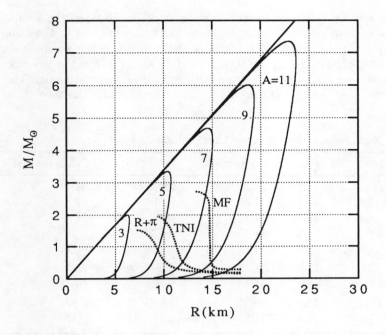

FIGURE 1 Semiempirical mass-radius relations. Mass-radius relations are shown for five different values of A. Dotted curves are the theoretical mass-radius relations for the mean field (MF), three-nucleon interaction (TNI), and Reid + pion-condensate ($R + \pi$) equations of state.[1]

5. DISCUSSION

The values of A have been listed in Table 1 for the X-ray pulsars having clear cyclotron line features in their spectra, except for the last two sources. In these sources cyclotron line energies are estimated from the cutoff energies in the X-ray spectrum following Makishima et al.[4] In Table 1 P, L_X, $\dot{P}$ and ΔE_∞ are also listed. Unfortunately, the spin-up rate $\dot{P}$ is known only for Her X-1 and 1626-67, since the other pulsars have exhibited no smooth spin-up rate but have sometimes exhibited spin-down episode. For Cen X-3 the averaged spin-up rate over ~19 years is listed in Table 1.

For Her X-1 the value of y_0 is close to the equilibrium value, providing $A = 5$. The semiempirical mass-radius relation is consistent with the theoretical neutron star models. For the Be-transient, 4U0115+63 the spin-up rate during outburst has yet to be observed. If the equilibrium value is adopted, the value of A is 13. The necessary spin-up rate to reduce A to about 5 is 10 - 20 times the secular spin-up rate. The spin-up rate during an outburst has been observed in the same class of the X-ray transient, A0535+26, which is about 30 times the secular value. Thus, for 0115+63 it may be possible to reduce A to about 5. For the low mass binary, 1E2259+586, necessary quantities are all determined observationally. The value of A is somewhat larger by a factor of 1.5 - 2 to reconcile with the theoretical models.

TABLE 1 X-ray Pulsars with Cyclotron Line Features

Source[1]	P (s)	L_X'(ergs/s)	$-\dot{P}$ (s/s)	ΔE_∞ (keV)	A
Her X-1 (L)	1.24	2E37	[eq][2]	34[(1)]	5
0115+63 (Be)	3.61	1E37	[eq]	12[(2)]	13
0331+53 (Be)	4.37	2E37	[eq]	28[(3)]	11
2259+586 (L)	6.98	1E35	$-6.2E-13$	7.2[(4)]	9
1538−52 (M)	529	4E36	[eq]	29[(5)]	110
Vela X-1 (M)	283	5E36	[eq]	28[(6)]	73
GX 301-2 (Be)	683	3E36	[eq]	40[(6)]	87
Cen X-3 (M)	4.84	5E37	3.6E−11	20[(3)]	17
1626−67 (L)	7.68	1E37	5.3E−11	[35][(3)]	10

[1] L = low-mass binary;   Be = Be transient;   M = massive binary.
[2] The equilibrium value has been adopted.

References: (1) Mihara et al., *MNRAS* **346**, 250 (1990). (2) Nagase et al., *Ap. J. Letters* **375** (1991) L49. (3) Makishima et al., *Ap. J.* **365**, L59 (1990). (4) Koyama et al., *Publ. Astron. Soc. Japan* **41**, 461 (1990). (5) Clark et al., *Ap. J.* **353**, 274 (1990). (6) Makishima, this volume (1992).

4U1538–52, Vela X-1 and GX 301-2 are characterized by their long spin periods. The values of A, assuming the equilibrium value of y_0, are about 100. In order to reduce A by a factor of 10, it requires the value of $y_0 = 0.1$. This means from the definition that the inner boundary of the accretion disk penetrates deep into the magnetosphere and much smaller than the corotation radius, suggesting that these sources are too far from a equilibrium state on a balance of torque. In order to calculate more realistic mass-radius relations, the spin-up rates should be determined more precisely. The semiempirical mass-radius relations may be available for diagnosis of the interaction between the magnetosphere and the accretion disk by comparing with the theoretical models.

ACKNOWLEDGMENTS

I thank F. Nagase for directing my attention to the present work and for providing necessary data for X-ray pulsars before publication.

REFERENCES

1. I. Wasserman and S. L. Shapiro, *Ap. J.* **265**, 1036 (1983).
2. Y.-M. Wang, *Astron. and Astrophys.* **183**, 257 (1987).
3. P. Ghosh and F. K. Lamb, *Ap. J.* **232**, 259 (1979).
4. K. Makishima et al., *Ap. J.* **365**, L59 (1990).

R. D. Blandford
Theoretical Astrophysics, Caltech, Pasadena, CA 91125, USA

Gamma Ray Bursts as Neutron Starquakes

A brief overview, update, and critique of a particular model for γ-ray bursts developed most recently by Blaes et al. (1989, 1990) is presented. In this model, it was proposed that interstellar gas accretes onto old neutron stars and undergoes pycnonuclear and electron capture reactions, possibly developing an (elastic) Rayleigh-Taylor unstable density inversion. This mechanism, or some other cause, may trigger a "deep-focus starquake" which excites intense seismic waves which are transmitted into the magnetosphere as large-amplitude, relativistic hydromagnetic waves. These waves, in turn, develop a field-parallel electric field which accelerates a relativistic electron-positron plasma, from which the observed γ-rays eventually emerge. It is concluded that

(i) If interstellar accretion supplies the fuel for the bursts, the γ-ray energy release must exceed ~ 10 keV n^{-1}.

(ii) Accreting neutron stars should have rotation periods of several hours.

(iii) Slowly accreted interstellar gas will form a pure helium quantum liquid between densities $3 \times 10^8 \lesssim \rho \lesssim 6 \times 10^8$ g cm^{-3} lying on an oxygen crust which transforms to ^{16}C at a density $\rho \sim 2 \times 10^{10}$ g cm^{-3}.

(iv) The ^{16}C crust may form an unstable interface with the primordial crust, but this instability is unlikely to account for bursts because too little mechanical energy will be released and if the crust gets hot enough to initiate nuclear burning then as much as 6×10^{45} ergs of nuclear energy will be released in a burst that could be seen up to ~ 1 Mpc away.

(v) Seismic waves produced below the helium liquid will be transmitted slowly, mainly through the regions of the highest magnetic field, and can energize long-duration bursts.

(vi) Large horizontal magnetic fields may form as a consequence of horizontal spreading of the accreted gas, and bursts could be due to MHD instability.

(vii) High altitude particle acceleration can be responsible for the majority of the observed γ-rays but Compton recoil cyclotron photons may produce a soft X-ray excess.

1. INTRODUCTION

Despite 18 years of observation and theoretical work, there is still no widely accepted explanation for γ-ray bursts. The discovery of apparent cyclotron lines,[1,2] isotropy,[3] periodicity,[4] and perhaps redshifted electron-positron annihilation lines[1] has, however, pointed to a local neutron star origin. One of the first neutron star models involved starquakes[5] and this has been discussed in several publications, most recently Blaes et al. (1989, 1990)[6,7] and references cited therein. In this contribution, I report on some recent observational and theoretical developments relevant to the particular models of Blaes et al., which will be discussed elsewhere by these authors, emphasizing some outstanding problems and future prospects.

More general recent discussions of γ-ray bursts include Refs. 8–11, to which reference should be made for more complete bibliographies.

2. INTERSTELLAR ACCRETION ONTO NEUTRON STARS

A typical faint burst fluence is $\sim 10^{-6}$ ergs cm^{-2} and the anticipated scale height of the old and slow pulsars in which we are interested is ~ 300 pc, corresponding to a velocity dispersion ~ 50 km s^{-1}.[12,13,14] The observed isotropy and source counts (now exhibiting some evidence for spatial inhomogeneity) indicate that the faintest bursts are located within two scale heights, implying a typical burst energy $E_\gamma \sim 10^{37}$ ergs. (Possible relativistic beaming will not affect this argument.) Now the maximum surface density of old neutron stars in the solar neighborhood is ~ 3 pc^{-2}, and this would would require a production rate averaged over the Galactic lifetime about 10 times the currently favored estimates.[15] We would therefore observe a changing population of $\lesssim 3 \times 10^5$ old neutron stars. As the burst rate appears to be $\gtrsim 100$ yr^{-1}, the recurrence time per neutron star must be $\lesssim 3000$ yr. Over a ~ 10 Gyr Galactic lifetime, each neutron star must produce $\gtrsim 3 \times 10^{43}$ ergs of γ-ray energy, independent of the neutron star scale height.

It was proposed that interstellar accretion at the Hoyle-Lyttleton rate [16]

$$\dot{M} \sim 10^{10} \left(\frac{v}{50 \text{ km s}^{-1}}\right)^{-3} \text{ g s}^{-1}. \tag{1}$$

would allow $\sim 3 \times 10^{27}$g of gas to settle onto the neutron star. A lower bound on the efficiency of γ-ray emission is $\epsilon_\gamma \sim 10^{16}$ ergs g^{-1} ~ 10 keV n^{-1}. It should be emphasized that this is a fairly strong lower bound if accretion provides the fuel. If pulsars are less numerous, move faster, or suppress accretion, then the efficiency must be even higher. Furthermore most mechanisms so far proposed only channel a fraction of the total energy release into γ-rays increasing ϵ_γ even more.

Kluzniak[17] has questioned whether or not accretion will even occur in the presence of an electromagnetic wind from the spinning neutron star. We can examine this issue assuming that the typical star speed is 50 km s^{-1}. The gas encountered by a neutron star is likely to be weakly ionized in its ambient state. However, the neutron star will have a surface temperature $\sim 3 \times 10^5$ K and will create ionizing photons at a rate $\sim 10^{40}$ s^{-1}, which will ionize the surrounding gas beyond the accretion radius

$$r_B \sim 2 \times 10^{13} \text{ cm}. \tag{2}$$

Now suppose that the neutron star has a dipolar field $\sim 10^{12} B_{12}$ G and a period P and slows down at the magnetic dipole rate. After a time t_9 Gyr, the period will be $P \sim 20 B_{12} t_9^{1/2}$ s. Accretion will proceed provided that the ram pressure of the interstellar gas exerted at the accretion radius exceeds the momentum flux emanating from the neutron star. Adopting the magnetic dipole formula for neutron star slow down, we find that the period P must satisfy

$$P \gtrsim P_B \sim 9 B_{12}^{1/2} \dot{M}_{10}^{-1/4} \text{s}. \tag{3}$$

This will happen after a time $t_9 \sim 0.2 B_{12}^{-1} \dot{M}_{10}^{-1/4}$. This is, perhaps, a conservative criterion as decay of the stellar dipole moment may further facilitate accretion.

Once accretion can proceed at the accretion radius, the ram pressure will increase $\propto r^{-5/2}$, whereas the pulsar momentum flux will increase $\propto r^{-2}$ so that the inflow will not be resisted. However, a new stress balance can be established within the neutron star light cylinder when the Reynolds stress exerted by the accreting matter balances the Maxwell magnetic stress of the dipolar field. The details of this interaction are still controversial. However, as long as the star spins faster than corotation at this radius, r_E, the matter stress can be approximated by $\sim \rho\Omega(GMr_E)^{1/2}$. Solving for r_E, we obtain

$$r_E \sim 2 \times 10^{10} B_{12}^{1/2} \dot{M}^{-1/4} \left(\frac{P}{P_B}\right)^{1/5} \text{ cm} \tag{4}$$

The torque acting on the star, $G_E \sim B_E^2 r_E^3$, will decelerate the star on a time scale $t_E \sim I\Omega/G_E \sim 6 \times 10^7 B_{12} \dot{M}_{10}^{-1/2}(P/P_B)^{-2/5}$ yr until corotation is achieved at a period

$$P_C \sim 4 \times 10^4 \dot{M}_{10}^{-3/7} B_{12}^{6/7} \text{ s} \tag{5}$$

when $r_E \sim 10^{11} \dot{M}_{10}^{-2/7} B_{12}^{4/7}$ cm. At this point, accretion onto the neutron star should definitely be able to proceed.

Periodic behavior is conspicuous by its absence in γ-ray bursts, expecially in those objects that exhibit cyclotron emission. Only the soft repeater in the LMC has an unambiguous period ($P = 8$ s), though a 2-s period[18] of questionable significance[19] has been reported. This may be consistent with the sources being slowly rotating, old neutron stars.

3. NUCLEAR REACTIONS

In the simplest model, accreted interstellar gas was supposed to settle uniformly on the surface. At these low accretion rates, typically some eight orders of magnitude smaller than those appropriate to the most luminous X-ray pulsars, the hydrogen will burn steadily to helium at such a low temperature that the helium will be unable to undergo thermonuclear reaction. (If the accreted gas is able to spread over the whole surface, the mass flux through the crust should be $\sim 10^{-3}$ g cm^{-2}s^{-1}, about four orders of magnitude smaller than the flux necessary to burn helium.) The helium will form a solid crust which melts to a quantum liquid (because of zero-point fluctuations) at a density $\rho \sim 3 \times 10^8$ g cm^{-3}.[20]

A major uncertainty in this scenario concerned the pycnonuclear reaction rates. New computations[21,22] have affirmed the Salpeter–Van Horn[23] binuclear rates for a solid lattice and give a 3α rate

$$r_{3\alpha} \sim 10^{22} \rho_9^{18} \text{ cm}^{-3}\text{s}^{-1} \tag{6}$$

for a pure-helium liquid with density $\rho = 10^9 \rho_9$ g cm^{-3} in the range $0.3 \lesssim \rho_9 \lesssim 1$. Once a carbon nucleus is formed, α capture follows in ~ 1 ns.[24] However further reaction takes over 1 Gyr to occur and in this time the oxygen and all the interstellar metals that have survived will precipitate out in $\sim 10^5$ yr to the bottom of the helium layer (about 5 fathoms) at a density of 5.6×10^8 g cm^{-3}. This is determined by requiring that the oxygen production rate balances the mass accretion rate. This interface is stable because the oxygen density is 1% larger than the helium density.

No further reactions will occur until the oxygen undergoes two-stage electron capture to form ^{16}C at a density $\rho \sim 2 \times 10^{10}$ g cm^{-3}. At our nominal mean accretion rate, the star must be $\sim 10^{10}$ yr old before this occurs, and many stars will not reach this stage. The last remaining possibility is that freshly formed ^{16}C can undergo binuclear reaction, which might result in the production of heavier species.

If the accretion proceeds in this fashion there will be considerable nuclear free energy available. The helium layer contains $\sim 3 \times 10^{25}$ g, which can release 2 MeV n^{-1} or $\sim 6 \times 10^{43}$ ergs if it is converted into the equilibrium nucleus, ^{64}Ni. Below this, there is $\sim 3 \times 10^{27}$g of oxygen, which could release a hundred times more energy, ample to account for the energy demands of all observed bursts. However, it is not clear how this energy can be released in small doses. If the temperature is raised to the values necessary to bring about a thermonuclear flash ($\sim 10^8$ K for helium, $\sim 5 \times 10^8$ K for oxygen), then the energy release is so great and the degeneracy so extreme, that all the available fuel should be burnt. A burst that releases $\sim 3 \times 10^{45}$ ergs could be seen as far away as 1 Mpc.

4. ELASTIC RAYLEIGH-TAYLOR INSTABILITY

If ^{16}C is formed by electron capture, then it will rest on primordial crust, probably comprising ^{56}Ti, ^{62}Cr or ^{62}Fe. The relative density jumps at these interfaces are found to be 5%, 3%, and 12% respectively,[7] in the sense that the upper phase is the denser. The formal criterion for elastic Rayleigh-Taylor instability at an equilibrium interface is that the relative density contrast exceed $0.9 Z^{2/3}$ percent. However, large-scale stresses in the crust will probably result in a weaker criterion. Instability remains a genuine possibility.

However, if there is a Rayleigh-Taylor overturn, the maximum gravitational and internal energy releases correspond to overall efficiencies of ~ 5 keV per nucleon of accreted gas. Allowing for the fact that there is likely to be considerable inefficiency in the production of γ-rays, it seems very unlikely that gravitational energy release can be responsible for most γ-ray bursts.

The temperature that results from the overturn depends linearly upon the thickness of the unstable layer but can be as high as ~ 200 keV, which is adequate to initiate photodisintegration and a nuclear flash as outlined above. However, smaller amplitude instabilities will not initiate nuclear reactions.

5. SEISMIC WAVES

It was argued[6] that if energy is released impulsively deep below the surface of the neutron star, by whatever means, then a large fraction will be liberated predominantly in the form of elastic shear waves which will probably propagate around the star and be refracted toward the surface in the inner crust and converted into magneto-elastic waves near the surface. As a wave packet approaches the surface, its wavelength will diminish slower than the scale height, and so ultimately it will be reflected when the WKBJ approximation breaks down. The expected transmission coefficient in the presence of a 10^{12}-G field was found to be $\sim 10^{-2}$ at high frequency. This allows the energy to be released over $\gtrsim 100$ dynamical times and goes some way toward accounting for the large difference between the duration of a typical γ-ray burst and a neutron star dynamical time scale.

The discovery of cyclotron lines suggests that a significant amount of energy is dissipated over a small fraction of the surface where the magnetic field strength is comparatively uniform. This fits in fairly well with the notion that the energy originates within the star and is transported to the surface by magnetoelastic waves propagating through the outer crust, because the energy transmission coefficient increases rapidly with magnetic field strength. So, even if the waves excite the whole star, the high-field regions will be the hottest part of the surface.

One modification to the original model that is mandated by the nuclear physics discussed above is that there may be a liquid helium layer surrounded on either side by solid crust. This is analogous to the interior of the Earth, and shear waves preferentially generated by starquakes will not be able to cross the liquid layer and must be converted to pressure waves. This will reduce considerably the transmission of seismic energy from the inner crust to the magnetosphere and will increase the duration of the burst resulting from energy release within the inner crust.

6. MAGNETIC FIELDS

A major uncertainty concerns the role of magnetic field. If the surface fields are typically as strong as $\sim 2 \times 10^{12}$ G, then the accretion will be channeled along the field, and the flux of mass through the polar regions will be larger than the flux through the crust of an unmagnetized star by the ratio of the area of the stellar surface to the polar caps. We suspect that this is unlikely to raise the *mean* accretion rate to the level necessary to allow a helium flash. However, it might raise the *intermittent* rate—for example, within a molecular cloud—to a level whereby a helium flash could be initiated.

Whatever the nuclear history, magnetic confinement must occur as long as the magnetic pressure exceeds the electron pressure multiplied by the ratio of the depth to the size of the polar cap.[25] However, as the pressure increases under the weight of subsequent accretion, this condition will eventually be violated and the accreted material must spread horizontally to establish isostasy. This can either happen through resistive diffusion or magnetic reconnection. The electrical conductivity in the lattice is probably bounded above by $6 \times 10^{23} \rho_9^{2/3} T_6^{-1} \text{s}^{-1}$, although it might be significantly smaller in the presence of abundant lattice defects and impurities.[26] Adopting this value, a magnetic diffusivity $\eta \sim 10^{-4} T_6 \rho_9^{-2/3} \text{cm}^2 \text{ s}^{-1}$ is obtained. (Strong Hall effects modify the magnetic evolution in practice, though they do not alter the conclusion of this argument.) If the gas flows horizontally at a depth where the pressure is $p = 10^{24} p_{24}$ dynes cm^2, with speed v, and the radius of the polar cap is r, then

$$rv = \frac{\dot{M}g}{2\pi p} \sim 0.2 p_{24}^{-1} \dot{M}_{10} \text{ cm}^2 \text{ s}^{-1}. \tag{7}$$

The magnetic field will be strongly perturbed if the magnetic Reynolds number $R_M = rv/\eta \gtrsim 1$. Now if the polar field is just vertical, then the magnetic pressure will no longer be able to withstand the horizontal pressure gradient at a pressure $p_{24} \sim B_{12}^{8/5} r^{4/5}$. Therefore, $R_M \sim rv/\eta >> 100$, implying that the field is strongly perturbed, becoming predominantly horizontal, and that its magnitude must increase to $\sim 10^{14}$ G. In this case, full horizontal spreading will be inhibited until $p_{24} \sim 100$.[27]

At these depths, the lattice is well below its melting temperature and the sideways spreading of the magnetized solid is unlikely to be steady. We can idealize this situation by supposing that a column of accreted material is maintained at an overpressure Δp with respect to the ambient external pressure p. The net gravitational and internal energy release per unit mass of accreted material as it sinks and spreads can be computed to be

$$\epsilon_Q = \left(\frac{Z}{4A}\right)\left(\frac{\mu_e}{m_u}\right)\left(\frac{\Delta p}{p}\right) \text{ MeV n}^{-1} \sim 0.5 p_{24}^{1/2} \left(\frac{\Delta p}{p}\right) \text{ MeV n}^{-1}. \tag{8}$$

If the flow proceedes predominantly through crustal cracking rather than creep, then this represents an estimate of the energy release through quakes. We see that at an overpressure $\Delta p/p \sim 0.1$ and a pressure $p_{24} \sim 100$, the quake energy release is ~ 0.5 MeV n^{-1}, roughly 50 times the minimum efficiency for production of γ-rays, ϵ_γ, computed above. If, illustratively, we suppose that a fraction $f \sim 0.1$ of the stress is relieved in each quake and that the size of the polar cap is $r \sim 1$ km, then the energy released per burst is roughly

$$\Delta E_Q \sim \left(\frac{Z}{4A}\right)\left(\frac{\pi r^2 \mu_e p f}{m_u g}\right)\left(\frac{\Delta p}{p}\right)^2 \sim 10^{38} \text{ ergs}, \tag{9}$$

again comparable with the inferred burst energy release. The presence of magnetic field is vital to this burst mechanism because the rubber-like elastic lattice is likely to limit the overpressure to $\Delta p/p \lesssim 10^{-4}$ before yielding.[6]

It is of interest to consider somewhat analogous processes which might operate during the nonaccreting phase of a neutron star's lifetime. Both positive and negative charges have to be continuously supplied to the magnetosphere and outflowing currents. If, as now seems likely, ions are not tightly bound to the surface, then the positive ions will probably be iron group nuclei from the outer crust. An estimate of the current is $\sim 3 \times 10^{11} B_{12} P^{-2}$A, which requires mass to be mined from depths where the magnetic stress is not dominant at a rate $\sim 2 \times 10^8 B_{12} P^{-2}$g s^{-1}. Equivalently,

$$\frac{dM}{d\ln P} \sim -2 \times 10^{23} B_{12}^{-1} \text{ g} \tag{10}$$

independent of the period. The magnetic Reynolds number is estimated to be $R_M \sim P^{-2} T_6^{-1} p_{24}^{-1/2}$. So, at a pressure $p_{24} \sim 1$, we estimate that large horizontal fields will be created in fast ($P \lesssim 1$ s), cool ($T_6 \lesssim 1$) pulsars, but that when their periods increase to $P \gtrsim 1$ s, ohmic diffusion becomes efficient. This may be relevant to understanding the age and period distribution of radio pulsars.

7. GAMMA-RAY EMISSION

Most models of γ-ray bursts now posit that relativistic electrons (and possibly positrons) are accelerated within a neutron star magnetosphere. The affirmation of the presence of cyclotron lines implies that at least in a significant fraction of sources, the ~ 20 keV X-ray emission is localized to a magnetic pole where the field strength is fairly uniform and narrow lines can be formed. Indeed, the cooling X-ray tails from bursts have been associated with blackbody emission from an area that is ~ 1 km in linear size, comparable with the area of a polar cap.[28] As explained above, the starquake mode provides a simple explanation, because seismic waves will be preferentially transmitted to the magnetosphere at high-field regions.

It is less likely that the bulk of the energy, radiated as $\sim$MeV γ-rays, is produced near the surface. High-energy γ-rays will pair produce on a perpendicular magnetic field above an energy $\sim 2 B_{\perp 12}^{-1}$ MeV and on soft photons above an energy $\sim 250 E_s^{-1}$ MeV, where E_s keV is the maximum soft photon energy at which the source is optically thick to two photon production by γ-rays above threshold $\sim 300/E_s$ MeV. If we assume that the high-energy γ-rays originate within the soft-photon source, then this latter condition translates into a minimum source size $\sim 3 \times 10^7$ cm above ~ 50 keV, for a typical burst. An additional indication that the γ-rays are created at large radius is the relative paucity of X-rays, which requires that few γ-rays irradiate the star.

This lower bound on the source size can be relaxed if the emission is relativistically beamed away from the stellar surface.[29] For example, several authors have proposed that energy is released at a magnetic polar cap, heating the surface to temperatures ~ 5 keV (see Ref. 30 and references therein). X-ray continuum photons are absorbed at the cyclotron frequency and its low harmonics by electrons lying above the photosphere and maintained at a Compton equilibrium temperature of $\sim \hbar\omega_G/4k$. Most of the γ-rays are produced in a corona containing mildly relativistic, outflowing electrons (and possibly positrons) that inverse Compton scatter the X-ray photons. (The acceleration of these relativistic particles is an aspect of these models that has not received much attention.) Resonance of the Doppler-shifted incident photon energy with the cyclotron energy increases the cross section for this scattering. The observed, very hard X-ray spectrum, breaking at an energy ~ 0.5 MeV, can be reproduced even when the relativistic electrons are able to lose most of their energy to the γ-rays.[31]

However, despite its advantages (reduction of the energy radiated per burst and the X-ray fraction and easier escape of γ-rays), relativistic beaming is not a panacea. First, it requires that bursting neutron stars rotate slowly, in contrast to the predictions of some models; otherwise a modulation of the γ-ray emission will be seen. Second, it cannot alter the overall energetics of a population of bursting objects. Third, a broad distribution of ratios of X-ray to γ-ray flux is predicted and this is not seen.

Partly for these reasons, it was proposed that the energy be transported away from the surface of the star by relativistic Alfvén waves and that it be dissipated where the waves became either nonlinear or charged-starved.[6] Under either of these circumstances, large electric fields parallel to the magnetic field will be produced that can create electron-positron showers that short out the field and produce γ-rays. We can place an upper bound of $\sim 10^8$ cm on the source size if we suppose that no more than $\sim 10\%$ of the magnetic energy is dissipated in a light crossing time. Adopting the lower bound from above, we therefore suppose that most of the power is produced from a $\sim 3 \times 10^7$ cm size region where the magnetic field strength is $\sim 3 \times 10^7$ G.

When inverse Compton scattering takes place in a magnetic field of intermediate strength, the electron can be left in a highly excited state of gyrational motion from which it can subsequently decay to the ground state with the emission of many Doppler-shifted cyclotron or synchrotron photons, which may, in a steady cycle, constitute the soft-photon spectrum that is scattered. We suppose that the electron acceleration is sufficiently strong to produce hard enough γ-rays to pair produce on this same soft photon source. In other words, the electron energy is $\sim m_e^2 c^4 / E_s$. The recoil energy, measured in the initial electron rest frame, will be $\sim m_e c^2$. The radiation cooling times for transverse gyrational motion are quite short and so the electron will quickly decay to its ground gyrational state with the emission of several Doppler-shifted cyclotron photons of energy $E_s \sim (m_e c^2 \hbar \omega_G)^{1/2} \sim 0.5$ keV. Roughly, $m_e c^2 / \hbar \omega_G \sim 10^6$ soft photons will be created per Compton scattering. This large yield of soft photons facilitates shorting out the induced electric field.

The shape of the γ-ray spectrum emitted under these conditions depends upon the detailed assumptions of the model, but preliminary calculations indicate that it is significantly softer than typically observed. A more careful study is needed.

8. OBSERVATIONAL PROSPECTS

Recent and forthcoming γ-ray burst observations should resolve many of these uncertainties. *Ginga* and *Phobos* observations have confirmed the presence of cyclotron lines, rapid temporal variation, and high-energy and hard γ-ray emission. The position on the reality of annihilation features is still unclear. The BATSE experiment aboard *GRO* ought to improve our understanding of both the isotropy and the source counts. Neutron star models in general predict a Galactic anisotropy and a flattening of the source density distribution for faint sources, and there are indications already of this latter property.[19] So far, no counterparts to γ-ray bursts in other parts of the electromagnetic spectrum have been conclusively reported. This is one of the main targets of *HETE*. Optical afterglow is to be expected if, in contrast to the type of model discussed here, the bursting neutron stars are in close binary systems. Even *HST* may eventually be able to search for a UV afterglow from the neutron star itself. This promises to be a very exciting decade for the study of these objects.

ACKNOWLEDGMENTS

I am indebted to my collaborators, Omer Blaes, Peter Goldreich, Steve Koonin, Piero Madau, and Lin Yan, for their contributions and Richard Epstein, Don Lamb, Mal Ruderman, Stefan Schramm, Friedrich Thielemann, and Stan Woosley for constructive criticism. Financial support under NSF grants AST 86-15325 and 89-17765 is gratefully acknowledged.

REFERENCES

1. E. P. Mazets et al., *Nature* **290**, 378, (1981).
2. T. Murukami et al., *Nature* **335**, 234, (1988).
3. D. Hartmann and G. Blumenthal, *Astrophys. J.* **342**, 521, (1989).
4. C. Kouveliotou et al., *Astrophys. J. Lett.* **322**, L21, (1988).
5. F. Pacini and M. Ruderman, *Nature* **251**, 399, (1974).
6. O. M. Blaes, R. D. Blandford, P. Goldreich, and P. Madau, *Astrophys. J.* **343**, 839, (1989).
7. O. M. Blaes, R. D. Blandford, P. Madau, and S. Koonin, *Astrophys. J.* **363**, 612, (1990).
8. J. C. Higdon and R. E. Lingenfelter, *Ann. Rev. Astron. Astrophys.* **28**, 401, (1990).
9. K. Hurley and D. Q. Lamb, *Phys. Rep.* (1991, in press).
10. A. Harding, *Phys. Rep.* (1991, in press).
11. D. Hartmann, *Ann. N. Y. Acad. Sci.* (1991, in press).
12. B. Paczyński, *Astrophys. J.* **348**, 485, (1990).
13. D. Hartmann, R. Epstein, and S. Woosley, *Astrophys. J.* **348**, 625, (1990).
14. O. Blaes and M. Rajagopal, *Astrophys. J.* (1991, in press).
15. R. Narayan and J. P. Ostriker, *Astrophys. J.* **352**, 222, (1990).
16. R. Hunt, *Mon. Not. R. Astr. Soc.* **154**, 141, (1971).

17. W. Kluzniak, *Ann. N. Y. Acad. Sci.* (1991, in press).
18. C. Kouveliotou et al., *Astrophys. J. Lett.* **330**, L101, (1988).
19. G. Vedrenne, *Ann. N. Y. Acad. Sci.* (1991, in press).
20. D. G. Yakovlev and D. G. Shalybkov, *Astrophys. Sp. Phys. Rev.* **7**, 311, (1989).
21. S. Schramm and S. E. Koonin, *Astrophys. J.* **365**, 296, (1990).
22. S. Schramm, K. Langanke, and S. E. Koonin, *Astrophys. J.* (1991, in press).
23. E. E. Salpeter and H. M. Van Horn, *Astrophys. J.* **155**, 183, (1969).
24. J. M. Hameury, S. Bonazzola, J. Haeverts, and J. Ventura, *Astron. Astrophys.* **111**, 242, (1982).
25. J. M. Hameury, S. Bonnazzola, J. Haeverts, and J. P. Lasota, *Astron. Astrophys.* **128**, 369, (1983).
26. D. G. Yakovlev and V. A. Urpin, *Soviet Astr.* **24**, 303, (1980).
27. R. D. Blandford, J. Applegate, and L. Hernquist, *Mon. Not. R. Astr. Soc.* **204**, 1025, (1983).
28. A. Yoshida et al., *Publ. Astr. Soc. Japan* **41**, 509, (1989).
29. J. Krolik and E. A. Pier, *Astronom. J.* **373**, 277, (1991).
30. J. M. Hameury and J. P. Lasota, *Astron. Astrophys.* **211**, L15, (1989).
31. C. Dermer, *Astrophys. J.* **360**, 197, (1990).

C. J. Pethick
Nordita, Blegdamsvej 17, DK-2100 Copenhagen Ø, Denmark; and Department of Physics, University of Illinois, 1110 W. Green St., Urbana, IL 61801, USA

Topics in the Physics of High Magnetic Fields

The strong magnetic fields expected to exist in neutron stars affect strongly the thermodynamic and transport properties of matter. Topics discussed here are TF type theories of matter in high magnetic fields, the general expression for magnetic forces in a nonlinear magnetic medium, and ambipolar diffusion as a mechanism for field decay in neutron star interiors.

1. INTRODUCTION

Magnetic fields at the surfaces of neutron stars are estimated to be as high as 10^{12}–10^{13} gauss. Close to the surface of a neutron star, the magnetic field completely dominates the properties of matter, which is quite different from matter in zero field (see, e.g., Ref. 1). In section 2 we describe some equilibrium properties of matter in high fields. First we discuss qualitative features of properties of atoms and of bulk matter calculated in the Thomas-Fermi and related theories. Then we describe the general expression for forces in a nonlinear magnetic medium. Section 3 is devoted to nonequilibrium phenomena—in particular, the decay of magnetic fields in the interiors of neutron stars. We show that the mechanism for rapid field decay proposed by Haensel, Urpin, and Yakovlev[2] may be understood as an example of ambipolar diffusion. We further argue that ambipolar diffusion may be hindered by the buildup of nonequilibrium composition gradients, thereby retarding the field decay.

2. THOMAS-FERMI TYPE THEORIES OF MATTER

Statistical theories of matter of the Thomas-Fermi (TF) type have been used extensively in calculations of the properties of dense matter because they are relatively simple computationally while at the same time sufficiently accurate for many purposes. Most studies to date have been for the case of zero magnetic field or for fields so high that electrons occupy the lowest Landau level for motion perpendicular to the field. Here we shall consider specific properties for the case of intermediate fields and some general properties of theories of the TF type for arbitrary fields.[3,4]

In TF-type theories of matter one solves simultaneously the Poisson equation and the condition that the electrochemical potential of an electron be constant everywhere the electron density is nonzero:

$$\mu_e[n_e(\vec{r})] + e\Phi(\vec{r}) = \text{const.},$$

where $\mu_e[n_e(\vec{r})]$ is the electron chemical potential as a function of the local electron density, $n_e(\vec{r})$, neglecting long-range electrostatic contributions, and $\Phi(\vec{r})$ is the electrostatic potential. In TF theory the electron chemical potential is taken to be that of a free electron gas, while in Thomas-Fermi-Dirac (TFD) theory the exchange energy is included in addition.

In a magnetic field, $\mu_e(n_e)$ for a free electron gas has a horizontal slope whenever a new Landau band begins to be occupied, that is when $\mu_e = n\hbar\omega_c$, where ω_c is the cyclotron frequency. This corresponds to an infinite compressibility for the electron gas, and it gives rise to the density profile in an atom (or matter) having a vertical slope.

When the exchange energy of the uniform electron gas is included in calculations of the equation of state, one finds that first-order phase transitions develop whenever new Landau bands begin to be populated.[3,4] These phase transitions are a consequence of the one-dimensional character of motion in a high magnetic field: The density of states in a band, α, varies as $1/k_{F\alpha}$, where $k_{F\alpha}$ is the Fermi wave number for the band. Since the density of electrons in a band varies as $k_{F\alpha}$, the density of states at the Fermi surface varies as $1/n_\alpha$, and consequently a system with an average electron-electron interaction that is negative and that vanishes less rapidly than $k_{F\alpha}$ for $k_{F\alpha} \to 0$ will be unstable to long-wavelength density fluctuations, because the effective interaction times the density of states at the Fermi surface, which is the dimensionless measure of the strength of the interaction, diverges as $n_\alpha \to 0$.

As a consequence of the phase transitions in the electron gas, the electron density distribution in atoms or matter will display steps whenever a new Landau band begins to be filled.

An interesting result for TF-type theories in zero magnetic field is that there is no molecular binding. Teller[5] enunciated this theorem following Sheldon's[6] numerical studies of the N_2 system in TF theory, which showed that the energy of the system increased as the distance between the two nitrogen nuclei decreased from large distances and, consequently, that binding of the N_2 molecule could not be accounted for in this theory. Subsequently the theorem was put on a rigorous footing by Lieb, Benguria, and Simon. (For a review, see Ref. 7.) TF-type theories give a good description of most of the electrons in an atom, but they fail close to nuclei and in the outer parts of atoms. Binding is a phenomenon associated with the outermost electrons, and the lack of binding in TF-type theories is due to the poor description they give of the outer parts of atoms.

TF-type theories have been used in attempts to determine whether or not matter in a high magnetic field can have a zero-pressure state at a density greatly in excess of typical terrestrial densities. This idea was proposed by Ruderman[1] and it has been the subject of extensive study over the intervening two decades. (For recent Hartree-Fock calculations, see Ref. 8.) It seems clear that at some level, there must be binding when all physical effects are included, but the binding energy of the matter and its equilibrium density have yet to be evaluated reliably.

An interesting question is whether TF theories of matter in a magnetic field can predict molecular binding. The answer is that they cannot. The theorems proved by Lieb, Benguria, and Simon are sufficiently general to demonstrate that the no-binding theorem holds, provided solutions of the TF equation exist. The result applies even if, as in the TFD approximation in a magnetic field, the equation of state for electrons displays first-order phase transitions. Thus TF-type theories cannot be used to estimate properties of a possible zero-pressure state of matter which is bound with respect to isolated atoms.

We note in passing that the theorem holds only if the relationship between the electron chemical potential and the electron density is a local one, and therefore it fails if one includes von Weizsäcker terms, proportional to $(\nabla n_e)^2$ in the energy functional. (see, e.g., Ref. 9 for TF and TFD calculations of properties of matter including a von Weizsäcker term for the case when all electrons occupy the lowest Landau band.) It should also be borne in mind that in the presence of a magnetic field, the von Weizsäcker term need not be isotropic.

2.1 FORCES IN A MAGNETIC FIELD

A related topic I wish to discuss is the nature of forces in high magnetic fields. In, for example, Ref. 10, one finds expressions for forces for the case when the relationship between the magnetization, $\vec{M}$, and the magnetic induction, $\vec{B}$, is linear and density dependent. In the surface layers of neutron stars, matter is a very nonlinear magnetic medium, and one may ask what the expression for the force is in the general case. Here I shall merely quote the result,[11] which is that the force, $\vec{F}$, per unit volume, neglecting gravitational terms, is given by

$$F_i = -\nabla_i p + M_j \nabla_i B_j + \frac{(\vec{j} \times \vec{B})_i}{c},$$

where $\vec{j}$ is the electrical current density and p is the pressure defined in the usual way as $p = n^2 \partial E(n, s, \vec{B})/\partial n$. Here n is the baryon density, and s the entropy per baryon.

An interesting feature of the second term in this equation is that its curl does not generally vanish. Thus there is a possibility that even in the absence of currents, flows can be generated in the surface layers of neutron stars. Another point is that at zero temperature, M exhibits discontinuities as a function of B. In laboratory situations, this leads to domain formation (see, e.g., Ref. 12), and in neutron stars there is a possibility that there could be density jumps associated with rapid dependence of the magnetization on density.

3. MAGNETIC FIELD DECAY IN NEUTRON STAR INTERIORS

The time dependence of magnetic fields in neutron stars has an important bearing on the evolution of neutron stars. In the standard picture, with normal neutrons, protons, and electrons in the interior of a neutron star the electrical conductivity is limited primarily by electron-proton collisions, and is estimated to be[13] $\sim 10^{27} T_9^{-2}$ s^{-1}. The characteristic ohmic decay time for the lowest mode of the magnetic field is $\sim 10^{10} T_9^{-2}$ yr, where T_9 is the temperature in units of 10^9 K. On the basis of this calculation it was argued that ohmic decay is an unimportant process.

Recently this conclusion has been questioned by Haensel, Shalybkov, Urpin, and Yakovlev,[2,14,15] who point out that there can be significant modifications to the above results if the magnetic field is sufficiently high that curvature of the trajectories of electrons and protons between collisions must be taken into account. They solve the coupled Boltzmann transport equations for the electrons, protons, and neutrons, allowing for the cyclotron motion of the charged particles and ensuring that momentum is conserved in collisions. Their results for the general case are rather complicated, but they become simple in the high field limit, when $\omega_{ce} \tau_e \omega_{pe} \tau_p \gg 1$. Here ω_{ci} is the cyclotron frequency, and τ_i the relaxation time of species i. With the usual estimates for collision rates, this condition is $0.1 B_{12} T_9^{-2} \gg 1$, and therefore finite Larmor radius effects are important for $T < 3 \times 10^8$ K if $B \sim 10^{12}$ gauss. Here B_{12} is the magnetic field in units of 10^{12} gauss. In this limit the effective conductivity is $\sim 10^{29} T_9^2/B_{12}^2$ s^{-1} and the characteristic decay time is $\tau \sim 10^{12} T_9^2/B_{12}^2$ yr. Notice that this time varies as T^2, whereas the standard result varies as T^{-2}, and that it varies inversely as B^2. (It is important to observe that neutron-proton collisions are the dominant ones for protons, while electron-proton collisions are dominant for electrons.)

The above results may be understood simply in terms of a calculation given for the case of star formation by Spitzer[16] for ambipolar diffusion in partially ionized plasmas.[17] We consider neutron star matter, with a magnetic field embedded in it, and estimate the forces acting on a unit volume of matter. We imagine that the magnetic field is so strong that the magnetic field and charged particles are effectively locked to each other. A typical magnetic force is of order $\nabla B^2/(8\pi) \sim B^2/(8\pi R)$, where R is a characteristic linear dimension, of order the stellar radius. This force acts on the charged components of the matter. If, for simplicity, one neglects the effects of gravity and pressure gradients, the other force acting on the charged particles must be the frictional force due to collisions between charged particles and neutrons, whose magnitude can be estimated to be $\sim mn_p(\vec{v}_c - \vec{v}_n)/\tau_{cn}$, where m is the nucleon mass, n_p is the proton density, $\vec{v}_c$ is the average velocity of the charged particles, and $\vec{v}_n$ is the average neutron velocity. The quantity τ_{cn} is a relaxation time for collisions between charged particles and neutrons, and it is due primarily to proton-neutron collisions. The drift velocity of the charged particles relative to the neutrons may be estimated by equating the magnetic and frictional forces and is given by

$$|\vec{v}_c - \vec{v}_n| \sim \frac{B^2 \tau_{cn}}{Rmn_p}.$$

Since the magnetic field may be regarded as being frozen to the charged particles, the time τ_D for the field to leave the neutron star is of order $\tau_D \sim R/|\vec{v}_c - \vec{v}_n|$, or

$$\tau_D \sim \frac{R^2 mn_p}{B^2 \tau_{cn}} \sim \frac{R^2}{\tau_{cn} v_A^2},$$

where v_A is the Alfvén velocity. Apart from numerical factors, this result agrees with the detailed calculations of Haensel, Shalybkov, Urpin, and Yakovlev.

Implicit in the above calculation is the assumption that matter is in beta equilibrium. Just how important such an assumption is may be judged by estimating the size of a nonequilibrium chemical potential difference that would give a force comparable to the magnetic force. The magnetic pressure is of order $B^2/(8\pi) \sim 10^{23} B_{12}^2$ ergs cm^{-3}, while the pressure associated with a nonequilibrium chemical potential difference is $\sim n_p(\mu_p + \mu_e - \mu_n)$. For proton densities typical of neutron stars, of order 0.1 fm^{-3}, one finds that a nonequilibrium chemical potential difference of order 1 meV or less produces forces similar in magnitude to a magnetic field of 10^{12} gauss. This underlines the importance of considering the evolution of the magnetic field and composition simultaneously.

We next consider how nonequilibrium composition gradients, and hence nonequilibrium chemical potential gradients, can build up in a neutron star with a magnetic field. Magnetic field lines will move outward in the star, carrying with them the charged particles. This will in turn give rise to a nonequilibrium composition. To what extent the nonequilibrium composition can grow will depend on the rate of beta processes, which tend to restore the composition to equilibrium and thereby allow a counterflow of neutrons and charged particles.

Let us now estimate the field decay rate, taking into account beta processes. For simplicity we assume that beta processes are so slow that the frictional force due to counterflow of neutrons and charged particles may be neglected. The departures of concentrations from their equilibrium values will be of order the magnetic energy density, $B^2/(8\pi)$, compared with a typical energy density of neutron star matter, $\sim nE_F$, where n is the baryon density, and E_F is a typical Fermi energy, that is, $\delta n_p/n_p \sim B^2/(8\pi n E_F)$. These departures from equilibrium will decay away on a typical beta decay time scale, τ_β. The time taken for the magnetic field to be expelled from a region must be of order the time for all protons and electrons in the neutron star to be converted to neutrons, and then converted back again. In a time $\sim \tau_\beta$, a fraction $\delta n_p/n_p$ of the charged particles are converted into neutrons, and therefore the time for all charged particles to be converted once is $\sim \tau_\beta n_p/\delta n_p \sim \tau_\beta 8\pi n E_F/B^2$, which is thus an estimate of the time for the field to decay. The weak interactions will determine the overall decay rate if τ_β exceeds $\tau_{Diff} \sim R^2/(v_F^2 \tau_{cn})$, the time for neutrons to diffuse relative to charged particles over a distance of order R. Here v_F is a typical Fermi velocity. (In making these order-of-magnitude estimates I have been rather cavalier in neglecting differences among the various Fermi energies and particle densities, but this should not affect results by more than a factor of ten at most.) If the relevant weak interaction processes are those for the modified URCA process, τ_β is $\sim 10^8 T_9^{-6}$ s. Since $\tau_{cn} \sim 10^{-19} T_9^{-2}$ s, nonequilibrium chemical potential gradients become important for $T \sim 3 \times 10^8$ K, which is a temperature of the same order as that at which finite Larmor radius effects become important for magnetic fields of order 10^{12} gauss. On the other hand, if some version of the direct URCA process occurs,[18] for which the characteristic time is of order $10^2 T_9^{-4}$ s, substantial nonequilibrium chemical potential gradients would build up only at much lower temperatures, and ambipolar gradients would be important only for $T < 3 \times 10^7$ K. This suggests that if the direct URCA process occurs, neutron stars will cool rapidly and their magnetic fields can also decay rapidly by ambipolar diffusion. On the other hand, if the modified URCA process is dominant, cooling is slow, and ambipolar diffusion is suppressed, thereby giving rise to a long decay time for magnetic fields. Another point to notice, is that if magnetic fields decay by processes other than ohmic dissipation, with a conductivity independent of the magnetic field, their time dependence will resemble a power law rather than an exponential.

In the above discussion we have not considered the possibility that the neutrons are superfluid. If they are, there will be little or no resistance to counterflow of neutrons and charged particles, and weak interaction rates will be modified. We also note that rates of weak interaction processes may be strongly dependent on density, because the direct URCA process has a sharp threshold as a function of density, in which case the evolution of the magnetic field may be strongly dependent on position in the star. The important lesson to be learned from these calculations is that minuscule nonequilibrium composition gradients can have a dramatic effect on the evolution of neutron star magnetic fields. More detailed studies are necessary to determine just how important nonequilibrium composition gradients are in practice. In particular, it is necessary to allow for the effects of the fact that in equilibrium the composition depends on density.

ACKNOWLEDGMENTS

The work described in this article is the result of collaborations with G. Baym, I. Fushiki, E. H. Gudmundsson, V. Urpin, D. G. Yakovlev, and J. Yngvason, to whom I wish to express my appreciation. This work was supported in part by U. S. National Science Foundation grants PHY86-00377 and DMR88-17613, and NASA grant NAGW-1583.

REFERENCES

1. M. A. Ruderman, *Phys. Rev. Lett.* **27**, 1306 (1971).
2. P. Haensel, V. A. Urpin, and D. G. Yakovlev, *Astron. Astrophys.* **229**, 133 (1990).
3. I. Fushiki, E. H. Gudmundsson, C. J. Pethick and J. Yngvason, *Phys. Lett. A*, **152**, 96 (1991).
4. I. Fushiki, E. H. Gudmundsson, C. J. Pethick and J. Yngvason, (in preparation).
5. E. Teller, *Rev. Mod. Phys.* **34**, 627 (1962).
6. J. W. Sheldon, *Phys. Rev.* **99**, 1291 (1955).
7. E. H. Lieb, *Rev. Mod. Phys.* **53**, 603 (1982); **54**, 311 (1982), (E).
8. D. Neuhauser, S. E. Koonin and K. Langanke, *Phys. Rev. A* **36**, 4163 (1987).
9. A. M. Abrahams and S. L. Shapiro, *Ap. J.*, (in press).
10. L. D. Landau and E. M. Lifshitz, *Electrodynamics of Continuous Media* (Pergamon, Oxford, 1960), p. 142).
11. This section is based on unpublished work by V. Urpin and the author.
12. D. Shoenberg, *Magnetic Oscillations in Metals*, (Cambridge University Press, Cambridge), Ch. 6).
13. G. Baym, C. J. Pethick and D. Pines, *Nature* **224**, 674 (1990).
14. D. G. Yakovlev, and D. A. Shalybkov, *Astrophys. & Spa. Sci.*, (in press).
15. D. G. Yakovlev, and D. A. Shalybkov, *Astrophys. & Spa. Sci.*, (in press).
16. L. Spitzer, *Physical Processes in the Interstellar Medium*, (Wiley, New York, 1978, p. 294).
17. This section is based on unpublished work by G. Baym, D. G. Yakovlev and the author.
18. J. M. Lattimer, C. J. Pethick, M. Prakash and P. Haensel, *Phys. Rev. Lett.* **66**, 2701 (1991).

Lee Lindblom
Department of Physics, Montana State University, Bozeman, MT 59717, USA

Instabilities in Rotating Neutron Stars

The maximum angular velocity of rotating neutron stars—and hence the minimum pulsation period of pulsars—is determined by the instabilities to which these objects are subject. This paper reviews the properties of the gravitational-radiation driven instability that is presently believed to limit the rotation of neutron stars. Numerical models of these instabilities are described along with estimates of the maximum angular velocities of rotating neutron stars.

1. INTRODUCTION

Numerous pulsars have been observed with millisecond pulsation periods, and more are being discovered each year. At present the shortest of these periods is 1.56 ms in pulsar PSR1937+21,[1] followed by 1.61 ms in PSR1957+20.[2] It is of considerable theoretical interest to understand what physical mechanism limits these periods, and to determine quantitatively what those limits are. The standard model of a pulsar is a neutron star whose pulsation period is determined by the star's rotation. This paper explores the instabilities that limit the rotation rates of neutron stars and hence, in the standard model, the pulsation periods of pulsars.

The angular velocities of rotating stars are limited by the nature of the equilibrium equations as well as by the existence of unstable solutions to the dynamical equations. The equilibrium states of neutron stars are expected to be rigidly rotating within a few years of their birth. The viscosity of neutron-star matter damps out differential rotation on a time scale of about[3] $\tau \approx 10^7 T_9^2$ s, where T_9 is the temperature in units of 10^9 K. Since the interior temperature is expected to drop to $T_9 \approx 1$ within a few years after its birth,[4] a neutron star is expected to be rigidly rotating after this time. Rigidly rotating stellar models can exist only if the angular velocity of the star does not exceed the "Keplerian" angular velocity of the equatorial circular orbit that coincides with the star's surface. Various studies indicate that this equilibrium limit on the angular velocity of a neutron star is given approximately by $\Omega < \Omega_{max} \approx 0.6 \sqrt{\pi G \bar{\rho}_o}$, where $\bar{\rho}_o$ is the average density of the non-rotating star of the same mass.[5,6] This limit appears to be rather insensitive to the equation of state of the stellar matter and applies to both Newtonian and to general-relativistic stellar models.

The angular velocities of rotating neutron stars may be limited further by the existence of unstable solutions to the dynamical equations. It appears that a gravitational-radiation driven instability is probably responsible for determining the maximum angular velocity of a neutron star. This instability was discovered by Chandrasekhar,[7] and was shown to be generic (i.e., it tends to make *all* rotating stars unstable) by Friedman and Schutz.[8,9] Therefore if no other physical mechanism operates, the maximum angular velocity of a neutron star is zero.

The action of the gravitational-radiation instability is not difficult to understand. Consider a neutron star that rotates with angular velocity Ω, and a small perturbation of this star having time dependence $e^{-i\omega t}$ and angular dependence $e^{im\phi}$, with m an integer. This perturbation propagates with angular velocity ω/m in the direction opposite the star's rotation (assuming $m > 0$) in sufficiently slowly rotating stars. (Note that with this sign convention, $\omega < 0$ for these perturbations.) Figure 1 depicts this situation schematically for $m = 4$. These perturbations create time-dependent mass-multipoles, which causes the star to emit gravitational radiation having negative angular momentum. The perturbation itself has negative angular momentum—since it propagates in the direction opposite the star's rotation—and so the gravitational radiation reduces the perturbation's amplitude in order to conserve angular momentum. Thus gravitational radiation damps out the perturbation. In sufficiently rapidly rotating stars these perturbations are forced to move in the opposite direction. The waves are in effect dragged along by the fluid in the star. In this case the star emits gravitational radiation having positive angular momentum. Since the angular momentum in the perturbation is negative—it still propagates against the rotational flow of the star—the perturbation's amplitude must grow in order to conserve angular momentum. Thus, any counter-rotating perturbation will become unstable when the star rotates rapidly enough to force it to corotate with the star.

This gravitational-radiation instability is quite generic. These perturbations are rather superficial and propagate much like waves on the surface of the ocean. Thus, the speed of the wave relative to the matter is rather independent of the rotation

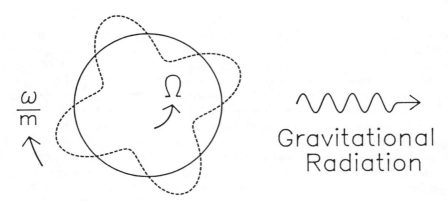

FIGURE 1 Representation of an $m = 4$ perturbation of a rotating neutron star

of the star. To a fairly good approximation, then, the angular-velocity dependence of the frequency of a perturbation is given by $\omega(\Omega) \approx \omega(0) + m\Omega$, where $\omega(0)$ is the frequency when the star is not rotating. These perturbations will reverse direction at the angular velocity where the frequency passes through zero, that is, when $\Omega \approx -\omega(0)/m$. It is easy to see why the instability is generic from this formula. The frequencies of the modes of nonrotating stars increase with m roughly as $\sqrt{m}$. Thus the angular velocity where a perturbation becomes unstable varies with m approximately as $\Omega \propto 1/\sqrt{m}$. An unstable perturbation can be found in any rotating star, therefore, simply by choosing m sufficiently large.

We know of course that all rotating stars are *not* unstable. The argument outlined above merely shows that some other physical mechanism must act to prevent gravitational radiation from driving these perturbations unstable. One such mechanism is internal dissipation in the stellar matter.[10] Viscosity and thermal conductivity quickly damp out any large gradients in the velocity or thermal perturbations. Those perturbations with angular dependence $e^{im\phi}$ have gradients that increase as m increases. Thus, the time scales for the internal dissipation mechanisms to damp out a perturbation tend to decrease as m increases. In contrast, the time scale for gravitational radiation to drive a perturbation unstable becomes very long as m gets large. This is because the radiation couples more weakly to the higher mass-multipole moments. Thus, for sufficiently large m, viscosity will suppress the gravitational-radiation instability.[11] As a consequence, sufficiently slowly rotating stars are stable.

While the presence of dissipation ensures the stability of some rotating stars, it also complicates considerably the analysis needed to determine which stars are actually stable. If the viscosity is large enough, for example, the gravitational-radiation instability can be suppressed in *all* rotating stars.[10] If the viscosity is very small, however, only the most slowly rotating stars will be stable. In order to determine which stars are stable, then, a detailed analysis of their perturbations must be carried out which includes the influences of gravitational radiation and viscosity. The remainder of this paper outlines the techniques that have been developed during the past several years to carry out this analysis, and some of the numerical results of that work are described.

2. THE THEORY OF STELLAR PULSATIONS

Consider the perturbations of a rotating star with time dependence $e^{-i\omega t}$ and angular dependence $e^{im\phi}$. (Since the analysis of such perturbations has only been completed to date in the context of Newtonian physics, the discussion here will be limited to that case.) All of the properties of such a perturbation are determined by two scalar potentials $\delta\Phi$ and δU.[12,13] The potential $\delta\Phi$ represents the perturbed Newtonian gravitational field, while δU is a potential related to the density perturbation $\delta\rho$ of the star by

$$\delta\rho = \rho \frac{d\rho}{dp}\left(\delta U - \delta\Phi\right), \tag{1}$$

where ρ and p are the density and pressure. Quantities not preceded by δ are equilibrium quantities. (While a more general formalism exists,[14] the equations given here apply only to barotropic perturbations, $\delta p = [dp/d\rho]\delta\rho$, of rigidly rotating stars.) The velocity of this perturbation δv^a is also determined by the potential δU. In neutron stars the dissipative forces (both viscosity and gravitational radiation) are weak in the sense that the dissipative time scales are much longer than the pulsation period $1/\omega$. Thus the dissipative effects may be ignored in the first approximation. In this case the perturbed Euler equation has a particularly simple form which determines δv^a in terms of δU:

$$\delta v^a = iQ^{ab}\nabla_b \delta U. \tag{2}$$

In this equation the tensor Q^{ab} is given by

$$Q^{ab} = -\frac{1}{\hat{\omega}}z^a z^b - \frac{\hat{\omega}}{\hat{\omega}^2 - 4\Omega^2}\left(g^{ab} - z^a z^b + \frac{2i}{\hat{\omega}}\nabla^a v^b\right), \tag{3}$$

where v^a is the velocity and Ω is the angular velocity of the unperturbed star, z^a is the unit vector parallel to the rotation axis, and $\hat{\omega} = \omega - m\Omega$. The Euclidean metric g_{ab} (i.e., the identity matrix in Cartesian coordinates) and its inverse g^{ab} are used to raise and lower tensor indices. The covariant derivative ∇_a associated with g_{ab} is just the partial derivative $\partial/\partial x^a$ in Cartesian coordinates.

The two scalar potentials δU and $\delta\Phi$ are determined by the perturbed mass-conservation and gravitational-potential equations. These form a system of second-order (in most cases elliptic) equations for the two potentials:

$$\nabla_a(\rho Q^{ab}\nabla_b \delta U) + \hat{\omega}\rho\frac{d\rho}{dp}(\delta\Phi - \delta U) = 0, \tag{4}$$

$$\nabla^a\nabla_a\delta\Phi + 4\pi G\rho\frac{d\rho}{dp}(\delta\Phi - \delta U) = 0, \tag{5}$$

where G is Newton's gravitation constant. The boundary condition $\delta\Phi \to 0$ must be imposed in the limit $r \to \infty$, where r is the spherical radial coordinate. The frequency of the perturbation ω plays the role of an eigenvalue in these equations. Although the most difficult step in the problem, it is reasonably straightforward to solve these equations numerically for the frequency ω and the eigenfunctions $\delta\Phi$ and δU even in rapidly rotating stars. The needed techniques are described in detail elsewhere[6,15] and will not be reviewed here.

It is easy to evaluate the effects of (weak) dissipation on the pulsation of a star once the frequency ω and the potentials δU and $\delta\Phi$ have been determined by the non-dissipative equations as outlined above. To this end, it is useful to introduce the following "energy" associated with the pulsations:

$$E(t) = \tfrac{1}{2}\int\left[\rho\delta v^a \delta v_a^* + \tfrac{1}{2}(\delta\rho\delta U^* + \delta\rho^*\delta U)\right]d^3x, \tag{6}$$

where * represents complex conjugation. This energy is conserved, $dE/dt = 0$, in the absence of dissipation. In general its time derivative is determined by the equations for the evolution of a viscous fluid coupled to gravitational radiation:

$$\frac{dE}{dt} = -\int\left(2\eta\delta\sigma^{ab}\delta\sigma_{ab}^* + \zeta\delta\sigma\delta\sigma^*\right)d^3x - \hat{\omega}\sum_{l=l_{\min}}^{\infty} N_l\,\omega^{2l+1}\delta D_l^m \delta D_l^{*m}. \tag{7}$$

In this expression $l_{\min}$ is the larger of 2 or $|m|$. The functions ζ and η are the bulk- and shear-viscosity coefficients, while $\delta\sigma^{ab}$ and $\delta\sigma$ are the shear and expansion of the perturbed fluid motion:

$$\delta\sigma^{ab} = \tfrac{1}{2}(\nabla^a\delta v^b + \nabla^b\delta v^a - \tfrac{2}{3}g^{ab}\nabla_c\delta v^c), \tag{8}$$

$$\delta\sigma = \nabla_a\delta v^a. \tag{9}$$

The gravitational-radiation energy loss is determined by the multipole moment

$$\delta D_l^m = \int \delta\rho\, r^l Y_l^{*m} d^3x \tag{10}$$

and the coupling constant N_l (with c the speed of light),

$$N_l = \frac{4\pi G}{c^{2l+1}} \frac{(l+1)(l+2)}{l(l-1)[(2l+1)!!]^2}. \tag{11}$$

When dissipation is present in the star, it is convenient to represent the time dependence of a perturbation in the form $e^{-i\omega t - t/\tau}$, where ω and τ are real. The energy $E(t)$ defined in Eq. (6) is a real function that is quadratic in the perturbation variables. Thus, its time derivative is given by

$$\frac{dE}{dt} = -\frac{2E}{\tau}. \tag{12}$$

This formula can be used to evaluate $1/\tau$. The integrals in Eqs. (6) and (7) that determine E and dE/dt may be evaluated to lowest order (in the strength of the dissipative forces) by using nondissipative values of the frequency ω and the potentials δU and $\delta \Phi$. Once evaluated, these integrals determine $1/\tau$ via Eq. (12). It is convenient to decompose the imaginary part of the frequency into contributions from each of the dissipative forces: $1/\tau = 1/\tau_\zeta + 1/\tau_\eta + 1/\tau_{GR}$. These individual damping times are defined—using Eqs. (7) and (12)—by the integrals

$$\frac{1}{\tau_\zeta} = \frac{1}{2E} \int \zeta \delta\sigma \delta\sigma^* d^3x, \tag{13}$$

$$\frac{1}{\tau_\eta} = \frac{1}{E} \int \eta \delta\sigma^{ab} \delta\sigma^*_{ab} d^3x, \tag{14}$$

$$\frac{1}{\tau_{GR}} = \frac{\hat{\omega}}{2E} \sum_{l=l_{min}}^{\infty} N_l \omega^{2l+1} \delta D_l^m \delta D_l^{*m}. \tag{15}$$

Consider a sequence of rotating stars—parameterized by the angular velocity Ω—of fixed mass and equation of state. A perturbation of one of these stars is stable whenever the imaginary part of the frequency of that perturbation $1/\tau(\Omega)$ is positive. Stars whose angular velocities do not exceed the smallest root of the equation $1/\tau(\Omega_c) = 0$ are stable (assuming that the nonrotating star in this sequence is stable). The problem of determining the maximum angular velocity of a neutron star has been reduced, therefore, to finding the values of the critical angular velocities Ω_c that are the roots of the equation

$$0 = \frac{1}{\tau(\Omega_c)} = \frac{1}{\tau_\zeta(\Omega_c)} + \frac{1}{\tau_\eta(\Omega_c)} + \frac{1}{\tau_{GR}(\Omega_c)}. \tag{16}$$

The integrals in Eqs. (6) and (13)–(15) are easily evaluated once the nondissipative pulsation problem has been solved. Then Ω_c is determined from Eq. (16) for each solution to the perturbation equations. The smallest of these critical angular velocities is, therefore, the maximum angular velocity of a stable neutron star.

3. A NUMERICAL EXAMPLE

In this section the techniques for analyzing the pulsations and stability of rotating neutron stars are illustrated with a numerical example. For simplicity, attention is limited here to neutron stars based on the idealized polytropic equations of state: $p = \kappa \rho^{1+1/n}$. The index n determines the "stiffness" of the equation of state. Realistic equations of state for neutron-star matter have $n \approx 1$, and so this discussion will focus on this value. Most of the results presented here are independent of the parameter κ in the polytropic equation of state. For numerical purposes its value is chosen to make the physical size of these models comparable to those based on more realistic equations of state.

The first task in analyzing the stability of rotating stellar models is to solve Eqs. (4) and (5) for the frequencies ω and the eigenfunctions δU and $\delta \Phi$ that describe the pulsations of a star in the absence of dissipation. As the discussion in section 1 indicates, the modes of primary interest here are those which propagate in the direction opposite the star's rotation. These are the modes which may become unstable via the emission of gravitational radiation in sufficiently rapidly rotating stars. The modes that are the most susceptible to this instability are those that reduce to the $l = m$ f-modes in nonrotating stars.[16] Table 1 gives the frequencies of these modes for $2 \leq l = m \leq 6$ in nonrotating $n = 1$ polytropes.[6] The frequencies are given here in units of $\Omega_o = \sqrt{\pi G \bar{\rho}_o}$, where $\bar{\rho}_o$ is the average density of these nonrotating stars. The ratios ω/Ω_o are independent of the parameter κ that appears in the polytropic equation of state. The angular velocity dependence of these frequencies is most conveniently expressed in terms of the dimensionless functions $\alpha(\Omega)$ defined by

$$\alpha(\Omega) = \frac{\omega(\Omega) - m\Omega}{\omega(0)}. \qquad (17)$$

These functions are displayed in Figure 2 for the $l = m$ f-modes of $n = 1$ polytropes.[6] The $\alpha(\Omega)$ are very slowly varying with $\alpha \approx 1$ over the entire range of angular velocities. This fact justifies the argument given in section 1 that the frequency of these modes is given approximately by $\omega(\Omega) \approx \omega(0) + m\Omega$. Also displayed in Figure 2 are the post-Newtonian versions of these functions.[17] These were computed for a reasonably relativistic ($GM/c^2 R \approx 0.2$) sequence of $n = 1$ polytropes with post-Newtonian mass $M = 1.4 M_\odot$. Thus, Figure 2 illustrates the errors that result from the neglect of general-relativistic effects.

Before the dissipation time scales τ_ζ and τ_η can be determined, expressions for the bulk- and shear-viscosity coefficients ζ and η must be given.[3] Bulk viscosity arises in neutron-star matter because the pressure and density perturbations become slightly out of phase due to the long time scale needed for the weak interactions to reestablish local thermodynamic equilibrium. Sawyer[18] calculates the bulk viscosity of neutron-star matter to be $\zeta = 6.0 \times 10^{-59} \rho^2 \omega^{-2} T^6$ in cgs units. Shear viscosity in neutron-star matter is primarily the result of neutron-neutron scattering (when the temperature exceeds the superfluid-transition temperature). Flowers and Itoh[19] calculate this form of shear viscosity to be approximately $\eta = 347 \rho^{9/4} T^{-2}$.

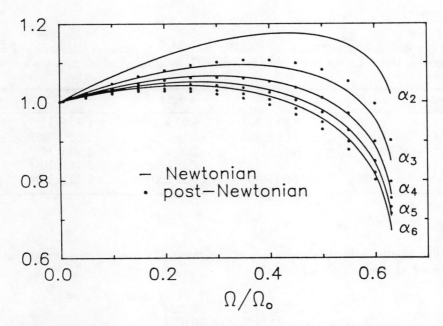

FIGURE 2 Frequencies of the $l = m$ f-modes are represented in terms of the functions $\alpha_m(\Omega)$ defined in Eq. (17)

Given these formulas for ζ and η, the frequency ω, and the eigenfunctions δU and $\delta\Phi$, it is straightforward to perform the numerical integrals needed to evaluate the expressions in Eqs. (13)–(15) for τ_ζ, τ_η, and τ_{GR}. These damping times are given in Table 1 for the nonrotating stellar models described in this study.[20] The viscous damping times τ_ζ and τ_η are given for neutron stars having the temperature $T = 10^9$ K. These damping times scale simply with temperature: τ_ζ as T^{-6} and τ_η as T^2.

TABLE 1 Frequencies and Damping Times of the $l = m$ f-modes of Nonrotating $n = 1$ Polytropes. These quantities are given in units of $\Omega_o = \sqrt{\pi G \bar{\rho}_o}$, where $\bar{\rho}_o$ is the average density.

$l = m$	$-\omega/\Omega_o$	$\tau_{GR}\Omega_o$	$\tau_\eta \Omega_o$	$\tau_\zeta \Omega_o$
2	1.415	2.43×10^2	6.00×10^{11}	1.85×10^{17}
3	1.959	1.06×10^4	6.15×10^{11}	2.66×10^{17}
4	2.350	5.21×10^5	7.00×10^{11}	4.90×10^{17}
5	2.667	2.95×10^7	7.98×10^{11}	9.14×10^{17}
6	2.939	1.92×10^9	8.99×10^{11}	1.65×10^{18}

The angular-velocity dependence of the damping times are most conveniently expressed as dimensionless functions:

$$\gamma(\Omega) = \frac{\omega(\Omega)}{\omega(0)} \left[\frac{\tau_\eta(0)}{\tau_{GR}(0)} \frac{\tau_{GR}(\Omega)}{\tau_\eta(\Omega)} \right]^{1/(2l+1)}, \qquad (18)$$

$$\epsilon(\Omega) = \frac{\tau_\zeta(0)}{\tau_\eta(0)} \frac{\tau_\eta(\Omega)}{\tau_\zeta(\Omega)}. \qquad (19)$$

These functions are independent of the temperature of the neutron-star matter and the parameter κ that appears in the polytropic equation of state. Figures 3 and 4 illustrate these functions for the $l = m$ f-modes of $n = 1$ polytropes.[20] These functions are very slowly varying except for the very highest angular velocities. The effects of bulk viscosity are suppressed in spherical stars because the nonradial pulsations have very little expansion $\delta\sigma$ associated with them. In very rapidly rotating stars, however, spherical symmetry is broken and the pulsations are no longer constrained to have small $\delta\sigma$. Figure 4 illustrates that as a consequence the τ_ζ are much shorter in rapidly rotating stars.

Having determined the angular-velocity dependence of the damping times $\tau_\zeta(\Omega)$, $\tau_\eta(\Omega)$, and $\tau_{GR}(\Omega)$, Eq. (16) can be solved for the critical angular velocities Ω_c where the perturbation becomes unstable. The numerical determination of Ω_c is made easier by transforming Eq. (16) into the form

$$\Omega_c = \frac{\omega(0)}{m} \left\{ \alpha(\Omega_c) + \gamma(\Omega_c) \left[\frac{\tau_{GR}(0)}{\tau_\eta(0)} + \frac{\tau_{GR}(0)}{\tau_\zeta(0)} \epsilon(\Omega_c) \right]^{1/(2l+1)} \right\}. \qquad (20)$$

This equation is easy to solve numerically because $\alpha(\Omega_c) \approx \gamma(\Omega_c) \approx \epsilon(\Omega_c) \approx 1$. Eq. (20) must be evaluated separately for each solution of the perturbation equations. The smallest of these Ω_c for a given sequence of stellar models is the maximum

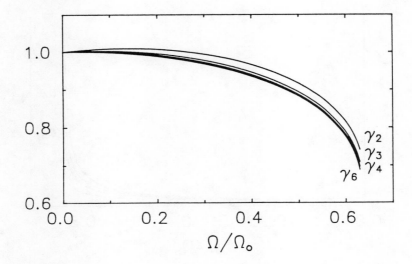

FIGURE 3 Angular-velocity dependence of the gravitational-radiation damping times as represented by the functions $\gamma_m(\Omega)$ defined in Eq. (3.2).

angular velocity with which a stable star may rotate. Since the viscosity of neutron-star matter depends on the temperature of the star, so too will these critical angular velocities. Figure 5 illustrates the smallest Ω_c for a range of neutron-star temperatures. These Ω_c are displayed as ratios, with $\Omega_{\max}$ the angular velocity above which mass shedding occurs. For the $n = 1$ polytropes considered here $\Omega_{\max} = 0.639\Omega_o$. Figure 5 shows that the gravitational-radiation instability is completely suppressed in neutron stars except for those with temperatures in the range 10^7 to about 10^{10}K. Shear viscosity suppresses the instability for lower temperatures while bulk viscosity suppresses it for higher temperatures. The analysis described here has not taken into account the superfluid nature of neutron-star matter at temperatures below about 10^9K. A preliminary investigation[21] indicates that dissipation in the superfluid state due to electron-vortex scattering completely suppresses the gravitational-radiation instability in all neutron stars cooler than the superfluid-transition temperature, $T \approx 10^9$K.

ACKNOWLEDGMENT

This research was supported by NSF grant PHY-9019753.

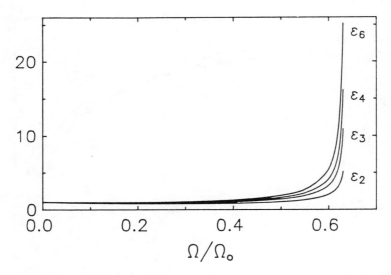

FIGURE 4 Angular-velocity dependence of the viscous damping times of the $l = m$ f-modes as represented by the functions $\epsilon_m(\Omega)$ defined in Eq. (19).

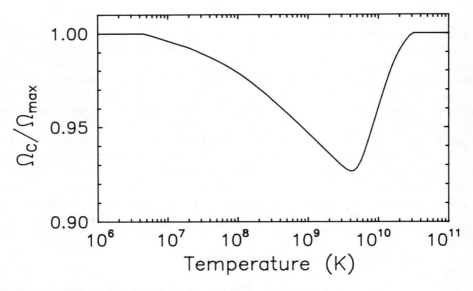

FIGURE 5 Critical angular velocities for rotating $n = 1$ polytropes.

REFERENCES

1. D. C. Backer et al., *Nature* **300**, 615 (1982).
2. A. S. Fruchter, D. R. Stinebring, and J. H. Taylor, *Nature* **333**, 237 (1988).
3. C. Cutler, L. Lindblom, and R. J. Splinter, *Ap. J.* **363**, 603 (1990).
4. K. Nomoto and S. Tsuruta, *Ap. J.* **312**, 711 (1987).
5. J. L. Friedman, J. R. Ipser, and L. Parker, *Ap. J.* **304**, 115 (1986).
6. J. L. Ipser and L. Lindblom, *Ap. J.* **355**, 226 (1990).
7. S. Chandrasekhar, *Phys. Rev. Lett.* **24**, 611 (1970).
8. J. L. Friedman and B. F. Schutz, *Ap. J.* **222**, 281 (1978).
9. J. L. Friedman, *Comm. Math. Phys.* **62**, 247 (1978).
10. L. Lindblom and S. L. Detweiler, *Ap. J.* **211**, 565 (1977).
11. L. Lindblom and W. A. Hiscock, *Ap. J.* **267**, 384 (1983).
12. J. R. Ipser and R. A. Managan, *Ap. J.* **292**, 517 (1985).
13. R. A. Managan, *Ap. J.* **294**, 463 (1985).
14. J. R. Ipser and L. Lindblom, *Ap. J.*, **379**, 285 (1991).
15. J. R. Ipser and L. Lindblom, *Phys. Rev. Lett.* **62**, 2777 (1989).
16. D. Baumgart and J. L. Friedman, *Proc. Roy. Soc. London A* **405**, 65 (1986).
17. C. Cutler and L. Lindblom, *Ann. N. Y. Acad. Sci.,* **631**, 97 (1991).
18. E. Flowers and N. Itoh, *Ap. J.* **206**, 218 (1976).
19. R. F. Sawyer, *Phys. Rev. D* **39**, 3804 (1989).
20. J. R. Ipser and L. Lindblom, *Ap. J.* **373**, 213 (1991).
21. L. Lindblom and G. Mendell, this volume (1992).

J. W. Clark, R. D. Davé, and J. M. C. Chen
McDonnell Center for the Space Sciences and Department of Physics, Washington University, St. Louis, MO 63130, USA

Microscopic Calculations of Superfluid Gaps

Revised variational calculations of the $^1S_0$ neutron pairing gap in the inner-crust regime of neutron stars have been carried out based on improved short-range correlations and more reliable methods for solving the gap integral equation. The results promise better agreement of medium-corrected microscopic predictions with semiphenomenological estimates and observational indications.

1. INTRODUCTION

Superfluidity of the nucleonic fluids contained in neutron stars exerts a profound influence on their thermal evolution and their internal dynamics.[1] A thorough understanding of the microscopic origin as well as the mesoscopic and macroscopic effects of nucleonic superfluidity is essential to the goal of exploiting neutron stars as cosmic hadron physics laboratories.[2]

Superfluidity of nucleonic matter may be attributed to pairing of nucleons having momenta near the relevant Fermi surfaces. The expected scenario[3-8] for pairing phenomena in neutron stars is as follows. In the inner crust of a cool neutron star, where a neutron gas coexists with a Coulomb lattice of neutron-rich nuclei and a sea of relativistic electrons, isotropic pairing of neutrons in the $^1S_0$ two-body state should prevail over a substantial density range, extending roughly to the depth of the crust-core boundary at about $(2-3) \times 10^{14}$ g cm^{-3}. The effect peaks at a density of about one-tenth the saturation density of symmetrical nuclear matter, which is usually taken as $\rho_o = 0.17$ fm^{-1} or 2.8×10^{14} gcm^{-3}. (A weaker singlet-S pairing of neutrons and protons may also occur inside the lattice nuclei.) In the quantum-fluid interior, or nucleonic core region, the dilute proton component reaches a partial density similar to that of the pair-condensed neutron gas of the inner crust, and the protons also pair in the $^1S_0$ state. Since the protons are charged, the proton superfluid is a superconductor. The strength of proton pairing may peak at densities somewhat beyond ρ_o. Finally, it is expected that neutron pairing in the $^3P_2 - ^3F_2$ partial wave may be favored in roughly the same density range of the quantum-fluid interior.

For modeling studies of cooling, glitch timing, and other observable behavior, the primary input from microscopic theories of nucleonic superfluidity is the energy gap $\Delta_{k_F} \equiv \Delta_F$ evaluated for the appropriate nucleon type and partial wave. As befits the origin of pair condensation in the forces between nucleons, the predicted values of Δ_F are on the scale of a hundred keV to a several MeV.

A few words are in order concerning the roots of the different types of pairing in different aspects of the basic nucleon-nucleon (NN) interaction. In neutron-star matter at low densities, S-wave collisions predominate and the interacting gas-phase neutrons of the inner crust sample mainly the longer-range attraction of the mutual $^1S_0$ potential. Hence the corresponding pairing interaction $V_{k_F k_F}$ at the Fermi surface is negative, favoring singlet-S pairing. As the density approaches ρ_o, the colliding neutrons see more and more of the strong short-range repulsion present in the NN force, $V_{k_F k_F}$ turns positive, and the $^1S_0$ neutron gap closes. An analogous situation holds for proton pairing in the quantum-fluid interior. The energetic advantage of neutron pairing in the $^3P_2 - ^3F_2$ state at the higher neutron densities in the quantum-fluid regime may be ascribed to the nature of the tensor and spin-orbit forces in this channel.

It is well known that both single-nucleon properties and the two-body interaction between nucleons are modified in a dense nucleonic medium, relative to free space. The above picture of the superfluid content of neutron stars is predicated on the assumption that these modifications are not extreme. The simplest effect is that the medium is dispersive, leading to an effective mass somewhat smaller than the bare mass and, correspondingly, a somewhat reduced value for the pairing gap. This rather trivial mean-field effect is small at the very low densities relevant to $^1S_0$ neutron pairing, but it is significant at the higher densities where $^1S_0$ proton pairing and $^3P_2 - ^3F_2$ neutron pairing may become important. Virtually all microscopic studies of nucleonic superfluidity include a simple dispersive modification of the single-particle spectrum, but it has generally been assumed that the pairing interaction in the medium suffers little modification relative to the bare interaction in free space, apart from a possible suppression or cutoff of the short-range repulsion at small distances. We shall refer to such treatments as "variational" gap calculations, even though some are based on other many-body methods such as fermion Green's function theory.[9,10] While the separation may not be very precise in some of the treatments, it is useful to think of the pairing interaction used in these calculations as the *direct* component of the actual pairing interaction, which in principle contains an additional part that is *induced* by the presence of the background medium and thus describes effects of higher-order processes such as exchange of density and spin-density fluctuations. Several investigations going beyond the variational stage have shown that the induced or "polarization" component can be quite significant.[11-14] Indeed, different approaches agree in the finding that higher-order medium processes act to quench the singlet-S neutron gap, reducing it by a factor something like 3–5. Qualitatively: The enhancement of the pairing interaction due to exchange of density fluctuations is overwhelmed by a suppression due to exchange of spin-density fluctuations.[11] Recent studies within the polarization-potential approach affirm this net quenching of the neutron gap in the inner crust[13] and indicate an even more emphatic suppression of the proton gap in the quantum-fluid region of the core.[14] Polarization and other higher-order medium effects must also be expected to alter the current theoretical status of $^3P_2 - ^3F_2$ neutron pairing in the quantum-fluid regime. There are preliminary indications[12,15] that the pairing interaction is enhanced in this case, possibly leading to a substantial increase of the energy gap over previous estimates. However, the difficult and important many-body problem of triplet-odd neutron superfluidity remains essentially open.

The present report will not be concerned in any detail with the issue of higher-order medium effects on pairing; instead, it will be devoted almost entirely to a reexamination of the singlet-S pairing problem at the variational level and an improved microscopic evaluation of the variational $^1S_0$ neutron gap in the range of Fermi momenta corresponding to the inner crust. This retreat to the simpler problem is necessitated by the quantitative inadequacy of earlier work and the fact that sound variational calculations provide the foundation for reliable microscopic determination of the properties of nucleonic superfluids.

2. CORRELATED BCS THEORY

The microscopic gap calculations to be discussed have been carried out within the method of correlated basis functions (CBF), applied to the superfluid ground state at the variational level. The relevant formal derivations may be found in Refs. 12, 16, and 17; specifically, we use the version of the theory outlined in Ref. 12, with some minor adaptations. This approach involves the use of a trial superfluid ground state formed by incorporating appropriate short-range spatial correlations as well as the usual BCS pairing correlations localized in momentum space. Essentially, the short-range correlations are cast in Jastrow form, being introduced for given particle number N through a factor $\prod_{1 \leq i < j \leq N} f(r_{ij})$, where $f(r_{ij})$ is a two-body correlation function, dependent on the pair separation $r_{ij} = |\mathbf{r}_i - \mathbf{r}_j|$. (In the calculations to be described, we invoke a simplification of the treatment of Ref. 12 that permits the

Microscopic Calculations of Superfluid Gaps 137

use of short-range correlations $f(ij)$ depending on the spin or the orbital angular momentum of the two-body state of particles i, j.)

In a variational treatment of pairing, the pivotal step is to vary the expectation value $<\hat{H} - \mu\hat{N}>$ with respect to the parameters characterizing the trial ground state of the superfluid. (Here $\hat{H}$ and $\hat{N}$ denote the Hamiltonian and number operators, respectively, and μ is the chemical potential.) For the unadorned BCS trial state $| \text{BCS} > = \prod_k[(1-h_k)^{1/2} + h_k^{1/2} a_{k\uparrow}^\dagger a_{-k\downarrow}^\dagger]| 0 >$, variation with respect to the real amplitude $h_k^{1/2}$ leads to the gap equation

$$\Delta_k = -\frac{1}{2}\sum_{k'} \frac{V_{kk'}}{[(\epsilon_{k'} - \mu)^2 + \Delta_{k'}^2]^{1/2}} \Delta_{k'} , \quad (1)$$

whose solution Δ_k yields the optimal amplitude $h_k^{1/2}$ through

$$h_k = \frac{1}{2}\left(1 - \frac{\epsilon_k - \mu}{[(\epsilon_k - \mu)^2 + \Delta_k^2]^{1/2}}\right) . \quad (2)$$

The quantity $E_k = [(\epsilon_k - \mu)^2 + \Delta_k^2]^{1/2}$ entering both Eqs. (1) and (2) is interpreted as the energy needed to create a quasi-particle of momentum k (and given spin projection σ) in the superfluid. The single-particle energy ϵ_k is given by

$$\epsilon_k = \frac{\hbar^2 k^2}{2m} + \sum_{l\sigma'} h_l < k\sigma, l\sigma'|v(12)|k\sigma, l\sigma' - l\sigma', k\sigma > . \quad (3)$$

For pairing in the $^1S_0$ channel, the pairing matrix elements entering Eq. (1) assume the form $V_{kk'} = < k \uparrow -k \downarrow |v(^1S_0)|k' \uparrow -k' \downarrow>$. The isotropy of S-wave pairing implies that these matrix elements depend only on the magnitudes $k = |\mathbf{k}|$, $k' = |\mathbf{k}'|$ of the momenta k, k'; likewise, the functions $\Delta_\mathbf{k} = \Delta_k$, $\epsilon_\mathbf{k} = \epsilon_k$, and $h_\mathbf{k} = h_k$ depend only on the magnitude k.

It is the usual practice to replace h_l in (3) by its normal-state limit $\Theta(k_F - l)$, converting the ϵ_k to Hartree-Fock (HF) single-particle energies. A similar replacement is made in the Lagrange-multiplier condition $<\hat{N}> = \sum_k h_k = N$, so that μ becomes the HF Fermi energy $\epsilon_{k_F} = \epsilon_F$. This approximation, which decouples the determination of ϵ_k and μ from solution of the gap equation and determination of h_k, is well justified in the present context (cf. Refs. 16, 18–20).

The variational CBF approach of Ref. 12 yields equations for the gap function Δ_k and amplitude $h_k^{1/2}$ that are identical in form with the results (1) and (2). However, this approach is predicated on the same kind of decoupling approximation as just described, so that ϵ_k and μ necessarily assume the normal-state, Hartree-Fock form.[16] A far more substantive distinction is that, in the CBF theory, the bare interaction $v(12)$ appearing in the pairing matrix elements and single-particle energies is replaced by a "dressed" CBF effective interaction $V(12)$, to be constructed in terms of the assumed short-range correlations by the normal-state CBF recipes given in Refs. 16 and 17. One simplification that should be mentioned is that the short-range dynamical correlations are not determined optimally in the presence of the superfluid correlations associated with the BCS *Ansatz;* rather, they are taken from a normal-state variational calculation. This simplification should not have an appreciable quantitative effect on the result for the gap.

For our numerical studies, we impose two further approximations:

1. The normal-state CBF matrix elements involved in the $V_{kk'}$ and ϵ_k are computed in two-body cluster order (or, more precisely, to leading order in the "smallness parameter" $\rho | \int [f^2(r) - 1] d^3 r |$). Such an approximation should be valid in the low-density regime under study, although its quality will deteriorate somewhat as the density ρ_o is approached. (In other words, fancy resummation of higher-order cluster diagrams by Fermi-hypernetted-chain [FHNC] techniques and related procedures is, for the most part, an unnecessary luxury [cf. Ref. 16].) Under "two-body" truncation, the effective interaction $V(12)$ takes the local form $w_2(ij) = (\hbar^2/m)[df(r_{ij})/dr_{ij}]^2 + f^2(r_{ij})v(ij)$.
2. The single-particle energies are parameterized in terms of an effective mass m^*, through the usual relation $(d\epsilon_k/dk)|_{k=k_F} = \hbar^2 k_F/m^*$. This approximation is tested and found to be generally quite satisfactory.

3. NUMERICAL RESULTS

We consider two model NN interactions: (1) the OMY4 potential,[21] acting in S-waves only, and (2) a central, but spin-dependent version of the Reid–soft-core interaction.[22] The OMY4 potential, which contains a hard core of radius 0.4 fm, gives reasonable fits of the low-energy NN data (effective ranges and scattering lengths) and of the $^1S_0$ phase shift out to intermediate energies. This choice coincides with one of the potentials considered in the early (and frequently cited) work of Refs. 3, 6, and 7 on superfluid nucleon gaps. Accordingly, we use the *same* one-parameter short-range correlations $f(12)$ as adopted in that work, to allow direct comparison of the old and new gap calculations.

The Reid–soft-core potential may be regarded as a realistic representation of the NN interaction in the low-density regime of neutron-star matter involved in the inner crust. Its v_6 operator version,[23,24] which omits spin-orbit and quadratic spin-orbit portions, should be adequate for our purposes; indeed, since the odd-state ($^3P_2$–$^3F_2$) tensor force acting in pure neutron matter is expected to be unimportant in this density range, we shall also omit the tensor component of v_6. The resulting potential, called Reid v_4, takes the following simple form in neutron matter: $v_4(12) = v^{(s)}(r_{12})P^{(s)}(12) + v^{(t)}(r_{12})P^{(t)}(12)$, where $P^{(s)}$ and $P^{(t)}$ are projectors onto singlet and triplet spin states. We assume a short-range correlation operator of the same form, whose singlet and triplet components have the common radial shape $f(r) = \exp\{-(b/r)^m \exp[-(r/b)^n]/2\}$, with the *spin-dependent* parameters $m^{(s,t)}$, $n^{(s,t)}$, and $b^{(s,t)}$, determined at each density $\rho = k_F^3/3\pi^2$ by minimization of the

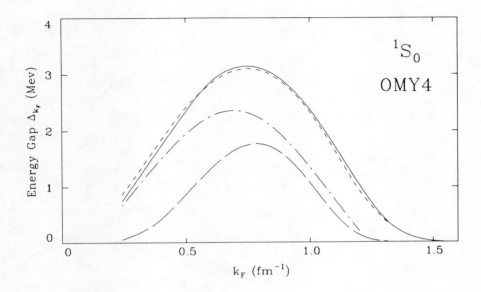

FIGURE 1 Variational results for $^1S_0$ pairing gap Δ_{k_F} in neutron matter based on OMY4 potential. *Solid curve:* straightforward iteration ($k_c = 16k_F$). *Short-dashed curves:* linearization-eigenvalue method ($k_c = 20k_F$). *Dot-dashed curve:* result from Yang and coworkers.[3,6,7] *Long dashed curve:* weak-coupling formula.

two-body cluster approximation to the normal-state variational energy of neutron matter.

In Figs. 1 and 2 we display and compare several different variational results for $\Delta_F(^1S_0)$ versus k_F in neutron matter, obtained for the OMY4 and Reid v_4 potentials, respectively. In a given figure, the various curves correspond to different numerical treatments of the gap integral equation (1), but are based on the same inputs for the CBF–variational pairing matrix elements and effective masses. Proceeding from the less to the more accurate, the four sorts of evaluation considered in our reappraisal of variational gap calculations are as follows:

(i) The familiar weak-coupling formula[25] $\Delta_F = 2\epsilon_F \exp[-1/N(0)g]$, where $\epsilon_F = \hbar^2 k_F^2/2m^*$, $N(0) = m^* k_F/2\pi^2 \hbar^2$, and $g = -V_{k_F k_F}$ (taking unit normalization volume for plane-wave states).

(ii) The prescription applied by Yang and coworkers,[3,6,7] involving determination of the parameters z and ζ in the *Ansatz* $\Delta_k = -zV_{\zeta k, \zeta k}$ for the gap function, by iterating the gap equation once with this form as input and matching value and slope of input and output at the Fermi surface.

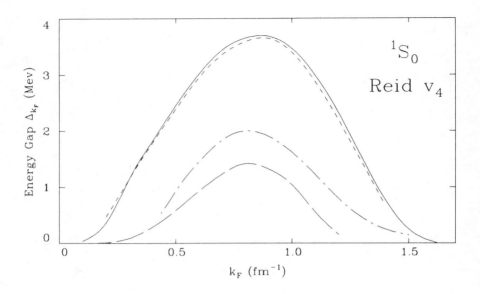

FIGURE 2 As in Fig. 1, except that calculations are based on Reid v_4 interaction, and dot-dashed curve is from straightforward iteration with $k_c = 2k_F$

(iii) The procedure introduced by Krotscheck[26] (cf. Ref. 27) and applied by Chen et al.,[12] wherein the insensitivity of the integrand of Eq. (1) to the behavior of $\Delta_{k'}$ away from k_F is exploited to convert the gap equation into a linear eigenvalue problem, which is solved recursively on a mesh of variable spacing.

(iv) Straightforward iteration of the gap equation, until the output and input agree to within a typical fractional tolerance $|\Delta_k(\text{out}) - \Delta_k(\text{in})|/\Delta_k(\text{in})$ of 10^{-5}.

When the pairing interaction is constructed directly from a bare NN interaction that contains a strong short-range repulsion (like the full Reid potential or the Argonne v_{14} interaction) or is highly nonlocal (like the Paris interaction), accurate solution of the gap equation requires that the upper limit on the k' integration of (1) be extended out to large values, *at least* to $\sim 10k_F$ (Ref. 9). The same is true for the variational-CBF effective pairing interactions of the present study, corresponding to OMY4 and Reid v_4 potentials. This fact is not reflected in method (i); nor has it been taken into account in the applications of method (ii).

The top pair of curves in Fig. 1 or 2, obtained by methods (iii) and (iv) with respective k' cutoffs k_c of 20 and 16 k_F, may be regarded as quantitatively reliable, within the assumed variational descriptions. The results from straightforward iteration are presumably the more accurate and are taken as standard. (Typically, 30 iterations are required to meet the chosen convergence criterion, if the process

Microscopic Calculations of Superfluid Gaps

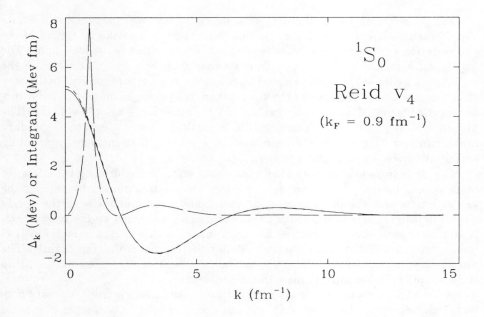

FIGURE 3 Gap function Δ_k in Reid v_4 neutron matter at $k_F = 0.9$ fm^{-1}. *Solid curve:* in effective-mass approximation. *Short-dashed curve:* with full normal-state single-particle spectrum. *Long dashed curve:* Integrand of gap equation.

is started with a vanishing gap function.) The OMY4 calculation of Yang and coworkers,[3,6,7] which yields the dot-dashed curve in Fig. 1, is seen to underestimate the gap by a significant margin; the variational gap is thus significantly larger than was hitherto believed. The weak-coupling formula (represented by long-dashed curves) is seriously in error for both potentials, predicting a gap that is too small by a factor roughly 0.5 in the peak region. The importance of integrating out to a large cutoff momentum is exemplified in Fig. 2, where the dot-dashed and solid curves show the results of straightforward iteration for $k_c/k_F = 2$ and 16, respectively.

Comparing the Δ_F curves obtained with the two model potentials by the standard iteration method (solid curves of Figs. 1 and 2), we see that, as expected, the energy gap is larger for the interaction with the softer repulsive core. Moreover, beyond its peak value the gap is quenched more rapidly with increasing k_F for the potential with the stiffer core. A necessary condition for stability of the normal trial state against pairing is $V_{k_F k_F} > 0$. We note that the pairing matrix element $V_{k_F k_F}$ changes sign from negative to positive as k_F increases through 1.5 fm^{-1} in the OMY4 case and 1.7 fm^{-1} for the Reid v_4 potential.

Fig. 3, referring to the Reid v_4 case at $k_F = 0.9$ fm^{-1}, serves to illustrate the characteristic behavior of the gap function Δ_k as well as the integrand of the gap equation (1), when a realistically strong NN interaction is used as input. The gap function is seen to fall rapidly from its maximum at $k = 0$, the width of the initial peak being $\sim 2k_F$. However, at larger k there appears a prominent oscillatory structure due to the presence of the strong short-range repulsion, with Δ_k showing non-negligible negative or positive values out to rather large $k \sim (10-15)k_F$. The plot of the integrand of the gap equation is particularly revealing: the secondary maximum and long tail would be missed in a scheme that includes momentum contributions only out to $2k_F$ (cf. Ref. 13).

Fig. 3 also provides information on the size of errors introduced by the effective-mass approximation. The short-dashed curve for Δ_k is obtained by solving the gap equation with the full normal-state variational ϵ_k as input. Use of the purely quadratic single-particle spectrum parameterized by m^* leads to the solid curve. A similar comparison (not shown) has been made for $\Delta_F = \Delta_{k_F}$ versus k_F. It is found that the effective-mass approximation underestimates the gap slightly. The discrepancy (amounting to about 3% around the peak of Δ_F) is unimportant on the scale of the other uncertainties in the problem.

A key approximation made in our variational calculations is truncation at "two-body" cluster order. To justify this simplification, we refer to an earlier numerical comparison[16] of two-body and FHNC results for the effective mass m^* and pairing matrix element $V_{k_F k_F}$ for the same OMY4 inputs as considered here. The two-body results are essentially bracketed by two different FHNC approximations to the higher cluster terms, the discrepancies between the three calculations remaining relatively small in the region where Δ_F is large. This optimistic assessment is accompanied by the caveat that the two-body approximation must break down as ρ increases past ρ_o. Accordingly, it is unlikely to yield a quantitatively reliable estimate of the gap-closure density.

We have verified that the Reid v_4 potential provides a quantitatively satisfactory representation of the NN interaction in the problem of $^1S_0$ neutron pairing by comparing gap results for this potential with those given by Baldo et al.[9] for the highly realistic Argonne v_{14} and Paris interactions. In this comparison, the required pairing matrix elements $V_{kk'}$ are constructed from the respective *bare* potentials, while the required normal-state single-particle energies ϵ_k are computed from the HF formula in terms of suitable *effective* interactions (a Brueckner G-matrix choice in Ref. 9 and the variational-CBF effective interaction w_2 in our treatment of the Reid v_4 potential). In the vicinity of the peak of Δ_F versus k_F, the bare-Reid v_4 results, calculated in effective-mass approximation, are only about 5%–7% smaller than the bare-Paris results quoted in Ref. 9 and in fact are quite close to the results cited there for the bare Argonne v_{14} potential. It may reasonably be concluded that the singlet-S gap in neutron matter is rather insensitive to the details of the NN interaction assumed; in particular, our neglect of tensor and spin-orbit forces is of little consequence as long as the density remains significantly below that of ordinary nuclear matter. The simple Reid v_4 choice should yield quantitatively useful

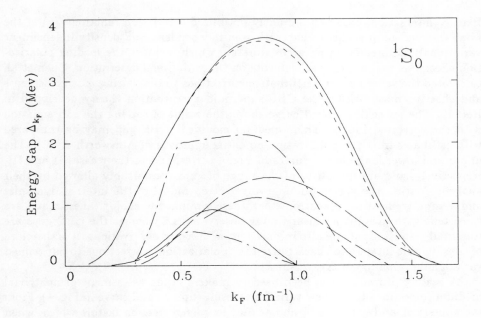

FIGURE 4 Predictions for $^1S_0$ pairing gap Δ_{k_F} in neutron matter. *Upper solid curve:* variational result for Reid v_4 interaction, with spin-dependent short-range correlations (same as solid curve of Fig. 2). *Short dashed curve:* corresponding result for spin-independent correlations. *Upper dot-dashed curve:* variational result of Chen et al.[12] for Reid v_6 interaction. *Lower dot-dashed curve:* second-order CBF result of Ref. 12. *Lower solid curve:* new CBF-scaled prediction. *Long-dashed curves:* results from polarization-potential analysis (cases II and III of Ref. 13).

results in this density regime, and the results for the OMY4 potential—although it contains an "outmoded" hard core—should be at least of semiquantitative value.

Finally, in Fig. 4 we consider some gap calculations that incorporate nontrivial, higher-order medium effects (generically, "polarization effects"). The results of three variational calculations are included in the figure, for reference. The topmost curve is the Reid v_4 result of the present investigation, obtained by straightforward iteration with $k_c = 16 k_F$. If the parameters b, m, and n specifying the short-range correlations are constrained to have the same values in singlet and triplet channels, the variational gap is reduced slightly, resulting in the short-dashed curve. The introduction of spin dependence into the short-range correlations is seen to have only a very minor effect, at least within this "two-body-cluster" evaluation. The upper dot-dashed curve is the older variational result of Chen et al.[12] for the

Reid v_6 interaction, based on a crudely optimized correlation function $f(r)$ of the same form as adopted here—but with spin-independent *and* density-independent variational parameters. In an approximation which includes the leading polarization effects, as well as other refinements on the variational description, Chen et al. calculated the second-order perturbative corrections to the pairing matrix elements and effective mass, within the CBF superfluid perturbation theory developed in Ref. 17. The lower dot-dashed curve shows the result of solving the gap equation with the corrected inputs. This "medium-modified" CBF gap may be compared with estimates from the polarization-potential approach of Ainsworth et al.[13] The upper and lower long-dashed curves of Fig. 4 correspond to their cases II and III, respectively, and are indicative of the range of gaps comfortably allowed by their analysis. Within the vortex-pinning–vortex-creep model of the internal dynamics and dissipative heating of neutron stars, these semiphenomenological estimates are consistent with postglitch pulsar timing observations,[1] whereas the CBF gaps are too small—and, indeed, are also inconsistent with X-ray observations. It is therefore of considerable interest to characterize the "polarization" corrections to our refined variational evaluation of Δ_F versus k_F.

A scaling procedure has been used to make a crude assessment of nontrivial medium corrections to the new variational result (upper solid curve in Fig. 4). From the numerical study of Chen et al. one may extract correction factors which, when applied to their variational results for m^* and $V_{k_F k_F}$, yield the calculated second-order CBF approximations to these quantities. (Although neither factor shows a gross deviation from unity, the exponential sensitivity of the gap to $m^* V_{k_F k_F}$ leads to a very strong quenching of Δ_F.) We have applied the *same* correction factors to the variational m^* and pairing matrix elements $V_{kk'}$ determined here for the Reid v_4 interaction. Solving the gap equation for these scaled inputs, we obtain the lower solid curve. To quote some definite numerical values, the variational (CBF-scaled) results for the gap are 3.03 (0.90), 3.69 (0.28), and 2.31 (0.0) MeV at k_F (ρ) points of 0.6 (0.0073), 0.9 (0.0246), and 1.2 (0.0584) fm^{-1} (fm^{-3}), respectively, with corresponding m^*/m values 0.96 (1.07), 0.93 (1.07), and 0.89 (1.05). Further details will be provided in a longer article.[20] The revised CBF gaps show improved agreement with the predictions of the polarization-potential approach, although significant discrepancies remain at the larger k_F values.

This improved agreement of microscopic and phenomenological accounts of $^1S_0$ neutron pairing comes about because the new calculation of the variational gap yields much larger values than were obtained by Chen et al.; while CBF corrections are again responsible for a dramatic reduction from the variational approximation, a substantial gap survives in the relevant density range. The improvements upon the work of Chen et al. include, primarily, the introduction of much better short-range correlations, which are more effective in taming the short-range repulsion, and, secondarily, more accurate treatment of the gap equation, including k' integration out to $k_c/k_F = 16$ or 20 rather than 8. The distinction between the assumed NN interactions (Reid v_4 and v_6) is quite unimportant, except possibly near ρ_o and beyond.

As remarked above, there is still a sizable discrepancy between microscopic CBF results and semimicroscopic polarization-potential estimates for the gap in the higher density range $k_F \sim (1-1.7)k_F$ where the gap closes. The behavior of $\Delta_{k_F}(^1S_0)$ in this range is of special interest for the physics of neutron-star interiors, especially in connection with the issue of magnetic-field decay.[28] Closure at a low density may imply the existence of a layer of normal neutron matter, in which the short-range repulsions and medium effects have extinguished singlet pairing, whereas the triplet gap, due to pairing in the $^3P_2-^3F_2$ partial wave, has not yet grown to a value exceeding $k_B T$, where T is the local temperature. Unfortunately, current many-body theories suffer from substantial uncertainties in the density regime $\rho \sim \rho_o$ and beyond, hampering definitive predictions.

The new CBF-scaled evaluation of Δ_F versus k_F may be faulted for the lack of an explicit calculation of CBF corrections with the short-range correlations actually assumed here, and for the neglect of higher-order cluster corrections to the CBF matrix elements. These omissions are likely to be most serious at the higher k_F values where the gap is closing. In particular, we expect that the scaling procedure leads to an *overestimate* of the CBF corrections and thus an underestimate of the gap, since the scaling factors extracted from Ref. 12 are based on the comparatively poor correlations assumed in that work. On the other side of the coin, the polarization-potential treatment of Ref. 13 may be faulted for its reliance on the weak-coupling formula and for its inclusion of particle-particle-reducible diagrams in the scattering amplitude. The former approximation, as we have seen, tends to underestimate the gap, while the overcounting due to the unwanted diagrams tends to increase the gap. A simple estimate based in part on the results of Ref. 9 suggests that the errors from these sources roughly cancel around $k_F = 0.6$ fm^{-1}, and presumably also around the peak of Δ_{k_F}. However, there is no reason to expect such a compensation to persist at higher densities, $\sim \rho_o/2$ to ρ_o, where the pseudopotential treatment of short-range correlations itself becomes less reliable.

4. CONCLUSIONS

In typical astrophysical considerations, decisive results can often be achieved on the basis of rough, order-of-magnitude estimates of key physical parameters. Unfortunately, this is not the case when we come to examine the consequences of nucleonic superfluidity for the dynamics and evolution of neutron stars. The qualitative physics can be distressingly sensitive to the pairing gaps, which provide the essential microscopic input to mesoscopic and macroscopic modeling. A factor of 2 can make all the difference! Moreover, precise determination of these energy gaps from fundamental microphysics and many-body theory is a very delicate matter, owing to (a) theoretical uncertainties associated mainly with the evaluation of polarization effects and other higher-order medium corrections to CBF, Green's function, and other many-body approaches and (b) numerical difficulties in solving the gap equation, for realistic nucleon-nucleon interactions. The "silver lining" of this cloud of gloom is that such heightened sensitivities to the microphysics and microphysical methods enhance the leverage of neutron stars as cosmic laboratories offering critical tests of fundamental nuclear theory.

The present work, in focusing on the important example of $^1S_0$ neutron superfluidity in the inner crust and updating variational estimates of the associated energy gap, has revealed several interesting computational aspects of the underlying many-body problem. We find that the variational gap is larger than reported in earlier work, due to improvements in the treatment of the short-range correlations and/or in the treatment of the integral equation for the gap function. The problematic nature of the gap equation is reflected in the fact that the integration over momenta must be extended at least to $\sim 10k_F$ to achieve a solution of satisfactory accuracy; consequently, the weak-coupling estimate of the gap is unreliable, undershooting the correct result by roughly a factor one-half. On the other hand, in dealing with the $^1S_0$ neutron gap, we have low density on our side—which means that (a) a "two-body" truncation of the relevant correlated matrix elements is acceptable and (b) predictions for the gap are not very sensitive to the choice of the bare NN interaction, noncentral components being relatively unimportant. These simplifications are lost in the much more challenging problems of $^1S_0$ proton pairing and $^3P_2 - ^3F_2$ neutron superfluidity, which await concerted microscopic investigation.

ACKNOWLEDGMENTS

This research was supported by the Division of Materials Research and the Physics Division of the U.S. National Science Foundation, under Grant No. DMR-9002863. We have benefited from discussions with T. Ainsworth, C. Pethick, D. Pines, B. Vonderfecht, and J. Wambach. JWC expresses his gratitude to the Japanese hosts of SENS '90 for their gracious hospitality and to the Japan Society for the Promotion of Science and the U.S. National Science Foundation for travel support.

REFERENCES

1. For up-to-date accounts of the observational consequences of nucleonic superfluidity, see contributions of D. Pines, R. Tamagaki, M. A. Alpar, R. Epstein, S. Tsuruta, and N. Shibazaki to this volume.
2. D. Pines and M. A. Alpar, *Nature*, **316**, 27, (1985); D. Pines, in *Proceedings of the Landau Memorial Conference on Frontiers of Physics*, edited by E. Gotsman, Y. Ne'eman, and A. Voronel (Pergamon, New York, 1990).

3. J. W. Clark and C.-H. Yang, *Nuovo Cimento Lett.*, **3**, 272, (1970); *Nuovo Cimento Lett.*, **2**, 379, (1971).
4. M. Hoffberg, A. E. Glassgold, R. W. Richardson, and M. Ruderman, *Phys. Rev. Lett.*, **24**, 775, (1970).
5. R. Tamagaki, *Progr. Theor. Phys.*, **44**, 905, (1970); T. Takatsuka and R. Tamagaki, *Progr. Theor. Phys.*, **46**, 114, (1971).
6. C.-H. Yang and J. W. Clark, *Nucl. Phys.*, **A174**, 49, (1971).
7. N.-C. Chao, J. W. Clark, and C.-H. Yang, *Nucl. Phys.*, **A179**, 320, (1972.)
8. T. Takatsuka, *Progr. Theor. Phys.*, **48**, 1517, (1972); *Progr. Theor. Phys.*, **50**, 1754, (1973); *Progr. Theor. Phys.*, **50**, 1755, (1973).
9. M. Baldo, J. Cugnon, A. Lejeune, and U. Lombardo, *Nucl. Phys.*, **A515**, 409, (1990).
10. B. E. Vonderfecht, C. C. Gearhart, W. H. Dickhoff, A. Polls, and A. Ramos, *Phys. Lett.*, **B253**, 1, (1991).
11. J. W. Clark, C.-G. Källman, C.-H. Yang, and D. A. Chakkalakal, *Phys. Lett.*, **B61**, 331, (1976).
12. J. M. C. Chen, J. W. Clark, E. Krotscheck, and R. A. Smith, *Nucl. Phys.*, **A451**, 509, (1986).
13. T. L. Ainsworth, J. Wambach, and D. Pines, *Phys. Lett.*, **B222**, 173, (1989).
14. T. L. Ainsworth, D. Pines, and J. Wambach, Nuclear Physics Divisional Meeting of the American Physical Society, Urbana, IL, October 25–27, 1990.
15. A. D. Jackson, E. Krotscheck, D. E. Meltzer, and R. A. Smith, *Nucl. Phys.*, **A386**, 125, (1982).
16. E. Krotscheck and J. W. Clark, *Nucl. Phys.*, **A333**, 77, (1980).
17. E. Krotscheck, R. A. Smith, and A. D. Jackson, *Phys. Rev.* **B24**, 6404, (1981).
18. R. C. Kennedy, *Nucl. Phys.*, **A118**, 189, (1968).
19. L. Amundsen and E. Østgaard, *Nucl. Phys.*, **A437**, 487, (1985).
20. R. D. Davé, J. W. Clark, and J. M. C. Chen, in preparation.
21. T. Ohmura, *Progr. Theor. Phys.*, **41**, 419, (1969).
22. R. V. Reid, Jr., *Ann. of Phys.*, **50**, 411, (1968).
23. V. R. Pandharipande and R. B. Wiringa, *Rev. Mod. Phys.*, **51**, 821, (1979).
24. J. W. Clark, *Progr. Part. Nucl. Phys.*, **2**, 89, (1979).
25. J. Bardeen, L. N. Cooper, and J. R. Schrieffer, *Phys. Rev.*, **108**, 1175, (1957); J. R. Schrieffer, *Theory of Superconductivity* (Benjamin, New York, 1964).
26. E. Krotscheck, *Z. Phys.*, **251**, 135, (1972).
27. V. J. Emery, *Nucl. Phys.*, **19**, 154, (1960).
28. P. Haensel, V. A. Urpin, and D. G. Yakovlev, *Astron. Astrophys.*, **229**, 133, (1960); and C. J. Pethick, private communication.

M. Ali Alpar
Physics Department, Middle East Technical University Ankara 06531, Turkey; and Loomis Laboratory of Physics, University of Illinois, 1110 W. Green St., Urbana, IL 61801, USA

Some Topics in Neutron Star Superfluid Dynamics

Accumulating timing observations of neutron stars pose several interesting questions and tests for models of superfluid dynamics in neutron stars. Three topics of interest are (1) implications of current observational limits on the crust-core coupling times of neutron stars; (2) the expected evolution of postglitch response in vortex creep models; (3) the relation between glitches and the long term dynamics of a pulsar.

1. INTRODUCTION

Observations of pulsar glitches, postglitch relaxation, and timing noise have stimulated considerable interest in the dynamics of neutron stars. Glitches and postglitch relaxation involve only a small fraction of the star's interior. As this requires, the bulk core superfluid is actually coupled to the observed crust quite rigidly. Recent observations provide tight limits on the crust-core coupling time. These are discussed in section 2. The glitches themselves are assumed to arise from the sudden unpinning of pinned vortex lines in the inner crust superfluid (Anderson and Itoh, 1975). The observed dynamics can be evaluated with a phenomenological approach that is independent of the yet unknown specific mechanism for this unpinning and the many interesting problems posed by the microscopic physics of pinned vortex lines. The dynamical behavior of a rotating superfluid is governed by an equation of motion of the form

$$\dot{\Omega} = -n\kappa V_r/r \qquad (1)$$

where n is the number of vortex lines per unit area, κ the vortex quantum, V_r the average radially outward velocity of the vortex lines, and r the distance from the rotation axis. This equation simply states that a superfluid spins down if vortex lines move outward. V_r depends on the lag $\omega = \Omega - \Omega_c$ between the superfluid and its container (the crust of the star); the exact dependence contains the physics of the vortex line–normal matter interactions and defines the model. Equations of the form (1) for each distinct superfluid region of the star are complemented by the equation of motion for the observed outer crust:

$$I_c\dot{\Omega}_c = N_{ext} - \sum I_i\dot{\Omega}_i \qquad (2)$$

Here N_{ext} is the external torque on the neutron star and the sum denotes the internal torques exerted on the crust by various components of the interior. A dynamical model in terms of a crust superfluid with pinned vorticity has been successfully applied to the postglitch dynamics from many pulsars. Section 3 discusses this vortex creep model, in which vortices move against a pinning potential by thermal activation. Postglitch behavior from pulsars of different ages can be interpreted with the same range of neutron star parameters when the different temperatures of the pulsars are taken into account. The link between the nature of dynamical response and the thermal evolution of the star is a crucial aspect of the vortex creep model. Vortex creep is also expected to determine the thermal evolution of old pulsars through energy dissipation in the interior, which makes these pulsars interesting for future X-ray observations. Section 4 discusses the relation between glitches and long-term dynamical relaxation. In the Vela pulsar, from which eight glitches have been observed, a large part of the discrete spin-up in each glitch does not subsequently relax, indicating that the glitches are themselves part of the long-term coupling of the crust superfluid to the crust.

2. THE CORE SUPERFLUID

In the neutron star superfluids, vortex lines experience two different types of environment. The superfluid in the inner crust of the neutron star coexists with a crystal lattice, and the motion of the vortex lines is constrained by pinning to the lattice. At densities above about 2×10^{14} g/cm^3 the core of the neutron star contains a homogeneous mixture of superfluid neutrons and superconducting protons which constitute a few percent of the baryons by number. Most of the mass and moment of inertia of the star reside in the neutron superfluid in the core. This superfluid is very tightly coupled to charged particles (electrons) and thereby to the crust of the star. Neutron-proton interactions drag superfluid protons along with the supercurrent around a neutron vortex line. The resulting magnetization of the vortex line leads to a drag force on it as electrons spinning down with the crust acquire a relative velocity with respect to the vortex lines (Alpar, Langer and Sauls, 1984; Alpar and Sauls 1988). The dynamical coupling time t is given by the approximate expression

$$\tau = 100(m/\delta m)^2 P \text{ s} \qquad (3)$$

where m and δm are the bare mass and the difference of effective from bare mass for protons or neutrons; P is the rotation period of the pulsar. An equilibrium lag between the rotation rates of the neutron superfluid and the crust provides the drag force necessary to set up a radially outward vortex flow at the rate required for the core superfluid to spin down at the same rate as the crust. After a perturbation to the crust, the core superfluids come into equilibrium with the crust within the time scale τ. On observationally available time scales after a glitch the core superfluid has already equilibrated with the crust. As far as postglitch timing observations are concerned the effective "crust" moment of inertia actually includes the core superfluid and is therefore almost the entire moment of inertia of the star. Only the crust superfluid with a few percent of the star's total moment of inertia is left to account for the observed postglitch relaxation phenomena, in agreement with the moment of inertia fractions deduced from the postglitch relaxation. The core superfluid is thus so tightly coupled to the crust as to be practically unobservable. The recent eighth glitch of the Vela pulsar was resolved down to 120 s after the event (McCulloch et al., 1990). Therefore τ is even shorter and $(\delta m/m)$ must be at least 0.3. A separate constraint comes from the timing noise from the accretion-powered X-ray pulsar Vela X-1. By virtue of the long period, $P = 283$ s, of this pulsar, τ could fall in the observable range of autocorrelation times. Analysis of timing data shows that at the 99% confidence level, at most 80% of the star's moment of inertia couples to the crust on time scales longer than 1 d (Boynton et al., 1984; Baykal, 1991). If a superfluid core is present, then its coupling times are shorter than 1 d, which implies that $(\delta m/m)$ is at least 0.6. If the 35 d cycle of Her X-1 reflects neutron star precession (Trümper et al., 1986), then the present dynamical coupling model implies that the core superfluid applies a torque $N = 1.5 \times 10^{37} I_{45} (\delta m/m)^2$ g cm^2 s^{-2} to damp the precession. External torques applied by the accretion disk on the neutron star must be of at least this magnitude to keep the precession going (Alpar and Ögelman, 1987). An upper limit on the external torques will translate into an upper limit on $(\delta m/m)$ and a lower limit on τ that may be in conflict with the above limits. It will be very interesting to observe the coupling time τ and to use it to obtain checks on astrophysical scenarios like the precession of Her X-1, or to measure the effective masses of nucleons in the neutron star core. The dynamical coupling of the core superfluids may alternatively be effected by interactions of neutron vortex lines and proton flux lines (Sauls, 1989). This possibility is important for linking the magnetic field decay in neutron stars to their spin histories (Srinivasan et al., 1990). The observational constraints discussed above also apply to the crust-core coupling times arising from this alternative mechanism. The primary significance of a future detection of a crust-core coupling time as predicted by some dynamical model of the core superfluid would of course be to demonstrate the superfluidity of the neutron star core, for which there is no observational evidence at present. Note however that the observations do limit us to very short coupling times and a demonstration that the observed coupling time cannot be furnished by normal matter would be in order.

3. THE CRUST SUPERFLUID

Vortex motion in the crust superfluid is constrained by pinning to the lattice. Vortex pinning by nuclei in the lattice has been estimated and several different regimes of pinning strength have been delineated (Alpar, 1977; Anderson et al., 1982; Alpar, Cheng, and Pines, 1989). Effects of vortex line tension and excitations on pinning were discussed by Epstein in this meeting (Epstein, Link, and Baym, 1992). Given a pinning energy E_p (the gain in energy when a vortex line is pinned), the rate of vortex flow by thermal activation over the energy barriers can be calculated. As a result of pinning the superfluid retains some excess vorticity and spins down at a lag with respect to the crust. A lag $\omega = \Omega - \Omega_c$ between the crust and the pinned superfluid will result in an average radially outward vortex velocity V_r:

$$V_r = 2V_o \exp(-E_p/kT) \sinh(E_p/kT.\omega/\omega_{cr}) \qquad (4)$$

where V_o is a microscopic vortex velocity and ω_{cr} is a critical lag at which vortices unpin. The vortex creep model obtained from Eqs. (1), (2), and (4) has been successfully applied to postglitch relaxation data from several pulsars (Alpar and Pines, 1989; Alpar, Cheng, and Pines, 1989; and references therein). There are two different creep regimes (Alpar et al., 1988; Alpar, Cheng, and Pines, 1989). In young and hot pulsars and for weak-pinning conditions a small lag ω suffices to establish steady state spin-down, and the theory can be linearized. In the opposite regime large lags must build up for vortex creep to spin the superfluid down at the same rate as the crust in a steady state, and the full nonlinear dependence of V_r on ω comes into play. Each glitch introduces a perturbation in ω. In the linear regime the response is linear in the initial perturbation and decays exponentially in time, and the model is formally identical to the two component theory (Baym et al., 1969). For a region i of the crust superfluid that is in the linear regime and experiences a perturbation $\delta\omega_i$ to the lag, the observed crust spin-down rate undergoes a sudden change

$$\Delta\dot{\Omega}_c/\dot{\Omega}_c = (I_i/I)(\delta\omega_i/\tau_i) \qquad (5)$$

which promptly starts to decay exponentially with a relaxation time τ_i. The total jump in the observed $\dot{\Omega}_c$ of the crust consists of contributions from many regions. For regions in the nonlinear regime, the contribution to the jump in $\dot{\Omega}_c$ either persists for a long time or decays very slowly and nonexponentially. The initial jump in spin-down rate directly equals the fractional moment of inertia of the nonlinear region involved,

$$\Delta\dot{\Omega}_c/\dot{\Omega}_c = I_i/I \qquad (6)$$

In either case one can identify the kind of response from its time dependence. A prompt exponential decay indicates linear response, whereas the recovery of the jump in spin-down rate is slow or nonexistent in the case of nonlinear response. Once the nature of each contribution to the postglitch response is thus diagnosed, the appropriate formula, Eq. (5) or Eq. (6), can be used to extract the moment of inertia involved. Remarkably, the amplitudes of both kinds of response have always given I_i/I of order 10^{-3}–10^{-2} in agreement with the moment of inertia fraction in the crust superfluid. It is particularly important to note that in the linear regime exponentially decaying jumps in the spin-down rate can have quite large amplitudes which do not imply large moments of inertia for the pinned superfluid (Alpar et al. 1988; Alpar, Pines, and Cheng, 1990). In the case of linear response, Eq. (5) can be used to test whether the perturbation $\delta\omega$ in the lag is simply the observed jump $\Delta\Omega_c$ in the crust rotation rate or whether it involves a larger $\delta\omega$ indicating that the rotation rate of the superfluid also underwent a reduction $\delta\Omega$. The latter is the case for exponential relaxation following the Crab pulsar glitches (Lyne and Pritchard, 1987). In terms of a generic two component model the Crab postglitch relaxation indicates that some internal component of the star has experienced a change in the fluid rotation rate. For the specific example of the vortex creep model this is evidence that vortex motion did occur at the time of the glitch.

The relaxation times can be interpreted in terms of the pinning energy and temperature. This is subject to some uncertainty in view of several possible pinning regimes. The vortex creep model predicts that old pulsars will display nonlinear response rather than simple exponential relaxation following glitches. As a pulsar cools, loss of thermal activation necessitates larger lags, and the nonlinear response regime prevails for most pinning energies. For such old pulsars the thermal evolution of the pulsar is in turn determined by energy dissipation in vortex creep (Alpar, Nandkumar, and Pines, 1985; Shibazaki and Lamb, 1989; Shibazaki, 1992). An alternative possibility for exponential relaxation comes from conditions of continuous scattering of vortex lines from the phonons of the lattice without going through a series of thermally activated pinnings and unpinnings (Jones, 1990). This alternative for linear response is basically temperature independent. The evolution of dynamical response with pulsar age can provide a valuable test to distinguish between the alternatives.

4. LONG-TERM RELAXATION OF THE VELA PULSAR

Timing observations of the Vela pulsar are of great importance as it has been followed through eight consecutive glitches. Extensive timing data exist after each of these glitches. The last four glitches have been caught within a day, and the eighth one actually occurred during an observing session. When the short time scale components of postglitch relaxation are over there is still a remaining offset both in frequency and in the spin-down rate with respect to an extrapolation of preglitch timing fits. A glitch at time $t = 0$ will have completely relaxed at some later time t if the integrated negative offset in the spin-down rate compensates exactly for the initial frequency increase:

$$\Delta\Omega_c(t) = \Delta\Omega_c(0) + \int_0^t \Delta\dot{\Omega}_c(t)dt = 0 \qquad (7)$$

Without integrating to test for this condition, one can examine the spin-down rate directly for signs that the relaxation process is completed. In the Vela pulsar the long time scale decrease in the spin-down rate heals with a linear dependence on time, so that there is a constant positive second derivative of the rotation frequency far in excess of that expected from pulsar braking. When the relaxation is over, the spin-down rate would itself settle to a constant equilibrium value, with only the second derivative due to pulsar braking left. In the vortex creep theory this equilibrium is reached when vortex creep has restarted in perturbed nonlinear response regions. Completion of the relaxation can involve a sudden increase in slope and even a steplike change in the spin-down rate, depending on the particular model assumptions. A search for such signatures was emphasized as an indicator of the relation between the healing of one glitch and the occurrence of the next (Alpar et al., 1984). Does a glitch heal before the next one comes along, does healing trigger the next glitch, or do unhealed remnants of glitches form part of the long-term dynamics of the pulsar? There is no sign of any change in the nature of the long-term relaxation following a glitch when the next glitch comes along (Cordes, Downs, and Krause-Polstorff, 1988). This question can be posed in a model independent way, as noted by Roger Blandford in this conference. Extrapolating from the rotation frequency prior to the first glitch, with the spin-down rate at that epoch and assuming that no glitches occurred, it is seen that the frequency obtained just prior to the sixth glitch is lower than the observed frequency at that epoch by a part in 10^5. Assessing the net observed offset with respect to a direct extrapolation from the situation before the first glitch is equivalent to doing the test in Eq. (7) over the span of five glitches, adding up the frequency jumps and the integrated spin-down offsets of the first five glitches and their postglitch timing observations. This means that the first five glitches did not relax back to zero frequency offset; rather, they collectively left a remnant frequency increase that is a significant part of the total of the initial frequency increases from these glitches. Notice that the glitches contribute a net spinup to the crust against the dominating background of spin-down by the external torque. Thus the glitches have the same sign as the internal torque on the right hand side of Eq. (2): The spin-down of an internal component is reflected as a spinup torque on the crust, compensating part of the external torque as the interior couples to the crust and shares in the overall spin-down of the star. This trivial but basic point means that the unrelaxed part of the glitches is simply a discrete complement to the continuous coupling torque between the interior and the crust. In the context of the vortex creep model the continuous outward flow of vortices is complemented by occasional unpinnings of large numbers of vortices, whose outward motion is manifested as the glitches. In the long run the remnants of the glitches are determined by the external torque on the system. The steady state vortex transport is such that the superfluid spins down at the same average rate as the crust through the sum of continuous vortex creep and the remnants of the discrete events. This interpretation of the steady state has been further explored to include the sixth, seventh, and eighth glitches in the analysis (Alpar et al., 1991). Elsewhere in these proceedings we provide a preliminary discussion of this model for evaluating the eight Vela pulsar glitches and the postglitch behavior (Pines and Alpar, 1992). The model is based on the hypothesis that a large part of the angular momentum transfer in each glitch is simply a lumped part of the internal torque coupling the crustal superfluid to the observed crust, and therefore this part of the spinup in the glitch does not relax back. The alternative possibility that the net unrelaxed part of each Vela glitch reflects a permanent reduction in moment of inertia due to a starquake can be excluded since the remnant offsets are comparable to the initial frequency jumps and it is already known that such large and frequent events cannot be supported by elastic energy storage and release. The more recently discovered glitches of PSR 1737-30 are similar and as frequent as those from Vela (McKenna and Lyne, 1988). With detailed postglitch analysis this pulsar can also provide a testing ground for investigating the long-term dynamical function of the glitches.

ACKNOWLEDGMENTS

I thank D. Pines and K. S. Cheng for many enjoyable discussions on these subjects over the years, the Aspen Center for Physics for opportunities to continue my collaboration with them, H. F. Chau for useful communications, and R. Blandford for his stimulating comments. This research was supported in part by the NSF International Grant NSF INT89-18665 and a NATO Collaborative Research Grant.

REFERENCES

1. Alpar, M. A., *Ap. J.* 213, 527 (1977).
2. Alpar, M. A., and Pines, D., "Vortex Creep Dynamics: Theory and Observations," in Timing Neutron Stars, H. Ögelman and E. P. J. van den Heuvel, eds., Kluwer Academic, Dordrecht (1989).
3. Alpar, M. A., Anderson, P. W., Pines, D., and Shaham, J., *Ap. J.* 278, 791 (1984).

4. Alpar, M. A., Langer, S. A., and Sauls, J. A., *Ap. J.* 282, 533 (1984).
5. Alpar, M. A., Nandkumar, R., and Pines, D. *Ap. J.* 288, 191 (1985).
6. Alpar, M. A., and Ögelman, H., *Astron. and Astrophys.* 185, 196 (1987).
7. Alpar, M. A., and Sauls, J. A., *Ap. J.* 327, 723 (1988).
8. Alpar, M. A., Cheng, K. S., Pines, D., and Shaham, J. *Mon. Not. Royal Astr. Soc.* 233, 25 (1988).
9. Alpar, M. A., Cheng, K. S. and Pines, D., *Ap. J.* 346, 823 (1989).
10. Alpar, M. A., Pines, D., and Cheng, K. S. *Nature* 348, 707 (1990).
11. Alpar, M. A., Chau, H. F., Cheng, K. S., and Pines, D. in preparation (1991).
12. Anderson, P. W., and Itoh, N., *Nature* 256, 25 (1975).
13. Anderson, P. W., Alpar, M. A., Pines, D., and Shaham, J. *Phil. Mag. A* 45, 227 (1982).
14. Baykal, A., unpublished Ph.D. thesis, Middle East Technical University (1991).
15. Baym, G., Pethick, C. J., Pines, D., and Ruderman, M. *Nature* 224, 872 (1969).
16. Boynton, P. E., Deeter, J. E., Lamb, F. K., Zylstra, G., Pravdo, S. H., White, N. E., Wood, K. S., and Yentis, D. J. *Ap. J. Lett.* 283, L53 (1984).
17. Cordes, J. M., Downs, G. S., and Krause-Polstorff, J., *Ap. J.* 330, 847 (1988).
18. Epstein, R., Link, B., and Baym, G., this volume (1992).
19. Jones, P. B. *Mon. Not. Royal Astr. Soc.* 243, 257 (1990).
20. Lyne, A. G., and Pritchard, R. S. *Mon. Not. Royal Astr. Soc.* 229, 223 (1987).
21. McCulloch, P. M., Hamilton, P. A., McConnell, D. and King, E. A., *Nature* 346, 822 (1990).
22. McKenna, J. and Lyne, A. G. *Nature* 343, 349 (1990).
23. Pines, D. and Alpar, M. A., this volume (1992).
24. Sauls, J. A., "Superfluidity in the Interiors of Neutron Stars," in Timing Neutron Stars, H. Ögelman and E. P. J. van den Heuvel, eds., Kluwer Academic, Dordrecht (1989).
25. Shibazaki, N., this volume (1992).
26. Shibazaki, N., and Lamb, F. K. *Ap. J.* 346, 813 (1989).
27. Srinivasan, G., Bhattacharya, D., Muslimov, A. G., and Tsygan, A. I., *Current Science* 59, 31 (1990).
28. Trümper, J., Kahabka, P., Ögelman, H., Pietsch, W., and Voges, W. *Ap. J. Lett.* 300, L63 (1986).

Richard I. Epstein, Bennett K. Link† and Gordon Baym‡

Los Alamos National Laboratory, Los Alamos, NM 87545, USA; †Astronomy Department, University of Illinois, Urbana, IL 61801, USA; ‡Loomis Laboratory of Physics, University of Illinois, 1110 W. Green St., Urbana, IL 61801, USA

Superfluid Dynamics in the Inner Crust of Neutron Stars

Interactions between the neutron superfluid and the neutron-rich nuclei in the inner crust of a neutron star may be responsible for glitches, post-glitch relaxations, and other timing irregularities. Here we describe these interactions as they relate to vortex pinning, vortex creep, and the dissipative coupling between the superfluid and the crust.

1. INTRODUCTION

In the inner crust of a neutron star a gas of free neutrons coexists with a lattice of neutron-rich nuclei. For stellar densities $\sim 10^{12}$–10^{14} g cm^{-3}, corresponding to free neutron densities $\sim 10^{-3}$–10^{-1} fm^{-3}, the free neutrons are believed to pair and form an isotropic s-wave superfluid. In a rotating star the neutron superfluid rotates by forming vortex lines, singular regions in which the superfluidity vanishes and around which the superfluid circulation is $\kappa = \pi\hbar/m_n$, where m_n is the neutron mass. The superfluid velocity in the neutron star is completely determined by the spatial arrangement of the vortex lines. A change in the velocity field requires a corresponding change in the vortex-line distribution, either a distortion of the configuration of the existing lines or the creation of new lines.

Superfluid Dynamics in the Inner Crust of Neutron Stars

Were the vortex lines in a region of the inner crust immobilized, for example by pinning to the crystal lattice, the angular velocity of the superfluid in this region would be largely fixed. As the angular velocity of the solid crust of the neutron star changes, due to magnetic braking or accretion, the difference between the angular velocities of the superfluid and the solid crust grows. This differential motion may be the source of free energy that powers glitches. A glitch may result from the sudden transfer of angular momentum from the superfluid to the crust caused by the catastrophic unpinning of many vortex lines or by cracking of the crust to which the vortex lines are pinned.[1,2] The post glitch relaxation may be understood in terms of variations in the slow outward vortex creep rate.[3]

To understand these variations in the rate of angular momentum transfer between the superfluid and the crust, we have been investigating the dynamics of vortex lines in the inner crust. Several aspects of our work, which we discuss here, include computing the interaction force between nuclei and vortex lines, the pinning energies that pin vortex lines to the nuclear lattice, the rates for vortex lines to unpin and creep through the stellar crust, and the dissipative coupling that transfers angular momentum from the superfluid to the crust. We now discuss each of these points.

2. VORTEX PINNING

We estimated the interaction of a single nucleus with a vortex line by computing the change in the superfluid pairing energy and the penetration of the superfluid into a nucleus by minimizing a Ginzburg-Landau energy functional.[4] To ensure that these estimates values were in accord with measured properties of nuclei, we adjusted the poorly known superfluid gaps so that the pairing energy of a laboratory nucleus would be of the order of 1 MeV. By including the classical Bernoulli force between a nucleus and a vortex line, we obtained the interaction energy $E_{\text{int}}(s)$ between a nucleus and a vortex line as a function of their separation s of the form

$$E_{\text{int}}(s) = \frac{E_S}{(1 + s^2/R^2)^4} + \frac{E_L}{1 + s^2/R^2}, \qquad (1)$$

plus terms independent of s. Here R is the effective nuclear radius, E_S is the energy related to the *short*-range part of the interaction, and E_L is related to the *long*-range repulsion; while E_L is always positive, E_S can be positive or negative depending on density ρ_* (see Table 1). The values of E_S we obtained are uncertain since a fully microscopic treatment of the short-range component of the vortex-nucleus interaction is not yet available. Furthermore, upper limits on the internal heating in neutron stars due to vortex creep, inferred from neutron star surface temperatures, suggest that our energy estimates are too large at high densities.[5] To allow for this possibility, we also consider the consequences of lower E_S. Since the long-range interaction is largely due to hydrodynamic effects, the values of E_L are more certain.

In the high-density ($\gtrsim 10^{13}$ g cm^{-3}) part of the inner crust, the vortex-nucleus interaction is attractive ($E_S < 0$), and the lines strongly pin on nuclei. At lower densities the interaction is repulsive ($E_S > 0$); the vortex lines tend to avoid nuclei and can weakly pin to the spaces between the nuclei. We refer to these two density regimes as the *nuclear-pinning* and *interstitial-pinning* regions, respectively.[4,6]

TABLE 1 Vortex Pinning Parameters

log ρ_* (g cm^{-3})	R (fm)	E_S (MeV)	E_L (MeV)	log ρ_s (g cm^{-3})	log v_B (cm s^{-1})	log v_* (cm s^{-1})
11.83	5.7	2.5	0.037	11.21	5.15 (5.15)	8.55 (7.14)
11.99	5.8	3.3	0.090	11.63	5.20 (5.20)	8.16 (6.80)
12.18	5.9	4.2	0.16	11.90	5.28 (5.28)	7.97 (6.63)
12.41	6.1	6.3	0.34	12.30	5.32 (5.32)	7.69 (6.39)
12.79	6.5	7.5	0.72	12.70	5.43 (5.43)	7.29 (6.08)
12.98	6.7	-1.3	0.94	12.89	7.20 (6.20)	5.88 (5.33)
13.18	6.9	-7.7	1.3	13.11	8.20 (7.20)	6.82 (5.27)
13.53	7.2	-16.4	1.4	13.48	8.23 (7.23)	6.94 (5.40)
13.89	7.3	-10.0	1.0	13.86	7.72 (6.72)	6.41 (4.86)
14.12	7.2	-7.8	0.49	14.09	7.58 (6.58)	6.24 (4.74)

When a vortex line is pinned to the crust, the superfluid streams past the line, producing a Magnus force per unit length of vortex line:

$$\vec{f}_M = \rho_s \vec{\kappa} \times (\vec{v}_v - \vec{v}_s), \qquad (2)$$

where $\vec{\kappa}$, whose magnitude is κ, is aligned with the vortex line; $\vec{v}_v - \vec{v}_s$ is the relative velocity of the vortex line with respect to the superfluid; and ρ_s is the superfluid density. A sufficiently large Magnus force breaks the pinning bonds and enables the vortex lines to move in the star. The characteristic relative velocity, v_B, at which vortex lines must break free is found by equating the magnitude of the Magnus force with the maximum pinning force per unit length; this balance is approximately given by

$$\rho_s \kappa v_B \simeq \frac{F_{\max}}{d_p}, \qquad (3)$$

where $F_{\max}$ is the maximum pinning force per site, and d_p is the minimum distance between pinning sites on a given line. We characterize the relative strengths of the pinning force and the Magnus force by the dimensionless parameter

$$\Delta \equiv \frac{v_B - |\vec{v}_v - \vec{v}_s|}{v_B}. \qquad (4)$$

The pinning force dominates for $\Delta \to 1$, and the Magnus force dominates for $\Delta \to 0$. Table 1 lists v_B for the values of E_S listed in the Table and for these values reduced by a factor of 10; the latter values are shown in parentheses. The abrupt change in v_B near a stellar density $\rho_* \sim 8 \times 10^{12}$ g cm^{-3} marks the transition from the interstitial-pinning region to the nuclear-pinning region. In the interstitial-pinning region v_B depends only on E_L and is unaffected by reductions in E_S.

3. ACTIVATION ENERGIES FOR VORTEX UNPINNING

The nature of vortex unpinning is determined by the relative importance of tension in the vortex line and the pinning forces acting upon it. If the vortex is relatively flexible, only one or a few bonds break simultaneously, whereas if it is relatively stiff, many pinning bonds break. In the nuclear-pinning region the vortex line is comparatively flexible, while for the interstitial-pinning regions the vortex is stiff.

We have modeled the vortex unpinning process by considering a vortex line pinned to a linear sequence of pinning sites with spacing l. When superfluid streams past a pinned vortex line, the Magnus force f_{Mag} bows the vortex line out and holds the line away from the pinning sites (see Figure 1). If the tension is not too large, the equilibrium configuration in the vicinity of one site is nearly independent of the configuration near other sites, and vortex unpinning occurs one site at a time; this is *single-site breakaway*. If the tension of the vortex line is much larger, the line may be too stiff to disengage from one pinning site while still pinned to nearby sites, and several pinning bonds must break simultaneously. We approximate this unpinning process as a *continuous breakaway*.[6]

For a static vortex line subjected to pinning, tension, and Magnus forces, the total energy E of a line of length L is[6]

$$E = \int_L \left[\frac{T}{2} \left| \frac{\partial \vec{s}}{\partial z}(z) \right|^2 + \rho_p(z) E_p(s) - \rho_s (\vec{\kappa} \times \vec{v}_s) \cdot \vec{s} \right] dz. \tag{5}$$

where $\vec{s}(z,t)$ is the displacement of the line, z is the distance along the line, T is the tension and ρ_p is the density of pinning sites along the line. Vortex-vortex interactions [7] are neglected here because the vortex lines are so widely spaced in observed neutron stars. The tension has the form $T = \rho_s \kappa^2 \Lambda / 4\pi$, where Λ is a logarithmic factor which is typically 3–10 in a neutron star.[6,8]

We now quantify what is meant by the vortex line's *stiffness* by defining

$$\tau \equiv \frac{T r_0}{F_{\text{max}} l}, \tag{6}$$

where r_0 is the separation at which $F_p = F_{\text{max}}$. A flexible vortex has τ much less than unity, while a stiff vortex has τ much greater than unity. In the low-density

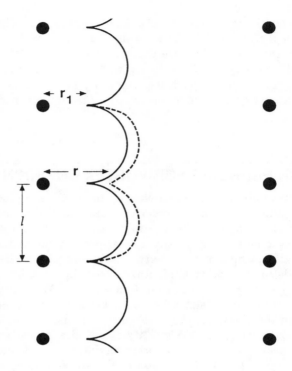

FIGURE 1 A pinned vortex under the influence of the Magnus force. The pinning sites are shown as filled circles. The equilibrium spacing between the vortex and a pinning nucleus is r_1. The perturbed spacing is r.

($\rho_* \lesssim 10^{13}$ g cm^{-3}) interstitial pinning region of the crust, pinning is so weak compared with the tension that a vortex is very stiff. In this region, unpinning necessarily involves many pinning bonds, and the continuous breakaway approximation is appropriate. In the higher-density nuclear-pinning region, a pinned vortex is flexible and can bend on a length scale comparable to the pinning spacing. Here the single-site breakaway approximation may be useful; however, for the reduced values of E_S, unpinning tends to involve 3–300 pinning sites, depending on density and the star's temperature and slowing rate.

To compute the activation energy for flexible vortex lines in the single-site breakaway approximation, we consider the energy change associated with moving the central pinning point to $s = r$ while the endpoints at $z = 0$ and $2l$ are held fixed at r_1. The vortex energy as a function of separation is shown in Figure 2. For this calculation we used a parabolic approximation for the pinning force.[6] The

Superfluid Dynamics in the Inner Crust of Neutron Stars

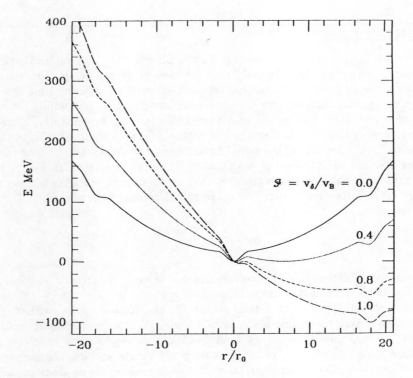

FIGURE 2 Vortex energy as a function of separation from the central pinning site in the single-site approximation, accounting for the attractive potential of a nearby nucleus. Curves for different values of $\mathcal{F} = 1 - \Delta$ are shown for $F_{max}r_0 = 11.3$ MeV, $l/r_0 = 18.4$, and $\tau \simeq 0.03$.

activation energy is the energy difference between the minimum in the pinned state and the peak to the right of it. We find that this energy is $A \simeq 0.75 F_{max} r_0 \Delta^{3/2}$. Figure 2 illustrates that a vortex line can unpin from one sight and move into a free configuration between pinning sites or, for the appropriate Δ, slip into an adjacent pinning site at lower Magnus energy shown here on the right (the vortex line can never unpin by hopping into a pinning site at larger pinning Magnus energy).

When breakaways involve many pinning sites the discrete nature of the pinning is unimportant; we treat the pinning as continuous, with a pinning density $\rho_p(z) = 1/l$. For stiff vortex lines and low Magnus force (i.e., $\Delta < 1/4$), the activation energy is $A \simeq 1.8 F_{max} r_0 \Delta^{3/2} j_*$ where $j_* \simeq 3.242 \tau^{1/2} \Delta^{-1/4}$ is the characteristic number of pinning sites involved in the unpinning event. For larger Δ, the forms of A_c and j_* are more complicated (see Ref. 6).

4. VORTEX CREEP

If the vortex lines in the inner crust are largely pinned, they can still move through the crystal lattice by the relatively slow process of *vortex creep*. The elemental process in vortex creep consists of a segment of vortex line unpinning from one set of pinning sites, moving to a set of sites at lower energy, and finally repinning. Since widely separated segments of a vortex line move independently, a creeping vortex line does not remain straight but rather advances sinuously. To obtain an expression for the average of the radial component of the creep velocity, we consider a vortex line pinned at a total of N sites. Let R_j be the rate at which $j \ll N$ bonds break simultaneously. After a segment with j bonds unpins, it moves a mean radial distance $\ell_r(j)$ before repinning. The mean radial vortex creep rate is

$$v_{\text{creep}} \simeq \sum_{j=1}^{N} R_j \ell_r(j). \qquad (7)$$

The problem of estimating the vortex creep rate thus reduces to obtaining the individual rates R_j and the mean translations ℓ_r.

When a vortex line is at a temperature T, the Kelvon mode oscillations, or "kelvons," are excited. A line segment of length jl has a fundamental oscillation frequency ν_j and minimum energy level spacing $\epsilon_j = h\nu_j$. At low temperature, a vortex has a high probability of being in its ground state and unpins primarily by quantum tunneling. At high temperature, the higher energy levels are occupied and the vortex can unpin by classical thermal activation. In the classical limit the line segment unpins by changing its shape as its energy rises to some activation barrier A_j and then declines. If we treat the line segment as having one degree of freedom and being thermally well coupled to the rest of the vortex lattice, the unpinning rate is[9,10]

$$R_{j,\text{class}} \simeq \nu_j e^{-A_j/k_B T}. \qquad (8)$$

At low temperatures, quantum tunneling is the dominant rate process, and the rate becomes nearly temperature independent. There is no well-established theory for computing the unpinning for an extended object such as a vortex line. We have estimated R_j at by employing a truncated parabolic model for the effective potential well for the modes of a pinned vortex line.[9] An approximate rate expression useful at all temperatures is

$$R_j \simeq \nu_j e^{-A_j/k_B T_{eff}}, \qquad (9)$$

where

$$k_B T_{eff} \equiv \frac{\epsilon_j}{2} \coth \frac{\epsilon_j}{2k_B T}. \qquad (10)$$

For high temperatures, R_j reduces to (8). Figure 3 shows the behavior of the rate (9) as a function of temperature; it is constant at low T, where quantum tunneling dominates, and increases with temperature at high T, where thermal activation is dominant. There is a "crossover temperature" $T_0 \sim \epsilon_j/k_B$ such that when $T < T_0$,

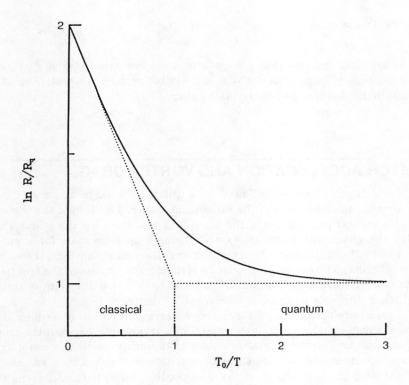

FIGURE 3 The rate as a function of temperature. At high temperatures, where motion past the barrier occurs classically, $\ln R$ is linear in $1/T$. At low temperatures, where quantum tunneling dominates, the rate is nearly constant. The transition between the two regimes is defined by the crossover temperature, T_0.

quantum tunneling dominates, while classical thermal activation dominates when $T > T_0$.

The radial creep rate, eq. (7), as evaluated in Ref. 9 is

$$v_{\text{creep}} \simeq v_0 e^{-A_*/k_B T_{\text{eff},*}}. \tag{11}$$

For stiff vortex lines the summation in eq. (7) is evaluated by steepest descent to give

$$v_0 = \nu_s \ell_{\min} j_* \left(\frac{2\pi k_B T_{\text{eff},*}}{A_*} \right)^{1/2}, \tag{12}$$

and for flexible vortices,

$$v_o = \nu_f \ell_{\min}. \tag{13}$$

Here the asterisks indicate that the quantities are evaluated at $j = j_*$; ν_s and ν_f and the oscillation frequencies for stiff and flexible vortices, respectively; and $\ell_{\min}$ is the minimum distance between pinning sites.

5. GLITCH ACCELERATION AND VORTEX DRAG

In the "catastrophic unpinning" model for glitches,[1] a large number ($\gtrsim 10^{10}$) of vortex lines unpin suddenly and the subsequent outward motion of the vortex lines lowers the angular momentum of the superfluid and spins up the crust. The rate at which the glitch instability grows and the spin-up time scale for a glitch are determined by the mechanisms that transfer angular momentum from the superfluid to the solid crust. The importance of understanding the physics of the transfer time scale was underscored by the recent determination[11,12] that the spin-up time scale for a glitch in the Vela pulsar did not exceed 2 minutes.

A physical explanation of pulsar timing observations entails describing the coupling mechanism by which angular momentum is transferred between the crust and the superfluid. Among the several processes that contribute to this coupling,[13–16] the most important one appears to be the excitation of waves on a vortex line by interaction with nuclei in the crust. As a vortex line moves past nuclei, the vortex-nucleus interaction bends and twists the vortex line and pulls the nuclei away from their equilibrium positions in the crystal lattice of the crust. These perturbations excite kelvons on the vortex line and phonons in the crystal. Interactions between phonons in the crystal lattice and kelvons tend to bring the vortex excitations into thermal equilibrium with the solid crust.

The power dissipated per length of vortex line due to scattering with nuclei is[16]

$$p_d = \rho_s \kappa |\vec{v}_v - \vec{v}_s|^{1/2} v_*^{3/2}. \tag{14}$$

Table 1 gives computed values of v_* in the inner crust for the values of E_S given in Table 1 and these values of E_S reduced by a factor of 10 (the latter values of v_* are shown in parentheses).

The power p_d is related to the motion of the unpinned vortex lines by

$$p_d = \rho_s \kappa \tan\theta_d |\vec{v}_v - \vec{v}_s|^2, \tag{15}$$

where θ_d is the *dissipation angle* which characterizes the trajectory of a vortex line. In a frame comoving with the crust a vortex moves at an angle θ_d with respect to the superfluid velocity, that is, the radial component of the vortex velocity is $|\vec{v}_v - \vec{v}_s| \sin\theta_d$.[13]

For $v_B \gtrsim v_*$ the spin-up time scale is found to be[16]

$$t_{su} \lesssim \frac{(v_B/v_*)^{3/2}}{2\Omega_s}, \qquad (16)$$

where Ω_s is the angular velocity of the superfluid n the inner crust. For smaller v_B the maximum dissipation angle is reached and

$$t_{su} \lesssim (\Omega_s)^{-1}. \qquad (17)$$

Among the entries in Table 1 the largest value of v_B/v_* is ~ 85, corresponding to an upper limit on the spin-up time scale of $t_{su} \lesssim 60\,P$, where P is the rotation period. The spin-up time scale for the nuclear scattering process is thus $\lesssim 5$ s for the Vela pulsar, more than adequate for explaining the observed upper limit on the glitch spin-up time. However, within the catastrophic unpinning model for glitches, the instability that forces many vortex lines to unpin rapidly remains an important unanswered question, and the rapid coupling between the crust and the core is not fully understood.

ACKNOWLEDGMENTS

This work was carried out under the auspices of the Department of Energy and supported in part by NSF Grant DMR 88-18713. We have had useful discussions with Ali Alpar, Lars Bildsten, Fred Lamb, Guy Miller, David Pines, Mal Ruderman, Edouard Sonin, and Noriaki Shibazaki about various aspects of this work.

REFERENCES

1. Anderson, P. W., and Itoh, N. 1975, *Nature*, **256**, 25.
2. Ruderman, M. 1976, *Ap. J.*, **203**, 213.
3. Alpar, M. A., Anderson, P. W., Pines, D. and Shaham, J. 1984, *Ap. J.*, **276**, 325.
4. Epstein, R. I., and Baym, G. 1988, *Ap. J.*, **328**, 680.
5. Shibazaki, N., and Lamb, F. K. 1989, *Ap. J.*, **346**, 808.
6. Link, B., and Epstein, R. I. 1991, *Ap. J.*, **373**, 592.
7. Baym, G., and Chandler, E. 1983, *J. Low Temp. Physics*, **50**, 57.
8. Thomson, W. (Lord Kelvin). 1880, *Phil. Mag.*, **10**, 155.
9. Link, B., Epstein, R. I., and Baym 1991, *Ap. J.*, forthcoming.
10. Hänggi, P., Talkner, P., and Borkovec, M. 1990, *Rev. of Mod. Phys.*, **62**, 251.
11. Flanagan, C. 1989, *IAU Circular* No. 4695.

12. Hamilton, P. A., King, E. A., and McCulloch, P. M. 1989, *IAU Circular* **No. 4708**.
13. Bildsten, L., and Epstein, R. I. 1989, *Ap. J.*, **342**, 951.
14. Feibelman, P. J. 1971, *Phys. Rev. D*, **4**, 1589.
15. Jones, P. B. 1988, *M. N. R. A. S.*, **235**, 545.
16. Epstein, R. I., and Baym, G. 1991, *Ap. J.*, submitted.

R. Tamagaki
Department of Physics, Kyoto University, Kyoto 606, Japan

Various Phases of Hadronic Matter in Neutron Stars and Their Relevance to Pulsar Glitches

Aspects of multihadronic phase structures in neutron stars are shown to be strongly dependent on neutron star masses, M. We attempt to correlate the pion condensation and glitch phenomena by taking into account the combined (charged and neutral) pion condensation, which provides a core "solid" and a rapid-cooling mechanism. From the viewpoint of a new extended corequake model for neutron stars with pion-condensed core, where the Vela neutron star with $M \simeq 1.6\ M_\odot$ and a "solid" core and the Crab neutron star with $M \simeq 1.2\ M_\odot$ and only a fluid core are assumed, the remarkable differences between the Vela and Crab glitches are explained. Postglitch timing behavior of the two pulsars is well described by a new three-component starquake model which takes into account the two kinds of neutron superfluids ($^1S_0$-one in the crust and $^3P_2$-one in the core) and the pion-condensed core. Comments on future problems are given.

1. MULTIHADRONIC PHASES IN NEUTRON STARS

In neutron stars involving a wide range of density including the nuclear density $\rho_0 \simeq 2.8 \times 10^{14}\ \text{g/cm}^3 \simeq 0.17\ \text{fm}^{-3}$, characteristic features of nuclear forces directly manifest themselves as various hadronic phases in neutron star interior. This manifestation appears through the state dependence of nuclear forces that their signs and strengths change state by state.

1.1 NUCLEON SUPERFLUIDS IN NEUTRON STAR INTERIOR

The attractive potentials of medium range in the $^1S_0$ and $^3P_2$ states bring about the pairing correlation in these states. They are strong enough to realize the superfluid states in neutron star matter, overwhelming the countereffects due to the short-range repulsion. The two superfluids ($^1S_0$-one and $^3P_2$-one) are different with respect to ingredients and density regions of realization. Here the discussion is restricted to these points.

The driving force to realize the $^1S_0$ **superfluid** is the intermediate-range attraction of the central potential originating from the exchange of two pions with the same quantum number as a neutral scalar, which is often simulated by the exchange of a single meson "σ." Many calculations hitherto made[1] have shown the existence of the neutron $^1S_0$ superfluid in the inner-crust region ("crust superfluid") at $\rho_1 \leq \rho \lesssim \rho_2$, where $\rho_1 \simeq 3 \times 10^{11}\text{g/cm}^3 \simeq 10^{-3}\rho_0$ is the neutron drip density and $\rho_2 \simeq 2 \times 10^{14}\text{g/cm}^3 \simeq 2\rho_0/3$ the transition density to the uniform matter. Since such pairing mechanism at subnuclear density is also operative for the low-density protons mixed with several percent at the fluid core; the proton $^1S_0$ superfluid appears at $\rho_2 \leq \rho \lesssim 3\rho_0$.[2] The disappearance at the high-density side is due to the decrease of the proton effective mass. Recent studies on medium effects on the $^1S_0$ superfluidity[3] have shown a reduction of the neutron $^1S_0$ gaps, compared with the ones previously obtained without such medium effects. For recent advances in microscopic studies of superfluid gaps, see the report given by Clark.[1]

The attractive interaction leading to the neutron $^3P_2$ superfluid[4,5] originates mainly from the spin-orbit force due to the exchange of the vector mesons (ρ,ω) and the neutral scalar meson ("σ"). In addition to this, the appearance of the neutron $^3P_2$ superfluid is assisted by the outer part of the tensor force dominated by the one-pion (1π) exchange. In this state where the coupling to the $^3F_2$ state due to the tensor force arises, in addition to the effect of the $^3P_2$ diagonal part, the supplementary attractive effect through this coupling is also important when we apply the realistic nuclear interaction, as shown by Takatsuka.[6] Because of the P-wave pairing, the energy gaps start to appear when the relative momentum of the pair neutrons becomes high enough to overcome the centrifugal effect. This onset density is near ρ_2 and the superfluid exists up to about $2.8\rho_0$, where the neutron effective mass decreases to about $0.7m_n$ due to the influence of the single-particle potential in neutron star matter. The regions of the neutron $^3P_2$ superfluid and the proton $^1S_0$ superfluid almost coincide where $\rho_2 \leq \rho \lesssim 3\rho_0$, the "core superfluid" region.

Some examples of the appearance of these hadronic superfluids in neutron stars, which is strongly dependent on M, are shown in Fig. 2(a).

1.2 PION CONDENSATION IN NEUTRON STAR MATTER

Pion condensation[7,8] is one of the characteristic phases in high-density hadronic matter, and is brought about by the pion-nucleon P-wave interaction as the driving interaction, $\mathcal{H}_{\pi NN} = f_{\pi NN}\ \overline{N}\vec{\sigma}\tau N \cdot \vec{\nabla}\pi$, the well-known Yukawa coupling between the nucleon (N) and pion (π), where $\vec{\sigma}$ (τ) is the spin (isospin) operator. The pion condensates have many order parameters because of three isospin components and the P-wave nature. Here we give the most plausible version, which is employed in studying pulsar glitches.

In the neutral pion condensation (abbreviated to NPC),[9,10] the ground-state expectation value of π^0 field develops along the condensed momentum $\vec{k}_0$ as a standing wave like $<\pi^0(\vec{r})> = A_0\ \sin \vec{k}_0 \cdot \vec{r}$. The NPC field $<\pi^0(\vec{r})>$ provides nucleons with the potential field $U_\pi = (f/m_\pi)\tau_3 \vec{\sigma} \cdot \vec{\nabla} <\pi^0(\vec{r})>$. Under U_π, neutrons ($\tau_3 = -1$) locate at the points of the steepest gradient of $<\pi^0(\vec{r})>$, with their spin directions alternately changing layer by layer. This neutron configuration is the structure which utilizes most efficiently the attractive effect of the 1π-exchange tensor potential. In other words, the NPC is nothing but the phase realized by the dominance of the 1π-exchange tensor force. The NPC phase can be described by such a specific localized spin structure, which we call the alternating-layer-spin (ALS) model.[11] It simultaneously possesses a spin order like the ferromagnetic substance and a spatial correlation like the smectic A-phase of liquid crystal. When protons ($\tau_3 = +1$) are mixed, they enter the layer with spins opposite to the neutron spin, enforcing the source function of $<\pi^0(\vec{r})>$. The phase transition to the NPC is considered to be first order.

In neutron star matter, since the neutron chemical potential μ_n is much larger than the proton one μ_p, the charged pion condensation (CPC) dominated by the π^- component starts to develop at the density where its chemical potential μ_c becomes equal to $\mu_n - \mu_p$.[8,12] A typical feature of the CPC field is of the running-wave type as $<\pi^c(\vec{r})> \exp(-i\mu_c t) = A_c \exp[i(\vec{k}_c \cdot \vec{r} - \mu_c t)]$, with the condensed momentum $\vec{k}_c$ and $\mu_c = \mu_n - \mu_p$. Correspondingly its source (nucleons) becomes the Fermi gas state of the quasi-neutrons which are described by a superposition of the neutron and proton states. The phase transition to the CPC is considered to be second order.

In the realistic situation, various effects come into play and largely modify the aspects given by the driving interaction only, such as the low values of the critical densities for the NPC and CPC predicted near ρ_0.[10,12] Suppressive effects come from the ρ-meson, the short-range correlation, and the antisymmetrization, whereas enhancing effects come from the mixing of the isobar, Δ, at 1232 MeV. In the NPC, for example, it increases the πNN coupling constant $f_{\pi NN}$ to an effective one $f_{\text{eff}} \simeq 1.7 f_{\pi NN}$ for the Δ-mixing probability $\simeq 0.1$.[10] In this case, neutrons (protons) become quasi-neutrons $\hat{n}$ (quasi-protons $\hat{p}$) due to the Δ-mixing. As a result of these effects, the critical densities for the NPC and CPC are in the range $\rho_t(\text{CPC}) \simeq (1.5-2)\rho_0$ [13] and $\rho_t(\text{NPC}) \simeq (2-4)\rho_0$,[14] where $\rho_t(\text{NPC})$ means the density where the energy curves of the NPC phase and the normal one cross between two transition

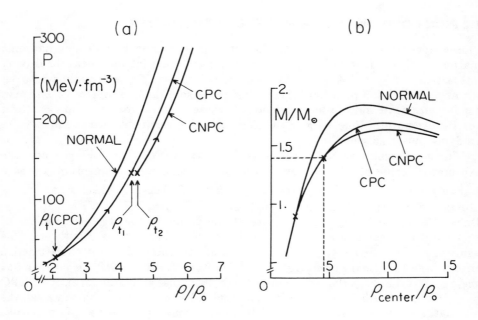

FIGURE 1 The EOS and the resulting mass profile of pion-condensed neutron stars. (a) Pressure (P) versus density (ρ), where the crosses denote the points of phase transitions. (b) Neutron star mass (M) versus central density (ρ_{center}). The dashed lines indicate the correspondence, $M = 1.4\,M_\odot$ versus $\rho_t(\text{CNPC}) = 4.5\rho_0$.

densities of the first-order phase transition. It seems reasonable to assume that $\rho_t(\text{CPC}) \lesssim \rho_t(\text{NPC})$.

Since both types of pion condensation are realized at the density $\rho > \rho_t(\text{NPC})$, we can consider a possible occurrence of a combined (charged and neutral) pion condensation (CNPC),[15] where the CPC and NPC coexist without serious interference in the sense that the total energy gain is almost additive. Such a structure, which is simple but seems most realistic, is the version where the NPC-momentum ($\vec{k}_0$) and the CPC-momentum ($\vec{k}_c$) are perpendicular. In this structure, the CPC develops in the layers of the baryonic ALS structure; $\hat{n} \uparrow$ and $\hat{p} \downarrow$ or $\hat{n} \downarrow$ and $\hat{p} \uparrow$ are superposed to make new quasi-nucleon states. In the following, the CNPC means the phase described by this model.

1.3 STRUCTURE OF PION-CONDENSED NEUTRON STARS

Neutron stars with the pion-condensed phase in their core regions (pion-condensed neutron stars) have such multiphase structure that the CNPC phase exists at $\rho \gtrsim \rho_t(\text{CNPC}) \simeq (3\text{--}5)\rho_0$,[16] the pure CPC phase at $\rho_t(\text{CNPC}) > \rho > \rho_t(\text{CPC}) \simeq 2\rho_0$, the core superfluid region and then the crust region follow, going from inner to outer. To model such a neutron star, which is used in studying glitch problems, we adopt the following equation of state (EOS). To get the EOS in the pion-condensed phase, we use the energy E, which is the sum of $E(\text{BJ–IH})$[17] in the normal phase (and also in the superfluid phase because of its small condensation energy) and the energy gain ΔE_π due to the pion condensation that arises from the Fermi gas state. Because the transition to the CNPC phase from the CPC phase is of the first order, the critical density is denoted by two values, $\rho_{t_1}(\text{CNPC})$ and $\rho_{t_2}(\text{CNPC})$. We give the energy of the system as follows:

$$E = \begin{cases} E(\text{BJ–IH}) + \Delta E_\pi(\text{CPC}) & \text{for} \quad \rho_t(\text{CPC}) \leq \rho \leq \rho_{t_1}(\text{CNPC}) \\ E(\text{BJ–IH}) + \Delta E_\pi(\text{CNPC}) & \text{for} \quad \rho > \rho_{t_2}(\text{CNPC}) \end{cases} \quad (1)$$

From the energy per baryon, $\mathcal{E} = E/N_B$, the pressure at zero temperature is obtained; $P = \rho^2 \partial \mathcal{E}/\partial \rho$. We use the EOS-I given in Ref. 18, which is brought about by a moderate pion condensation. The critical densities are $\rho_t(\text{CPC}) = 2.1\rho_0$, $\rho_{t_1}(\text{CNPC}) = 4.4\rho_0$, $\rho_{t_2}(\text{CNPC}) = 4.6\rho_0$. Since the last two are very close, we use $\rho_t(\text{CNPC}) = 4.5\rho_0$ for convenience of explanation. The resulting EOS, P–ρ plot, is shown in Fig. 1(a) and the neutron star masses, M, versus the central density, ρ_{center}, are shown in Fig. 1(b). The most noticeable feature of such pion-condensed neutron stars is that the transition density from the CPC to the CNPC around $(4\text{--}5)\rho_0$ is near the central density of neutron stars with the standard mass, $M \simeq 1.4\ M_\odot$.

2. RELEVANCE OF VARIOUS HADRONIC PHASES TO PULSAR GLITCHES

Here, our approach to pulsar glitches is given from the viewpoint of an extended starquake model which fully utilizes the effects due to various hadronic phases in neutron star interiors. (In this respect, it differs from the approach of the vortex creep model,[19] in which glitch phenomena are described as inherently related to the properties of the crust phase, especially the crustal superfluid.) This work has been performed in collaboration with Takatsuka. Details can be found in Refs. 18 and 20, so only motivations and main results are mentioned here.

2.1 MOTIVATIONS

We have three motivations in this work:

1. To understand pulsar glitch (PG) phenomena by extending the starquake model and taking into account the rich aspects of internal structure due to pion condensation.
2. To understand the remarkable differences between the Crab glitches and the Vela glitches, which is a long-standing problem, by virtue of the structural difference depending on M for pion-condensed neutron stars. The key point is

$$\rho_{\text{center}}(M \simeq 1.4\ M_\odot) \simeq \rho_t(\text{CNPC}) \simeq (4\text{--}5)\rho_0, \tag{2}$$

   as emphasized at the end of section 1.3. If we take $M(\text{Crab}) = (1.1\text{--}1.3)M_\odot$, its core consists of the fluid phase only, even for the realization of the CPC. If we take $M(\text{Vela}) = (1.4\text{--}1.6)M_\odot$ its core possesses the CNPC as well as the fluid phase. Typical aspects of the structures of two such neutron stars are shown in Fig. 2(a) for $M = 1.2\ M_\odot$ and $M = 1.6\ M_\odot$. For the Crab pulsar only a crustquake with a magnitude $\Delta\Omega_0/\Omega \sim 10^{-8}$ is possible, but the Vela pulsar can undergo a corequake with magnitude $\Delta\Omega_0/\Omega \sim 10^{-6}$, where $\Delta\Omega_0$ is the discontinuous jump of the pulsar angular velocity Ω at glitch epoch. As is well known, to overcome the difficulty in explaining the Vela glitches by means of the crustquake model successful for the Crab glitches, Baym, Pethick, Pines, and Ruderman[21] proposed a corequake model, but its shortcomings were soon noticed. A baryonic solid in the usual sense is unlikely, and the accumulated heat from the repeated corequakes leads to an increase in surface temperature, in contradiction to observations. We can remove such shortcomings by considering the CNPC; the NPC provides a kind of "solid," the ALS "solid," and the rapid cooling due to the CPC removes the energy released by the corequake before the next glitch. We have shown that $\Delta t(\text{cooling time}) \lesssim 1\ \text{yr} < \text{glitch interval} \simeq (2\text{--}4)\text{yr}$.[18]
3. Data accumulation and analyses have shown that the postglitch timing behavior is described by at least two exponential terms with a long relaxation time, $\tau_1 = $ months to years, and a short one, $\tau_2 = $ days to a week. A natural view is to consider that two different neutron superfluids are associated with this behavior. They have the long and short relaxation times, τ_s and τ_p, intrinsic in the $^1S_0$-one and the $^3P_2$-one, respectively, which are closely related to the τ_1 and τ_2 observed.

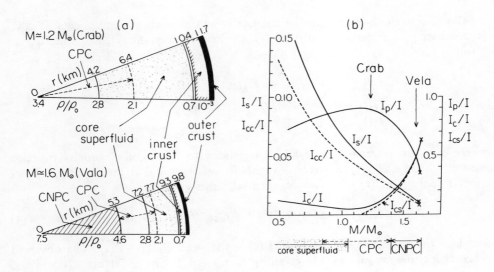

FIGURE 2 (a) Profile of multi hadronic phase structure of pion-condensed neutron stars for Crab (*upper*) and Vela (*lower*). (b) Moment-of-inertia ratio of each component versus neutron mass, where the corresponding hadronic phase at the center is shown at the bottom.

2.2 EXTENSION TO THREE-COMPONENT STARQUAKE MODEL

According to the neutron star structure shown in Fig. 2(a), three components are associated with PGs. The first one is the tightly coupled component consisting of the crust, charged particles, and normal fluids, denoted by c. The second one is the neutron $^1S_0$ superfluid denoted by s. The third one is the neutron $^3P_2$ superfluid denoted by p. Because the internal structure largely depends on M, the moments of inertia of these three components change largely as M increases, as shown in Fig. 2(b). The typical values of I_i ($i = c, s$ and p) relative to the total moment of inertia ($I = I_c + I_s + I_p$) are

$$I_c/I : I_s/I : I_p/I \simeq \begin{cases} 0.05 : 0.05 : 0.90 & \text{for} \quad M = 1.2\ M_\odot \\ 0.48 : 0.01 : 0.50 & \text{for} \quad M = 1.6\ M_\odot. \end{cases}$$

Here I_c consists of two parts, I_{cc} and I_{cs}, where I_{cc} (I_{cs}) is the part coming from the crust solid (the ALS solid) including the neighboring charged and normal-fluid components.

Basic equations describing the angular velocities of three components, $\Omega \equiv \Omega_c$ (the subscript c is omitted only for Ω_c, the observed quantity), Ω_s, and Ω_p, are given as follows:

$$I_c\dot{\Omega} = -\alpha - [I_c(\Omega - \Omega_s)/\tau_s] - [I_c(\Omega - \Omega_p)/\tau_p],$$
$$I_s\dot{\Omega}_s = I_c(\Omega - \Omega_s)/\tau_s,$$
$$I_p\dot{\Omega}_p = I_c(\Omega - \Omega_p)/\tau_p, \qquad (3)$$

where α is the external torque which is related to the angular velocity function without glitches, $\Omega^{no}(t) = -\alpha t/I + \text{const}$. By setting $t = 0$ at the glitch epoch, we have the solution $\Omega(t)$ with two exponential terms given by

$$\Omega(t) = \Omega^{no}(t) + \Delta\Omega_0[Q_1 e^{-t/\tau_1} + Q_2 e^{-t/\tau_2} + (1 - Q_1 - Q_2)] \qquad (4)$$

and also the angular velocity lags, $\Omega_s(t) - \Omega(t)$ and $\Omega_p(t) - \Omega(t)$. The observed relaxation times, τ_1 and τ_2, are given by τ_s, τ_p, and I. The Q-factors, Q_1 and Q_2, which are closely related to properties of the neutron superfluids, are given by these factors and the fractional change of Ω at the glitch epoch, $\Delta\Omega_{io}/\Omega_i$, which is fixed by a reasonable assumption that the angular momentum conservation would hold for each component at the glitch epoch, $\Delta\Omega_{io}/\Omega_i \simeq -\Delta I_i/I_i$ ($i = c, s,$ and p). The detailed expressions have been given in Ref. 20).

2.3 EXTRACTION OF STRUCTURAL INFORMATION

By fitting $\Delta\Omega(t) \equiv \Omega(t) - \Omega^{no}(t)$ and its time derivative $\Delta\dot{\Omega}(t)$ with observations, we can extract information on structure and dynamical behavior of neutron star interiors.

We have analyzed the giant glitches of the Vela pulsar with $\Delta\Omega_0/\Omega \sim 10^{-6}$, for which $\Delta\dot{\Omega}$ fitting can be done with reference to the analysis by Cordes et al.[22] We can fit $\Delta\dot{\Omega}(t)$ for five giant glitches by using the relaxation times and Q-values in the range $\tau_1 = 20$–100 d, $Q_1 = (0.3$–$1.7) \times 10^{-2}, \tau_2 \simeq 3$ d, and $Q_2 = (0.3$–$1.4) \times 10^{-3}$. The fittings to the first, second, third, fourth, and sixth glitches are given in Ref. 20. For the fifth glitch,[23] τ and Q inferred from the observed $\Delta\dot{\Omega}(t)$ are within these ranges. The smallness of Q_2 (signifying the dominance of the long term for the Vela glitches) and a rather large variation in τ_1 remain to be explained.

For the Crab macro glitches with $\Delta\Omega_0/\Omega \sim 10^{-8}$, only the short term is visible. From the fitting to the observed $\Delta\Omega(t)$, we have $\tau_2 = $ a few days, $Q_2 \simeq 0.9$. It is difficult to see the long-term behavior if $Q_1 \lesssim 0.1$ even for $\Delta\dot{\Omega}$, because $Q_1/\tau_1 \ll Q_2/\tau_2$.

Structural information is extracted from the τ and Q thus obtained. The intrinsic relaxation times (τ_s and τ_p) can be extracted from τ_1 and τ_2. Because of $I_s/I \sim 10^{-2}$, we have

	Crab	Vela
$\tau_1 \simeq \tau_s(I_s/I_c)$:	$\tau_1 \sim \tau_s$	$\tau_1 \simeq 0.02\tau_s$
$\tau_2 \simeq \tau_p(I_p/I)$:	$\tau_2 \simeq 0.9\tau_p$	$\tau_2 \simeq 0.5\tau_p$.

Therefore, we need $\tau_s = (2\text{--}11)$ years and $\tau_p \sim$ days.

The Q-factors infer the portion of the neutron superfluids. The properties of $Q_1(\text{Vela}) \sim 10^{-2}$ and $Q_2(\text{Crab}) \simeq 0.9$ are given directly by our model, because $Q_1(M = 1.6\ M_\odot) \simeq I_s/I \simeq 0.02$, $Q_2(M = 1.2\ M_\odot) \simeq I_p/I \simeq 0.90$. The small $Q_1(\text{Crab})$ can be understood as $Q_1(\text{Crab}) \simeq (-1 \text{ to } +1)I_s/I \sim (-10^{-2} \text{ to } + 10^{-2})$. For the very small $Q_2(\text{Vela}) \lesssim 10^{-3}$, we need another explanation besides that based on I_i/I. A reasonable account is that the neutron $^3P_2$ superfluid immediately reflects the shape change of the core solid just beneath it. Therefore the smallness of $Q_2(\text{Vela})$ means that more than just the crust superfluid with $I_s/I \sim 10^{-2}$ is associated with the Vela glitches.

2.4 SUMMARY AND FUTURE PROBLEMS

The main results are summarized as follows.

1. The combined (charged and neutral) pion condensation in the center of a neutron star with $M \simeq 1.6\ M_\odot$ enables us to apply an extended corequake model to the Vela glitches.
2. Postglitch timing behavior for both Vela and Crab glitches can be described in a three-component starquake model, in which the Vela giant glitches are brought about by the corequake and the Crab macroglitches by the crust-quake.
3. The Q-factors (Q_1 and Q_2) provide us with important information on structure and dynamical behavior of neutron star interiors, as well as the relaxation times (τ_1 and τ_2). From the results on Q we have obtained, the neutron $^3P_2$ superfluid is responsible for the postglitch behavior of short (but still macroscopic) time.

The following problems remain unsolved and must be studied in future.

1. In our model, $\tau_p \sim$ days are needed. We can get $\tau_p \sim$ hours by taking into account the large coherent length of the neutron $^3P_2$ superfluid $\gtrsim 250$ fm, following Alpar, Langer and Sauls.[24] Still, the calculated time is less than that necessitated by observations by one order of magnitude.
2. Estimated glitch intervals are not unreasonable for the Crab case but are too large for the Vela case. To solve this long-standing problem inherent in the starquake model we need the introduction of some mechanism that acts as a trigger for the corequakes.

3. Rather large variation of τ_1 observed in the Vela glitches must be explained. It may be attributed to the time variation of internal temperature T_{in} due to corequakes, because the relaxation time τ_s of the crust superfluid due to the Feibelman mechanism[25] is very sensitive to T_{in} and ρ as $\tau_s \propto \exp\left[\Delta_F^2(\rho)/\epsilon_F T_{in}\right]$, with the Fermi kinetic energy ϵ_F and the neutron $^1S_0$ energy gap $\Delta_F(\rho)$.

We have not yet analyzed the eighth giant glitch of Vela, which was observed on December 24, 1988.[26] The very short speedup time of less than 2 min indicates that the starquake picture and its extended versions must be studied further.

REFERENCES

1. J. W. Clark, this volume.
2. N. C. Chao, J. W. Clark, and C. H. Yang, *Nucl. Phys.* **A179**, 320 (1972); T. Takatsuka, *Prog. Theor. Phys.* **50**, 1754 (1973).
3. T. L. Ainsworth, J. Wambach, and D. Pines, *Phys. Lett.* **222B**, 173 (1989).
4. R. Tamagaki, *Prog. Theor. Phys.* **44**, 905 (1970).
5. M. Hotteberg, A. E. Glassgold, R. W. Richardson, and M. Ruderman, *Phys. Rev. Lett.* **24**, 775 (1970).
6. T. Takatsuka, *Prog. Theor. Phys.* **48**, 1517 (1972).
7. A. B. Migdal, *Zh. Eksp. Teor. Fiz.* **61**, 2210 (1971); ibid. **63**, 1933(1972); idem, *Rev. Mod. Phys.* **50**, 107 (1978).
8. R. F. Sawyer, *Phys. Rev. Lett.* **29**, 382 (1972); D. J. Scalapino, *Phys. Rev. Lett.* **29**, 386 (1972); R. F. Sawyer and D. J. Scalapino, *Phys. Rev.* **D7**, 953 (1972).
9. A. B. Migdal, *Phys. Lett.* **52B**, 264 (1974).
10. As review articles, e.g., R. Tamagaki, *Nucl. Phys.* **A328**, 352 (1979); and idem, in *Nucleon-Nucleon Interaction and Nuclear Many-Body Problems*, S. S. Wu and T. T. S. Kuo (eds.) (World Scientific, Singapore, 1984), p. 97.
11. T. Takatsuka, K. Tamiya, T. Tatsumi, and R. Tamagaki *Prog. Theor. Phys.* **59**, 1933 (1978).
12. As review articles, e.g., G. E. Brown and W. Weise, *Phys. Reports*, **27**, 1 (1976); and G. Baym and D. K. Campbell, in *Mesons in Nuclei III*, M. Rho and D. Wilkinson (eds.) (North-Holland, Amsterdam, 1979). p. 1031.
13. W. Weise and G. E. Brown, *Phys. Lett.* **48B**, 297 (1974); ibid. **58B**, 300 (1975); T. Tatsumi, *Prog. Theor. Phys.* **68**, 1231 (1982).
14. K. Tamiya and R. Tamagaki, *Prog. Theor. Phys.* **66**, 948 and 1361 (1981); T. Takatsuka, Y. Saito and J. Hiura, *Prog. Theor. Phys.* **67**, 254(1982); O. Benhar, *Phys. Lett.* **106B**, 375 (1983); idem, *Nucl. Phys.* **A437**, 590 (1985).
15. K. Tamiya and R. Tamagaki, *Prog. Theor. Phys.* **60**, 1753 (1978).
16. T. Muto and T. Tatsumi, *Prog. Theor. Phys.* **78**, 1405 (1987).

17. R. C. Malone, M. B. Johnson and H. A. Bethe, *Astrophys. J.* **199**, 741(1975).
18. T. Takatsuka and R. Tamagaki, *Prog. Theor. Phys.* **79**, 274 (1988).
19. M. A. Alpar, P. W. Anderson, D. Pines and J. Shaham, *Astrophys. J.* **276**, 325 (1984); ibid. **278**, 791 (1984); M. A. Alpar, R. Nandkumar, and D. Pines, *Astrophys. J.* **288**, 191 (1985); M. A. Alpar, this volume.
20. T. Takatsuka and R. Tamagaki, *Prog. Theor. Phys.* **82**, 945 (1989).
21. G. Baym, C. Pethick, D. Pines, and M. Ruderman, *Nature* **224**, 872 (1969).
22. J. M. Cordes, G. S. Downs, and J. Krause-Polstorff, *Astrophys. J.* **330**, 847 (1988).
23. P. M. McCulloch, P. A. Hamilton, G. W. R. Royle, and R. N. Manchester, *Nature* **302**, 319 (1983).
24. M. A. Alpar, S. A. Langer, and J. A. Sauls, *Astrophys. J.* **282**, 533 (1984).
25. P. J. Feibelman, *Phys. Rev.* **D4**, 1589 (1971).
26. C. Flanagan, IAU Circular No. 4695 (1989); C. S. Flanagan, *Nature* **345**, 416(1990); P. A. Hamilton, E. A. King, D. McConnel, and P. C. McCulloch, IAU Circular No. 4708 (1989).

Toshitaka Tatsumi†and Takumi Muto‡
†Department of Physics, Kyoto University, Kyoto 606, Japan; and ‡Yukawa Institute for Theoretical Physics, Kyoto University, Kyoto 606, Japan

Static and Dynamic Properties of Pion-Condensed Neutron Stars

In this report we shall try to give an account of our recent studies on pion condensations and neutron star phenomena. In particular our main concern here will be some consequences of the first-order phase transition, which results from the neutral pion condensate, and their relevance to the anomalous gamma-ray burst of 5 March 1979.

1. INTRODUCTION

Pion condensations in high-density matter should play a significant role in neutron star physics. In the last few years we have been studying the pion-condensed phases in a realistic situation and exploring their implications in the various phenomena of neutron stars.[1,2,3]

As a consequence of these studies we have obtained a phase structure in the core region of neutron stars which includes a combined condensation of neutral and charged pions (CPC) surrounded by a pure charged-pion (π^c) condensate. In the CPC phase, charged-pion condensate develops on the basis of baryonic alternating-layer-spin (ALS) structure,[4] which accompanys neutral-pion (π^0) condensate as well. This phase enjoys two outstanding features inherent in pion condensations:

(i) drastic change of baryon basis due to π^0 condensate like a liquid crystal, which leads to a large softening of the equation of state (EOS); (ii) extra weak-process such as a quasi-particle URCA process, which leads to, for example, rapid cooling of neutron stars.

Another characteristic of the CPC phase, which is important as well, is that the transition to this phase is of first order. This should account for some of the dynamic behavior of neutron stars; for example, minicollapse may occur through the decay of metastable states and accompany a large energy release. Vibration of neutron stars, which may follow this phase transition, will damp mainly through viscosity due to the extra weak-process. We shall discuss the anomalous gamma-ray burst of 5 March 1979 from this viewpoint.

2. COMBINED $\pi^0 - \pi^c$ CONDENSED (CPC) PHASE

We shall pick out the essential points of our formulation[1,2] instead of going into detail: (i) Chiral symmetry, $SU(2)_L \times SU(2)_R$, which governs the low-energy regime of pion-nucleon dynamics, (ii) SU(4) constituent quark model, by which isobar degrees of freedom supplementing the p-wave $\pi - N$ interaction in the chiral symmetry approach are easily incorporated, (iii) minimal interactions between baryons to simulate the effects of the short-range correlations (s.r.c.),

$$\mathcal{H}_{min} = 1/2 g' \sigma_i^{q_1} \tau_\alpha^{q_1} \sigma_i^{q_2} \tau_\alpha^{q_2}, \tag{1}$$

in terms of quarks q_1, q_2; $g' = \tilde{g}'(f/m_\pi)^2$ ($= g'_N$) is the Landau-Migdal (LM) parameter which measures the strength of correlation in the $N - N$ channel.[5]

The ground state is composed of classical pion fields and quasi-particles ("η-particles"). η-particles are constructed in two steps. First we have defined quasi-baryon states $(\tilde{p}, \tilde{n}, \tilde{\Delta})$ to take account of the admixture of isobars $\Delta(1232)$ in the CPC phase. Then η-particle is given as a superposition of quasi-nucleons, for example, $|\eta\rangle = \cos\phi|\tilde{n}\rangle + i\sin\phi|\tilde{p}\rangle$. Results of energy gain in neutron matter, $\Delta E/N \equiv (\epsilon_{p.c.} - \epsilon_{Fermi\ gas})/\rho$, are shown in Fig. 1.

Thick lines show previous results[1] based on the universality assumption that all the strengths in the various baryon channels are equal; thin lines show our recent results obtained by relaxing this assumption and taking into account the implications of nuclear experiments.[6] Note that the latter results based on more refined considerations still lie between previous ones. Thus we could say that studies based on the universality and somewhat smaller value of $\tilde{g}'$ (=0.5–0.6) should be regarded as an *effective* substitute for a proper treatment of s.r.c. between baryons. Moreover we could regard these two curves, shown in thick lines, as boundaries of the allowed region within the ambiguities of the analysis.

From Fig. 1 we can see that pure π^c condensation first occurs in relatively low densities ($\rho_{cr}^{\pi^c}$ =[1.5–2.2]ρ_0, where $\rho_0 = 0.17$ fm^{-3} is the nuclear density), and further transition to CPC phase occurs in higher densities (ρ_{cr}^{CPC} =[3–5]ρ_0). The

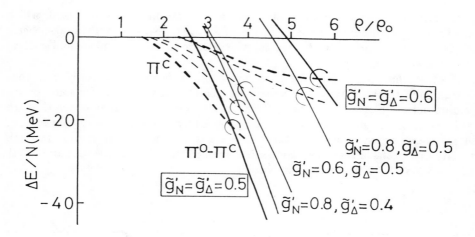

FIGURE 1 Energy gain per particle in MeV. Solid and dashed curves show those in CPC phase and pure π^c condensed phase, respectively.

latter phase transition is first order due to the drastic change of baryonic structure, and a major softening of the EOS should result, which will play an important role in the following discussion.

3. DECAY OF THE METASTABLE (SUPERCOMPRESSED) STATE AND ENERGY RELEASE

Since the transition from π^c to CPC phase of first order, metastable states, "supercompressed" and "superexpanded," will appear. The supercompressed states are the key ingredient in the energy release following the phase transition (see Fig. 2).

Then total released energy δE is given simply as a mass difference $-\delta M$ between a metastable neutron star and a stable one, $\delta E = -\delta M = -[M(S) - M(MS)]$ subject to baryon number conservation $\delta A = 0$. It consists of internal and gravitational energy differences, which largely cancel each other.

Decay of the metastable state goes through *nucleation*; droplets composed of a stable CPC phase spontaneously appear in the metastable π^c condensed state and they, once produced, grow to infinitely large size and finally fill the whole space. To estimate the lifetime of the metastable state we have employed a formulation given by Lifshitz and Kagan and Haensel and Schaeffer[7] based on hydrodynamics.[3] The system is characterized by the Gibbs free energy under constant pressure, P. The

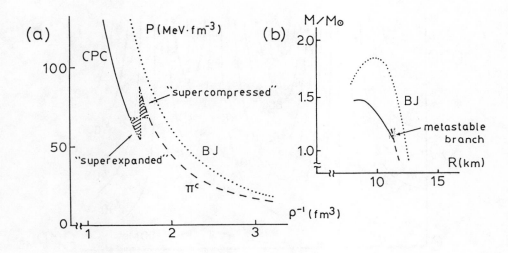

FIGURE 2 Pressure (a) and gravitational mass (b) for $\tilde{g}' = 0.5$. The metastable branch containing the supercompressed phase appears in (b).

relative Gibbs free energy $G(\rho, A)$ for A-baryon droplet with density ρ embedded in the metastable phase with density ρ_{MS} can be written as

$$G(\rho, A) = A\Delta g_{vol}(\rho) + A^{2/3} g_{surf}(\rho), \qquad (2)$$

where $\Delta g_{vol}(\rho) \equiv g_{vol}(\rho) - g_{vol}(\rho_{MS})$, $g_{vol}(\rho) = \epsilon_{p.c.}/\rho + P/\rho$, and $g_{surf}(\rho)$ is the surface contribution.[3] Contour plots of $G(\rho, A)$ are shown in Fig. 3.

The nucleation proceeds by quantum tunneling through the free energy barrier in the (ρ, A) plane. Instead of the proper treatment of the multidimensional tunneling problem we have considered two one-dimensional paths, the baryon number (A) fluctuation path (dashed line) and the density (ρ) fluctuation path (dotted line), for the sake of a rough estimate. Then lifetime can be estimated by the semiclassical formula for tunneling.[3] The released energy δE and the lifetime τ_{life} are given in Table 1 together with the radius reduction δR.

In actual situations, metastable stars will be compressed in a macroscopic timescale by accretion or cooling, so that a *macroscopic* lifetime is needed. Note that δE changes gradually and τ_{life} changes abruptly.

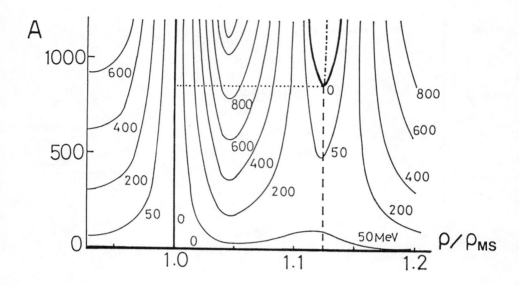

FIGURE 3 Contour plots of $G(\rho, A)$ in the (ρ, A) plane for a typical case of pressure $P = 72$ MeV fm^{-3} and $\tilde{g}' = 0.5$.

TABLE 1 Released Energy δE and the Radius Reduction δR Due to a Minicollapse from a Metastable Star with Radius $R(MS)$ and Central Density $\rho_c(MS)$ to a Stable One, and the Lifetime τ_{life} of the Metastable Star.

g' ($\tilde{f}^2$)	$\rho_c(MS)$ (fm^{-3})	$R(MS)$ (km)	$-\delta R$ (km)	δE ($\times 10^{49}$ ergs)	τ_{life} (s)
0.5	0.580	11.482	0	0	∞
	0.592	11.413	0.059	1.85	$7.04 \times 10^{+75}$
	0.593	11.407	0.065	2.20	5.14×10^{-17}
	0.594	11.401	0.072	2.60	—
	0.607	11.334	0.164	9.91	—
	0.617	11.283	0.246	18.9	0
0.6	0.937	10.714	0	0	∞
	0.941	10.703	0.002	0.0165	$7.65 \times 10^{+27}$
	0.943	10.697	0.005	0.0462	3.61×10^{-16}
	0.952	10.671	0.022	0.489	—
	0.961	10.647	0.052	1.48	—
	0.979	10.598	0.182	3.84	0

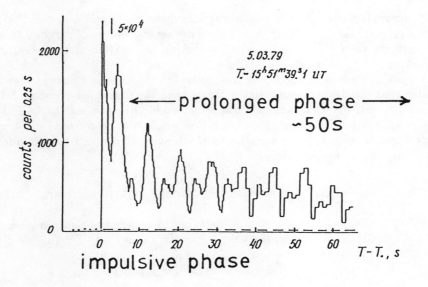

FIGURE 4 The time profile of the 5 March 1979 event taken from Ref. 8.

4. ANOMALOUS GAMMA-RAY BURST OF 5 MARCH 1979 REVISITED

In this section we reconsider the anomalous gamma-ray burst of 5 March 1979 from the viewpoint obtained in the preceding sections. This event is intriguing but unelucidated and requires some extraordinary mechanisms for the energy release. In Fig. 4 we show the time profile of the 5 March event taken from Mazets et al.[8]

Some characteristics of this event are summarized as follows;

1. Rapid rise time (< 0.2 ms) and a short duration time (~ 0.15 s) in the impulsive phase.
2. Large energy intensity at the peak flux $\sim 10^{-3}$ ergs cm^{-2} s^{-1}, which is an order of magnitude larger than ordinary ones.
3. Pulse structure with a period of 8.1 ± 0.1 s and a duration time of ~ 50 s in the prolonged phase.

The energy source of this gamma-ray burst should be a compact object, a neutron star, from (1) and (3), but its location is unclear. There is controversy about this point even now. Some locate the source in the supernova remnant N49 in the LMC. In the following discussion we will presume this is the case. Burst energy amounts to $\sim 5 \times 10^{43}$ ergs in the impulsive phase and $\sim 1 \times 10^{44}$ ergs in the prolonged phase. Note that these values are several orders of magnitude larger than ordinary ones (10^{38}–10^{39} ergs). A sufficient explanation of these features is still lacking.

Adopting the minicollapse scenario given by Ellison and Kazanas,[9] we have assumed that this large energy release comes from the first-order phase transition, induced by some external factors, from a metastable neutron star to a CPC one. Released energy following the phase transition would first be converted to thermal energy and mechanical energy such as radial vibration. The magnitude of vibration energy E_{vib} can be estimated as

$$E_{vib} \sim 1/2\omega^2 \int d^3r \tilde{\epsilon}(r)(\delta r)^2 \sim 2\pi\omega^2(\delta R/R)^2 \int_0^R r^4 \tilde{\epsilon}(r)dr, \qquad (3)$$

where $\tilde{\epsilon}(r)$ is the mass density, and ω the typical frequency of neutron stars, $\omega = 2\pi/\tau = (10^4$–$10^5)$ s^{-1}. Putting the values given in Table 1 into Eq. (3), we get

$$E_{vib} \simeq \begin{cases} 5.6 \times 10^{47} \text{ergs} & \text{for } \tilde{g}' = 0.5 \\ 1.1 \times 10^{45} \text{ergs} & \text{for } \tilde{g}' = 0.6 \end{cases}, \qquad (4)$$

which are much less than the total released energy δE, $\delta E \sim 2 \times 10^{49}$ (2×10^{47}) ergs for $\tilde{g}' = 0.5$ (0.6) (Table 1). Thus almost all the released energy is converted into thermal energy, so that the neutron star is heated; the change in internal temperature T_{in} is

$$T_{in} : \sim 10^8 \text{K} \longrightarrow \sim 10^{10}(10^9)\text{K} \qquad (5)$$

for $\tilde{g}' = 0.5$ (0.6) (see Table 2). It should be noted that the resultant temperature still is very low in the light of the strong-interaction energy scale.

It is, however, worthwhile to see that mechanical energy is a tiny fraction ($\sim 1\%$) of total released energy but it would be sufficient to produce such a gamma-ray burst from the viewpoint of energetics.

The conversion mechanism of the released energy to the gamma-ray burst is still unclear, but Ellison and Kazanas have proposed that prompt compression of the atmosphere near the surface by the radial vibration produces a shock wave, which in turn compresses and heats gas in the atmosphere successively. This process leads to the creation of e^+e^- pairs and finally to the gamma-ray burst. If we accept such a scenario we could say that minicollapse-vibration explains the 5 March event.

5. DAMPING MECHANISM OF THE RADIAL VIBRATION IN THE PROLONGED PHASE

Here we shall pursue further the minicollapse-vibration scenario in a somewhat different context. We shall discuss the prolonged phase with duration time ~ 50 sec by considering the damping of the radial vibration.

Ellison and Kazanas assumed that the so-called Σ^- process,[9] which is known to be dominant in the hyperon matter, could damp the radial vibration on a time scale of 50 s. However, vibrational energy should dissipate mainly through a nonequilibrium η-particle β-decay process,

$$\eta(\vec{p}) \longrightarrow \eta(\vec{p}') + e^- + \bar{\nu}_e \quad (a), \qquad e^- + \eta(\vec{p}) \longrightarrow \eta(\vec{p}') + \nu_e \quad (b), \qquad (6)$$

in a pion-condensed neutron star because of its large reaction rate. This process is inherent in the π^c condensed phase and closely related to the rapid cooling (η-particle URCA process). In the following we consider only the pure π^c condensed phase, because (i) formulation becomes clear in this case and (ii) the CPC phase exists in the core region so that the outer, π^c condensed phase would give a dominant contribution. Previously, Wang and Lu briefly discussed this process[10] but their treatment is not clear, so we think it deserves more elaborate study. Further, this process is also important in more general situations involving the vibration.

The mean dissipation rate per baryon of the vibration energy can be expressed as

$$\left(\frac{dW}{dt}\right)_{av} = -\tau^{-1} \int_0^\tau P(t) \frac{d}{dt} \delta\rho^{-1} dt = \tau^{-1} \int_0^\tau \frac{d}{dt}[\delta P(t)] \delta\rho^{-1} dt, \qquad (7)$$

where $\delta\rho$ and δP are the deviations from equilibrium values, $\rho(t) = \rho_{eq} + \delta\rho \cos[2\pi t/\tau]$ and $P(t) = P_{eq} + \delta P(t)$, respectively. The pressure deviation δP can be further expressed by using the density deviations of η-particles $\delta\rho(t)$, electrons $\delta\rho_e(t)$, and condensed pions $\delta\rho_\pi(t)$. These deviations from equilibrium are caused by the volume change and the nonequilibrium process (6), so that

$$\delta\rho_e \equiv \delta_v \rho_e + \delta_c \rho_e \quad \text{and} \quad \delta\rho_\pi \equiv \delta_v \rho_\pi + \delta_c \rho_\pi. \qquad (8)$$

Note that $\delta_v \rho_{\pi(e)} = (\rho^{eq}_{\pi(e)}/\rho_{eq})\delta\rho(t)$ and the relation $\delta_c \rho_e(t) = -\delta_c \rho_\pi(t)$ holds through the process (6). Hence we get

$$\delta P \equiv \delta_v P(t) + \delta_c P(t),$$

$$\delta_v P(t) = \left[\left(\frac{\partial P}{\partial \rho}\right)_{eq} + \left(\frac{\partial P}{\partial \rho_e}\right)^v_{eq} \frac{\rho^{eq}_e}{\rho_{eq}} + \left(\frac{\partial P}{\partial \rho_\pi}\right)^v_{eq} \frac{\rho^{eq}_\pi}{\rho_{eq}}\right] \delta\rho(t),$$

$$\delta_c P(t) = \left[\left(\frac{\partial P}{\partial \rho_e}\right)^c_{eq} - \left(\frac{\partial P}{\partial \rho_\pi}\right)^c_{eq}\right] \delta_c \rho_e(t) \equiv \tilde{\mu} \delta_c \rho_e(t). \qquad (9)$$

Substituting these formulas into Eq. (7) and using the time dependence of $\delta\rho$, we can see that $\delta_v P$ is not effective in calculating the dissipation rate. The time dependence of $\delta_c \rho_e(t)$ is given as a difference between forward (a) and backward (b) reaction rates of (6),

$$\frac{d}{dt}\delta_c\rho_e(t) = R(u_t) - R(-u_t), \qquad (10)$$

where $u_t \equiv [\mu_e(t) - \mu_\pi(t)]/T$ with chemical potentials μ_e, μ_π, which measures to what extent the system deviates from equilibrium ($\mu_e = \mu_\pi$). The reaction rate for (a), or (b) is easily calculated to find

$$R(u_t) = \epsilon T^{-6}\frac{I_2(u_t)}{I_3(0)}T^5, \qquad I_n(u) \equiv \int_0^\infty dx\, x^n \frac{\pi^2 + (x+u)^2}{1 + e^{(x+u)}}. \qquad (11)$$

Here we can see the close relation between the reaction rate R and corresponding emissivity ϵ.[11] By way of Eqs. (7), (9), and (10) we finally get

$$\left(\frac{dW}{dt}\right)_{av} = -\frac{\tilde{\mu}}{\tau}\frac{\delta\rho}{\rho_{eq}^2}\int_0^\tau [R(u_t) - R(-u_t)]\cos(\frac{2\pi}{\tau}t)dt. \qquad (12)$$

Then damping time τ_d can be estimated by way of Eqs. (4) and (12): $\tau_d \sim E_{vib}/D$, with the total dissipation rate

$$D = \int_{\text{pion core}} \left(\frac{dW}{dt}\right)_{av} \rho_{eq} d^3r.$$

We have also used further approximations such as small deviation for u_t and constant temperature during the one period. Detailed description and full discussions will be presented elsewhere.[12]

Some results relating to the 5 March even are given in Table 2.

TABLE 2 Radius reduction ratio $\delta R/R$, vibrational energy E_{vib} internal temperature T, the total dissipation rate D and the damping time τ_d.

$\tilde{g}'$	$\delta R/R$	E_{vib}(erg)	T(MeV)	D(erg·s^{-1})	τ_d(sec)
0.5	5.2×10^{-3}	5.6×10^{47}	0.67	6.26×10^{48}	$\sim 10^{-1}$
0.6	1.9×10^{-4}	1.1×10^{45}	6.9×10^{-2}	1.63×10^{42}	$\sim 10^{+3}$

We can see that our results include the observational value (~ 50 s) within the theoretical uncertainties.

6. SUMMARY

We have seen that first order phase transition to the CPC phase not only gives rise to a drastic change in the static properties such as mass-radius relation, but also induce a mini-collapse under some external conditions such as accretion. As a result a large energy release surely follows mini-collapse from a metastable neutron star with a *macroscopic* lifetime.

Vibration may also follow such mini-collapse. Its energy is a tiny fraction of the total released energy, but sufficient to produce anomalous gamma-ray burst from the viewpoint of the energetics. Damping mechanism of the vibration in the pion-condensed neutron stars has been formulated and as a application we discussed the prolonged phase of 5 March 1979 event. We have shown that dissipation time is consistent with data.

We would like to thank Professor R. Tamagaki and Professor T. Takatsuka for useful discussions.

REFERENCES

1. T. Muto and T. Tatsumi, *Prog. Theor. Phys.* **78**, 1405 (1987).
2. T. Tatsumi, *Prog. Theor. Phys. Suppl.* **91**, 299 (1987); T. Muto and T. Tatsumi, *Prog. Theor. Phys.* **79**, 461 (1988); ibid. **80**, 28 (1988).
3. T. Muto and T. Tatsumi, *Prog. Theor. Phys.* **83**, 499 (1990).
4. T. Takatsuka, K. Tamiya, T. Tatsumi and R. Tamagaki, *Prog. Theor. Phys.* **59**, 1933 (1978).
5. E.g., T. E. O. Ericson and W. Weise, *Pions and Nuclei* (Oxford Univ. Press, Oxford, 1988).
6. T. Tatsumi and T. Muto, *Nuclei in the Cosmos,* Proc. of the Int. Symp. on Nuclear Astrophys., forthcoming.
7. I. M. Lifshitz and Yu. Kagan, *Soviet Phys.* JETP **35**, 206 (1972); P. Haensel and R. Schaeffer, *Nucl. Phys.* **A381**, 519 (1982).
8. E. P. Mazets, S. V. Golenetskii, V. N. Il'inskii, R. L. Aptekar' and Y. A. Guryan, *Nature* **282**, 587 (1979).
9. D. C. Ellison and D. Kazanas, *Astron. Astrophys.* **128**, 102 (1983).
10. D. Q. Wang and T. Lu, *Phys. Lett.* **148B**, 211 (1984).
11. T. Tatsumi, *Prog. Theor. Phys.* **69**, 1137 (1983).
12. T. Muto and T. Tatsumi, in preparation.

Gordon Baym
Loomis Laboratory of Physics, University of Illinois, 1110 W. Green St., Urbana, IL 61801, USA

Ultrarelativistic Heavy-Ion Collisions and Neutron Stars

After reviewing possible states of matter in the cores of neutron stars, this talk describes the present and future laboratory program to create and study high-density matter in ultrarelativistic heavy-ion collisions.

1. INTRODUCTION

The outstanding fundamental problem in understanding the overall structure of neutron stars—for example, the mass density profile, $\epsilon(r)/c^2$, and hence the radius and moment of inertia of a star of given mass, and the maximum possible neutron star mass—is our lack of knowledge of the microscopic nature of the matter under the extreme conditions deep in the stellar interiors. Laboratory experiments on individual nuclei and low-energy collisions of nuclei with hadrons and other nuclei teach us the properties of nuclear matter at densities near normal nuclear matter density, $\rho_0 \simeq 2.8 \times 10^{14}$ g/cm$^3 \simeq 0.16$ baryons/fm^3 (1 fm = 10^{-13}cm). Descriptions of nuclear matter constructed from such experimental data are based on nucleons interacting via two-body potentials, of well-established form, supplemented with rather uncertain three-body potentials. The gross features of neutron stars are, however, determined essentially by the matter in the cores of the stars, which have

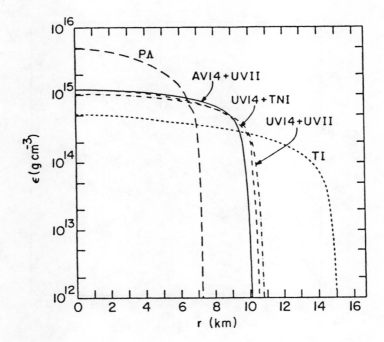

FIGURE 1 Density profiles of a $1.4 M_\odot$ neutron star calculated for stiff[2] (TI), soft[3] (PΛ), and modern neutron matter equations of state, based on two- and three-nucleon interactions. The corresponding energies per nucleon are shown in Fig. 2. From Ref. 1.

densities in the range ~ 5–$10\rho_0$, as illustrated in Fig. 1, which gives typical mass-energy density profiles of a $1.4 M_\odot$ neutron star calculated with a variety of nuclear matter equations of state described below.[1]

Inferring the nature of matter at higher densities represents a large theoretical extrapolation from laboratory nuclear physics. Beyond a few times ρ_0 the assumption that the forces between particles can be described in terms of static few-body potentials begins to break down, as the rich variety of hadronic and quark degrees of freedom in the nuclear many-body system comes into play. Further hadronic degrees of freedom include Δ's, hyperons, and higher baryon resonances, as well as mesonic degrees of freedom. At higher densities we can no longer continue to assume that the system can even be described in terms of well-defined "asymptotic" laboratory particles.

Various exotic states at higher densities have been suggested, including pion-condensed and strangeness-condensed matter and quark matter, not to mention strange quark matter. The presence of such states of matter in neutron stars, and

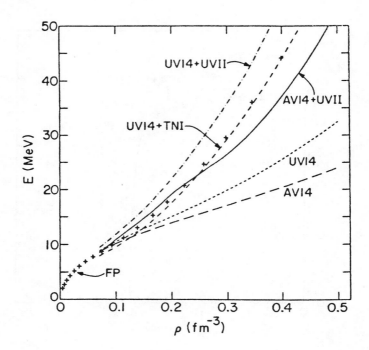

FIGURE 2 Energy per nucleon vs. baryon density in pure neutron matter, with and without three-body forces. From Ref. 1.

in addition limited information on their properties, could be inferred indirectly from their effects on stellar behavior. Exotic states modify the equation of state and hence the observable gross features of neutron star models. These states characteristically have larger neutrino emissivity than very neutron-rich nuclear matter, and so their presence would also lead to more rapid neutron star cooling.

Ultimately, we would like to learn directly about the states of matter at densities well above that in equilibrium nuclei through laboratory experiments. Over the past decade, nuclear and particle experimentalists have begun a program to create nuclear matter at high baryon and energy density by colliding large atomic nuclei together at ultrarelativistic energies in accelerators. One of the principal aims of these studies is to measure the properties of matter under extreme conditions, as well as to discover and examine new states such as the quark-gluon plasma. In this talk I would like to survey possible states of dense matter and then summarize how present and future heavy-ion experiments should enable one to map out the properties of matter at high energy and baryon density.

2. NUCLEAR MATTER REGIME

At densities near ρ_0, matter in neutron stars is well described in terms of interacting nucleons, accompanied by a sea of electrons to preserve charge neutrality. The equation of state in this regime is determined first by extracting the nucleon-nucleon interaction from pp and pn scattering experiments, at energies below $\sim$ 300 MeV, and the properties of the deuteron, and then solving the many-body Schrödinger equation to find the energy density as a function of baryon number.

In recent years considerable progress on both steps has brought the calculation of the equation of state near ρ_0 under good control. Modern fits to the interactions, described in Ref. 1, include the Paris, Urbana-v_{14} (UV14), and Argonne-v_{14} (AV14) two-body potentials (the 14's refer to the number of different components, such as central, spin-orbit, etc., included in the interactions). The Schrödinger equation can now be accurately solved numerically by powerful variational techniques.[1,4]

Two-body forces are not the whole story, however. While they lead to a reasonable binding energy of symmetric bulk nuclear matter, the predicted saturation density is about a factor 2 too high. They also tend to underbind light nuclei (e.g., ^{3}He and ^{4}He). Intrinsic three-body forces acting between nucleons—arising, in a hadronic description, from processes such as two nucleons scattering and becoming internally excited to intermediate Δ states while a third nucleon scatters from one of the Δ's—must be taken into account. The detailed structure of the three-body forces remains uncertain, and present fits are phenomenological. They must increase the binding in the neighborhood of ρ_0, but to avoid overbinding nuclear matter, they must also become repulsive at higher densities. This repulsion leads to a stiffening of the equation of state of neutron star matter at higher densities over that computed from two-body forces alone. Fig. 2 illustrates the stiffening effects on the energy per nucleon, $E = \epsilon/\rho$, of neutron matter versus baryon density ρ, from inclusion of three-body forces via the Urbana "Three-Nucleon" (TNI) and three-body Urbana VII (UVII) interactions. The uncertainties in three-body interactions are reflected in the equation of state, particularly at higher densities.

3. HADRONIC MATTER: PION AND KAON CONDENSATION

With increasing density, more massive baryons such as Δ's and hyperons, Λ, Σ, etc., can be present in the matter. How these particles interact with nucleons and among themselves is not well understood in detail, and the consequences of their presence for the dense matter equation of state can only be estimated from particular models of their interactions.[5,6] The meson fields that mediate the low-energy interactions between baryons also become dynamical degrees of freedom with increasing density. If mesons appear in matter in its ground state, they will, because they are bosons, macroscopically occupy the lowest available mode, in other words, form a condensate. A meson condensate is characterized by a macroscopic excitation of the meson field rather than a sea of mesons, as occurs for fermion particles. Two condensates that may exist in neutron stars are those associated with the lowest mass mesons, the pion and the strange K meson.

At first ignoring the interactions of pions with the medium, we would conclude that charged pions make their appearance once the neutron-proton chemical potential difference $\mu_n - \mu_p = \mu_e$ (the electron chemical potential) exceeds $m_\pi = 139.6$ MeV, the charged-pion mass, for it is then energetically favorable for a neutron at the top of the neutron Fermi sea to turn into a proton and a π^-:

$$n \to p + \pi^-. \qquad (1)$$

In neutron star matter in beta equilibrium, $\mu_n - \mu_p$ reaches ~ 110 MeV at nuclear matter density, and so we might expect π^- to appear at slightly higher densities.

However, the coupling of the nucleon particle-hole states and Δ-nucleon hole states to the pionic degrees of freedom produces an important mixing of these states and leads to a somewhat different physical picture of the onset of condensation. (The theory of condensation and numerical calculations are reviewed in Refs. 7–10; see also Ref. 11.) Essentially, nuclear and neutron star matter has a collective mode with the quantum numbers of the charged pion—an oscillation of the matter with spatially-varying nucleon spin (S = 1) and isospin (I = 1)—which at a certain critical density, ρ_π, goes to zero frequency at a critical wavevector, k_c. This "softening" of the collective mode causes the nucleon eigenstates to become rotated in isospin space; instead of the states being pure neutron $|n\rangle$ and proton $|p\rangle$, the eigenstates spontaneously undergo a [chiral SU(2)⊗SU(2)] rotation to become linear superpositions of neutron and proton states of the form

$$\begin{aligned} |N'\rangle &= \cos\theta |n\rangle + \sin\theta |p\rangle \\ |P'\rangle &= \cos\theta |p\rangle - \sin\theta |n\rangle, \end{aligned} \qquad (2)$$

where the condensation angle θ grows from zero as the density increases above ρ_π. To conserve charge as the nucleon eigenstates are rotated, the system develops a macroscopic spatially varying (p-wave) pion field, $\langle \pi \rangle \sim e^{i\mathbf{k}\cdot\mathbf{r}}$—the condensate—of net negative charge. The magnitude of the condensation wavevector k begins as k_c at ρ_π. An analogous neutral-pion condensed state can also be formed through softening of the neutron particle-hole collective mode; the neutral condensed state is characterized by a spatially varying finite expectation value of the neutral-pion field.[12] Neutral and charged-pion condensates can in principle coexist.

Early estimates predicted the onset of charged pion condensation at $\rho_\pi \sim 2\rho_0$. However, these estimates are very sensitive to the strength of the effective nucleon particle-hole repulsion in the I = 1, S = 1 channel (described by the Landau Fermi-liquid parameter[13] g'), which tends to suppress the condensation mechanism. Recent measurements in nuclei tend to indicate that the repulsion is too strong to permit condensation;[14] however, see Ref. 15.

Pion condensation would have two important effects on neutron stars. It would soften the equation of state above the critical density for onset of condensation, and hence reduce the maximum neutron star mass and increase the maximum central density.[16] It would have a neutrino luminosity[17] $\sim T^6$, where T is the temperature, enhanced over that of the normal modified-URCA process $\sim T^8$.

A further and perhaps more attractive form of condensation involves spontaneous formation of K mesons (kaons). The underlying chiral SU(3)⊗SU(3) symmetry of strong interactions, exact in the limit that the up (u), down (d), and strange (s) quark masses vanish, implies that K mesons have an effective attractive interaction with nucleons of the form $H_{eff} \sim -\rho \bar{K} K$, where ρ is the baryon density, and K is the K-meson field. This interaction acts as a density-dependent term in the kaon effective mass which lowers the kaon energy in the matter. As first noted by Kaplan and Nelson,[18] the energy of a K^+ falls below $\mu_p - \mu_n$, the chemical potential for K^+, at a critical density $\rho_K \sim 2.5$–$3\rho_0$; above this density the system should form a kaon or strangeness condensate, with a macroscopic expectation value of the charged K field.

In kaon condensation the nucleons undergo a chiral SU(3)⊗SU(3) rotation, analogous to the rotation of the nucleon eigenstates in pion condensation, in which a neutron state becomes a linear superposition of a neutron and Σ^-,

$$|N'\rangle = \cos\theta_K |n\rangle + \sin\theta_K |\Sigma^-\rangle, \quad (3)$$

while a proton state is rotated into a linear superposition of a proton, Σ^0, and Λ. From the point of view of the underlying quark structure of the baryons (n = udd, p = uud, Σ^- = dds, Σ^0 = uds, Λ = uds), the u and s components are rotated into each other by the angle θ_K. The rotation leads to a nonzero field expectation value in matter, $\langle \bar{s}u \rangle$, with the quantum numbers of the K^+ (where u and s are here the quark fields); the condensed state spontaneously breaks the chiral SU(3)⊗SU(3) symmetry. The matter can also form an η meson condensate, a state with a nonvanishing expectation value, $\langle \bar{s}s \rangle$, with the quantum numbers of the η meson.

Kaon condensation would also soften the equation of state and enhance the neutrino luminosity.[19,20] However, present uncertainties in the parameters of the effective interaction of kaons and the effective strange-baryon nucleon-hole couplings in neutron star matter lead to large uncertainties in the condensation angle expected at a given density and hence the magnitude of the effects of kaon condensation on neutron stars.

4. QUARK MATTER

At lower densities, the basic degrees of freedom of matter are hadronic; at very high densities quark degrees of freedom become dominant. Since nucleons are each made of three quarks, we expect that nuclear matter squeezed to sufficiently high density (or sufficiently heated) will turn into a liquid of uniform quark matter, composed of quarks, and at finite temperature, a *quark-gluon plasma* containing antiquarks and gluons as well. While it is impossible to remove an isolated quark from a hadron, in the plasma state the quarks and gluons are no longer confined in individual hadrons but are free to roam over macroscopic distances in the matter. Quark matter at low temperatures consists of Fermi seas of degenerate u and d quarks.

The only reliable approach at present to determining the transition from hadronic to quark matter is through Monte Carlo calculations of lattice gauge theory (for recent reviews see Refs. 21 and 22). These calculations have been successful so far only for the case of zero baryon density at finite temperatures. The results depend strongly on the masses assumed for the quarks. For infinitely heavy quarks, only gluons play a dynamic role; lattice calculations in this case predict a sharp first-order phase transition associated with deconfinement. The energy in the confined phase required to separate a test quark-antiquark pair, $\bar{Q}Q$, far apart grows linearly with separation, while it remains finite in the deconfined phase.

With finite-mass quarks one can always separate a test quark-antiquark pair, $\bar{Q}Q$, with finite energy, since at sufficient separation, it becomes energetically favorable to create a $\bar{q}q$ pair in the system, which screens out the interaction between the test pair; the q binds to the $\bar{Q}$, and the $\bar{q}$ to the Q, creating effectively a pair of mesons which can be separated to infinity with finite energy. The point is that once light quarks are in the system there no longer exists a good measure of whether the system is in a confined or deconfined state, and there need not be a sharp transition between the confined and deconfined phases. The transition between the two phases can be smooth, as occurs, for example, in ionization of a gas as it is heated, where the system goes gradually from gas molecules to electrons and nuclei; the two states are qualitatively different and there is a reasonably rapid onset of ionization, but it is not sharp. Alternatively, the transition may be first order, as in the boiling of water.

For massless quarks, on the other hand, one again finds a first-order phase transition, associated now with the spontaneous breaking of the SU(3)⊗SU(3) chiral symmetry of strong interactions. In the hadronic world chiral symmetry is spontaneously broken, analogous to the breaking of rotational symmetry in a ferromagnet, while in the deconfined phase it is restored. Since chiral symmetry is not exact for finite mass quarks, the situation for realistic quarks (u and d are light with $m_u \sim m_d \sim 10$ MeV, and the strange quark has $m_s \sim 150$ MeV) is not fully understood; present calculations indicate that the transition is likely first order, at a critical temperature ~ 200 MeV (or 10^{12} K), with a latent heat of order a few GeV/fm^3.

Unfortunately, lattice gauge calculations at nonzero baryon density are beset by technical problems; to date we do not have a reliable estimate of the transition density at zero temperature from nuclear to quark matter or even compelling evidence that there is a sharp phase transition. One can roughly estimate the location of the deconfinement transition for neutron star matter by asking whether one's favorite theory of nuclear matter or quark matter has a lower energy per baryon as a function of baryon density.[23] Such an approach necessarily implies a first-order transition with a discontinuity in the baryon density; typically one finds the onset of deconfinement at $\rho \sim 5 - -10\rho_0$. Unlike in lattice gauge theory, however, phenomenological theories of nuclear matter and of quark matter are generally based on inequivalent physical descriptions of the two phases and thus cannot be expected to describe the transition accurately.

Although the density range of the deconfinement transition is possibly above the central density found in models of neutron stars with masses $M \sim 1.4 M\odot$ based on nuclear equations of state, the question of whether neutron stars can have quark matter cores remains open. In the absence of information about the equation of state at very high densities, the issue of whether a separate family of quark stars with higher central densities than neutron stars can exist also remains open.

At this point we can ask whether the lowest energy state of quark matter at given baryon density contains strange quarks. Such strange quark matter is a distinct state from strangeness-condensed hadronic matter discussed above. Conversion of a u or d quark in quark matter to a strange quark, s, via the interactions

$$d + u \to s + u, \quad e^- + u \to s\,(+\nu_e), \tag{4}$$

creates further Fermi seas for the same number of fermions, thus lowering the Fermi energy of the system. The cost of the conversion, neglecting interaction effects for the moment, is the increased mass of the s quark, $m_s \sim 150$ MeV. If the gain in Fermi energy exceeds the mass cost, as it always will at very high density, conversions would lower the energy of the quark matter. At sufficiently high densities, strangeness-containing quark matter should be the lowest energy state. Strange quark matter can possibly become the favorable state of matter before ordinary quark matter. It is even conceivable that quark matter with a finite net strangeness per baryon could be self-bound with an energy below that of normal nuclear matter and thus be the absolute ground state of matter.[24,25]

If the baryon density ρ_s above which strange quark matter becomes the most favorable state is less than the central density of a neutron star made of ordinary matter then the central regions of the star at some point would begin to burn from normal to strange quark matter producing a strange core at densities above ρ_s. Were strange quark matter, the absolute ground state, the burning would proceed to the surface, leaving a *strange star*.[26,27] The gross structure of stars made entirely of strange quark matter would not be very different from that of normal neutron stars.[25] The crucial difference is the absence of a normal solid crust; a crust is needed for pulsar timing stability and is as well the most likely site for the mechanism of sudden pulsar speedups or glitches. Unless strange stars can have substantial crusts, they cannot be pulsars as we observe them.[28] If, however, the critical density ρ_s is above ρ_0, then strange stars would have normal crusts.

5. ULTRARELATIVISTIC HEAVY-ION COLLISIONS

As the above musings on possible states of matter at high densities make very evident, knowledge of the deep interiors of neutron stars is strongly limited by the lack of experimentally based information on the nature of nuclear matter beyond a few times ρ_0. With discovering properties of matter at high density and temperature as one of its goals, a comprehensive program of ultrarelativistic nucleus-nucleus collisions—with beam energies per nucleon much greater than the nucleon rest mass—has been under way at CERN and Brookhaven since the mid-eighties. (Useful general references are the recent proceedings of the conferences on quark matter.[29,30]) I would now like to describe how such experiments enable one to create temporarily regions of dense, hot matter and indicate the ways in which one can probe the collisions to learn about the matter created.

The initial phase of the program began in the autumn of 1986 with lighter nuclear projectiles on fixed nuclear targets. The CERN experiments have been carried out in the SPS accelerator at lab energies of 60 and 200 GeV per nucleon (GeV/A) with ^{16}O and ^{32}S beams, as well as proton beams. (The nucleon rest mass is $\simeq 1$ GeV.) The Brookhaven program has been carried out in the AGS accelerator at lab energies of 10–14.5 GeV per nucleon with ^{16}O, ^{28}Si, and p beams. The second generation of the lighter projectile experiments at both laboratories is currently being implemented and will be running over the next few years. The next step is to employ heavier beams: A new booster ring adjacent to the AGS will allow injection of fully stripped ions of arbitrarily large mass into the AGS; similarly, CERN is designing a "Pb Injector" for the SPS. Experiments with heavy beams, at ~ 12 GeV/A at the AGS and ~ 170 GeV/A at the SPS, should begin within the next few years.

Looking further into the future, the Brookhaven Relativistic Heavy Ion Collider (RHIC), on which construction has just begun, and which is expected to have beams by 1997, will provide the capability of colliding nuclei as heavy as Au on Au at 100 GeV/A in the center of mass (equivalent to 20 TeV/A in a fixed target experiment). CERN is also discussing building the Large Hadron Collider (LHC) in the LEP tunnel, with the possibility of injecting heavy-ion beams from the SPS, thus enabling heavy-ion collisions at ~ 4 TeV/A in the center of mass; at least one intersection region would contain heavy-ion detectors. If built, the LHC could have heavy-ion beams as early as 1998.

The physics of ultrarelativistic heavy-ion collisions changes with increasing energy. In the lower energy AGS regime we can picture the two colliding nuclei as effectively stopping each other (in their center of mass), to a crude first approximation forming a high-density fireball. (In reality, of course, parts of the nuclei pass through the collision volume rather than remaining in a fireball. With light projectiles employed so far in the AGS and SPS experiments, matter in the collision volume, which is small transverse to the beam direction, ought not to achieve thermal equilibration necessary for a fireball description.) Collisions between heavy nuclei in this regime may reach energy densities of order a few GeV/fm^3—possibly high enough to reach a quark-gluon plasma—and baryon densities several times normal nuclear matter density, ρ_0. The high-density matter produced in such collisions is relatively baryon rich and should be a good testing ground for neutron star matter. One important problem is that the matter is, in fact, produced in a highly excited state and disperses over times at most tens of fm/c, some 10^{-22} s, rather than cooling to the ground state; one will therefore have to extrapolate its properties to the low temperatures of matter in neutron stars.

The proposed PS-Collider at KEK would be ideal for neutron star equation of state studies, since it would be able to focus on producing baryon-rich high density matter. This machine would use the present 12-GeV proton synchrotron as an injector to produce colliding beams of heavy nuclei at $\simeq 5$ GeV/A, directly in the region where stopping should be maximal.

As the beam energy increases, the colliding nuclei begin to pass through each other, becoming highly excited internally and at the same time, leaving the vacuum between them in a highly excited state, containing matter of low baryon concentration. Such "nuclear transparency" begins to be important in the regime of the CERN fixed target experiments and should be completely manifested at RHIC and LHC collider energies. The highly excited fragments of the original nuclei, which recede from each other at the speed of light, contain essentially all the initial baryons and can be used for studying matter at high baryon density. To a first approximation, the central region between the nuclei has no baryon excess and would resemble the hot vacuum of the early universe; the matter in this region is entirely manufactured in the collision. After its production and thermalization, the matter cools and expands, and if having initially reached a quark-gluon plasma state, it undergoes a hadronization transition (the deconfinement transition in reverse).

Experiments at CERN and BNL have initially concentrated on exploring the global structure of events—the degree of energy deposition and particle formation in collisions—as well as exclusive behavior, including strangeness production (K mesons, Λ's) and charm production (J/Ψ mesons).[29,30] The development of effective probes of the collision volume, such as Hanbury-Brown Twiss interferometry between pion pairs, and now kaon pairs, as a tool for learning its geometry, has been one of the important outcomes of the present round of ultrarelativistic heavy-ion experiments.

6. ENERGY DEPOSITION IN COLLISIONS

The crucial measure of whether experiments will succeed in producing nuclear matter under extreme conditions is the amount of energy deposited in collisions. The degree to which colliding nuclei transfer energy and stop each other is determined most directly from measuring the transverse energy, E_T, of the event, defined as

$$E_T = \sum_i E_i \sin\theta_i, \qquad (5)$$

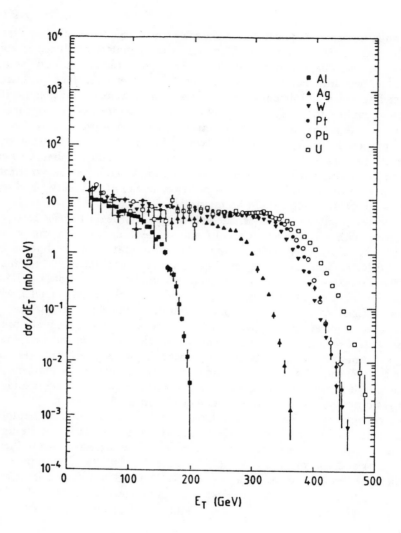

FIGURE 3 Transverse energy distribution $d\sigma/dE_T$ for ^{32}S at 200 GeV/A on various targets. From Ref. 31.

where E_i is the energy from final particle i deposited in the calorimeter in which it lands (mesons deposit their total energy, and baryons their total energy minus the nucleon rest mass), θ_i is the angle that particle i makes with respect to the beam axis, and the sum is over all final particles emerging from the collision. (While the transverse *energy* is not Lorentz invariant, as is the transverse momentum, it has the great advantage of being directly measurable without needing to identify the type of particles detected.)

Measurements of E_T reveal strikingly large energy deposited in collisions. Fig. 3 shows a characteristic total transverse energy distribution, $d\sigma/dE_T$, for 200-GeV/A ^{32}S beams on Al, a light target, on Ag, an intermediate-mass target, and on W, Pt, Pb, and U, heavy targets, as measured by the CERN Helios collaboration (NA34).[31] The angular coverage is essentially the entire forward hemisphere (in the beam direction) in the lab. Events in the tail of the distribution have nearly 500 GeV deposited in the collision volume!

This energy is approximately half the kinematic limit on the energy deposition, which equals the total kinetic energy in the center of mass of the nucleons participating in the collision. For a central collision of a projectile of nucleon number B on a target of nucleon number A ($> B$) the limit is

$$E_T^{max} = \sqrt{s_p} - m(B + A_p), \qquad (6)$$

where A_p is the number of participating target nucleons, and the participant cm energy squared is

$$s_p = 2mBA_p E_{lab} + (B^2 + A_p^2)m^2. \qquad (7)$$

Here m is the nucleon mass, and E_{lab} the lab projectile energy per nucleon. At ultrarelativistic energies the projectile drills a tube of radius equal to the projectile radius, $R_B \simeq 1.2B^{1/3}$ fm, through the target nucleus, leaving the spectators in the target nucleus initially undisturbed. For a central collision the number of target participants in the tube is $A_p \simeq (3/2)B^{2/3}A^{1/3}$ when $B \ll A$. In the ^{32}S on heavy target collisions shown in Fig. 3, the available energy is $\simeq 1$ TeV, half of which is deposited in the low-probability events in the tails of the distribution. Experiments at lower beam energies at the AGS indicate a high degree of stopping in central collisions.[32]

The general structure of $d\sigma/dE_T$ in Fig. 3 primarily reflects the geometry of the collision. The large cross section at low E_T arises from peripheral interactions of the projectile and target, which occur with large probability but produce only little transverse energy. The plateau arises from events in which the entire projectile strikes the target, while the tail is produced by central collisions. The Gaussian structure of the tail arises from fluctuations in transverse energy production in central collisions; these fluctuations provide a fertile probe of the production processes.

Let us turn now to estimating the energy density, ϵ, in the collision volume. The energy density decreases as the system evolves; thus, in order to estimate it, we must know the underlying space-time evolution of the collision volume. Eventually measurements from pion and kaon interferometry will be able to give detailed information on the evolving collision geometry. At present we must rely on various approximate model descriptions; the resulting estimates of the energy density are considerably model dependent.

The simplest estimate is to assume that the entire E_T is deposited uniformly in the collision volume in the participant center of mass, a cylinder with axis along the beam direction, of transverse area πR_B^2, where R_B is the projectile radius. The cylinder lies between the Lorentz contracted projectile and target nuclei, which recede from each other at the speed of light, and is thus of length ct, where t is the time since the initial collision of the nuclei, measured in the participant center of mass. This back-of-the-envelope estimate of the energy density gives

$$\epsilon = \frac{E_T}{\pi R_B^2 t}, \qquad (8)$$

which for ^{32}S depositing 500 GeV in collision with a heavy target is $\sim (10/t_{\text{fm/c}})$ GeV/fm^3, decreasing in time.

We are interested in the energy density only after the excitations in the collision volume, whether hadrons or quarks and gluons, have had time to form. Characteristic formation times, measured in the frame comoving in the beam direction with the excitation, are $\lesssim 1$ fm/c. Furthermore, we would like to study the system after it has had time to approach local thermodynamic equilibrium. A measure of the equilibration time is the viscous relaxation time, τ_η, which has recently been calculated exactly[33] to leading logarithmic order in the weak-coupling limit in a quark-gluon plasma; the result is $\tau_\eta = 0.24/T\alpha^2 \log(1/\alpha) \gtrsim 1$ fm/c, where T is the plasma temperature and $\alpha = g_c^2/4\pi$ is the qcd fine-structure constant. Stopping times for two interpenetrating plasmas are very similar.[33,34] The equilibration time in a strongly interacting plasma, for which we do not yet have a good description, could be considerably shorter than the weak-coupling result.

Taking $t \sim 1$ fm/c in the estimate (8), we would infer that we are seeing events with energy densities an order of magnitude beyond that in equilibrium nuclear matter, $\epsilon_0 \simeq 0.15$ GeV/fm^3 (essentially the rest mass of the nucleons). Such large values augur well for producing locally equilibrated matter at high energy and baryon densities in future experiments with heavier, more energetic beams.

7. CONCLUSION

Once one begins to make high-density matter in laboratory collisions, the important question is, "what is the next step after we find it?"[35], that is, how does one go about learning its properties? To do so will be a nontrivial problem, involving a considerable interplay of many different types of measurements and theoretical simulations of collisions. Unlike in solid-state experiments, the matter produced in heavy-ion collisions evolves dynamically from its initial formation through equilibration and final freeze-out to the noninteracting particles that are detected. Measurements of strongly interacting particles, including Hanbury-Brown Twiss interferometry of meson pairs, directly reveal only conditions at the time of freeze-out, analogous to how photons from a star provide direct information only about conditions at the stellar surface. Electromagnetically produced particles such as electron-positron pairs, muon-antimuon pairs, and direct photons interact negligibly with the matter after they are made and thus are particularly well suited to probe conditions in the interior of the interaction volume, in the way that solar neutrinos can provide direct information about the deep interior of the Sun. Detecting onsets of exotic phases of matter will likely not be signaled by the appearance of "smoking guns" but rather will involve correlating experiments carried out with a wide range of nuclear sizes and energies to search for changes in observed quantities such as final particle distributions. Despite difficulties in deciphering the properties of the matter in experiments, ultrarelativistic heavy-ion collisions are a unique way to study nuclear matter under extreme conditions in the laboratory.

ACKNOWLEDGMENT

This work has been supported in part by National Science Foundation grants DMR88-18713 and PHY89-21025.

REFERENCES

1. R. B. Wiringa, V. Fiks, and A. Fabrocini, *Phys. Rev.* **C38**, 1010 (1988).
2. V. R. Pandharipande and R. A. Smith, *Nucl. Phys.* **A237**, 507 (1975).
3. B. Friedman and V. R. Pandharipande, *Nucl. Phys.* **A361**, 502 (1981).
4. V. R. Pandharipande and R. B. Wiringa, in *The Nuclear Equation of State*, W. Greiner and H. Stöcker (eds.) (Plenum Press, New York, 1989), p. 585.
5. V. R. Pandharipande and V. K. Garde, *Phys. Lett.* **B39**, 608 (1972).
6. J. I. Kapusta and K. A. Olive, *Phys. Rev. Lett.* **64**, 13 (1990).
7. G. E. Brown and W. Weise, *Phys. Rept.* **27**, 1 (1976).
8. A. B. Migdal, *Rev. Mod. Phys*, **50**, 107 (1978).
9. G. Baym, in *Nuclear Physics with Heavy Ions and Mesons, Les Houches session XXX*, R. Balian, M. Rho, and G. Ripka (eds.) (North-Holland, Amsterdam, 1978), p. 745.
10. G. Baym and D. K. Campbell, in *Mesons in Nuclei*, vol. 3, M. Rho and D. Wilkinson (eds.) (North-Holland, Amsterdam, 1979), p. 1031.
11. T. Muto and T. Tatsumi, *Progr. Theoret. Phys.* **78**, 1405 (1987).
12. T. Takatsuka, K. Tamiya, T. Tatsumi, and R. Tamagaki, *Progr. Theoret. Phys.* **59**, 1933 (1978).
13. S.-O. Bäckman and W. Weise, in *Mesons in Nuclei*, vol. 3, M. Rho and D. Wilkinson (eds.) (North-Holland, Amsterdam, 1979), p. 1095.
14. G. E. Brown, K. Kubodera, D. Page, and P. Pizzochero, *Phys. Rev.* **D37**, 2042 (1988).

15. T. Tatsumi and T. Muto, in *Nuclei in the Cosmos*, H. Oberhummer and C. Rolfs (eds.) (Springer-Verlag, Berlin, 1991); and this volume (1992).
16. T. Takatsuka, *Progr. Theoret. Phys.* **78**, 516 (1987).
17. O. V. Maxwell, G. E. Brown, D. K. Campbell, R. F. Dashen, and J. T. Manassah, *Astrophys. J.* **216**, 77 (1977).
18. D. B. Kaplan and A. E. Nelson, *Phys. Lett.* **B175**, 57 (1986).
19. T. Tatsumi, *Progr. Theoret. Phys.* **80**, 22 (1988).
20. D. Page and E. Baron, *Astrophys. J.* **354**, L17 (1990).
21. A. Ukawa, *Nucl. Phys.* **B** (Proc. Suppl.) **17**, 118 (1990).
22. S. Gottlieb, *Nucl. Phys.* **B20** (Proc. Suppl.), 247 (1991).
23. G. Baym and S. A. Chin, *Phys. Lett.* **62B**, 241 (1976).
24. E. Witten, *Phys. Rev.* **D30**, 272 (1984).
25. C. Alcock and A. Olinto, *Ann. Rev. Nucl. Part. Sci.* **38**, 161 (1988).
26. G. Baym, R. Jaffe, E. W. Kolb, L. McLerran, and T. P. Walker, *Phys. Lett.* **160B**, 181 (1985); H. Heiselberg, G. Baym, and C. J. Pethick, *Nucl. Phys.* **B24** (Proc. Suppl.), 144 (1992).
27. A. V. Olinto, *Phys. Lett.* **192B**, 71 (1987).
28. M. A. Alpar, *Phys. Rev. Lett.* **58**, 2152 (1987).
29. G. Baym, P. Braun-Munzinger, and S. Nagamiya (eds.), *Quark Matter '88, Proc. 7th Intl. Conf. on Ultra-relativistic Nucleus-Nucleus Collisions, Nucl. Phys.* **A498** (1989).
30. J.-P. Blaizot, C. Gerschel, and A. Romana (eds.), *Quark Matter '90, Proc. 8th Intl. Conf. on Ultra-relativistic Nucleus-Nucleus Collisions, Nucl. Phys.* **A525** (1991).
31. T. Åkesson et al. (Helios collaboration), *Nucl. Phys.* **B353**, 1 (1991).
32. J. Stachel, *Nucl. Phys.* **A525**, 23c (1991).
33. G. Baym, H. Monien, C. J. Pethick, and D. G. Ravenhall, *Phys. Rev. Lett.* **64**, (1867) 1990; *Nucl. Phys.* **A525**, 415c (1991).
34. G. Baym, H. Heiselberg, C. J. Pethick, and J. Popp, *Proc. 9th Intl. Conf. on Ultra-relativistic Nucleus-Nucleus Collisions, Nucl. Phys.* **A** (1992), in press, and to be published.
35. K. Nakai, *Nucl. Phys.* **A418**, 377c (1984).

K. Iwasawa,† K. Koyama,† and J. P. Halpern‡
†Department of Astrophysics, Nagoya University, Japan; and ‡Columbia Astrophysics Lab, Columbia University, New York, NY, USA

Cyclotron Lines and the Pulse Period Change of X-ray Pulsar 1E2259+586

We report on two *Ginga* observations of the 7-s X-ray pulsar 1E2259+586, the central source in the supernova remnant G109.1−1.0.[1] The observations were made with the LAC aboard *Ginga*[2] on 1989 December 15−17, and 1990 August 8−10. We obtained both pulsar timing and energy spectral information. The pulse period has been continuously increasing at small rate. we found a hint of a cyclotron resonance feature in the energy spectrum.

1. PULSAR LUMINOSITY AND THE PULSE PERIOD CHANGE

The X-ray flux of 1E2259+586 on August 1990 was about two times larger than those of December 1989 and the other previous observations.[3] We also found that the pulse profile shows significant difference between two observations. The pulse period history (Fig. 1)[3,6] shows that 1E2259+586 has been continuously spinning

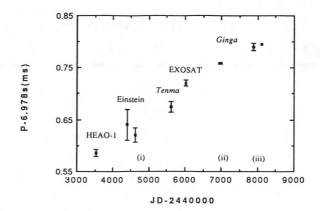

FIGURE 1 The pulse period history of 1E2259+586

down, but the spin down rate has been decreasing to a small rate of $\dot{P} \approx 3 \times 10^{-13}$ s s^{-1}.

From the very small spin-down rate, we suggested that 1E2259+586 is rotating with a period close to its "equilibrium" period with slightly smaller accelarating torque than decelarating torque.[4] The increase in luminosity may suggest an increase in accelerating torque. Therefore 1E2259+586 would become closer to its "equilibrium," resulting smaller spin-down rate. Thus, present observations give further support that 1E2259+586 is a binary X-ray pulsar.[3,4,5]

2. CYCLOTRON HARMONIC FEATURE

Energy spectrum of 1E2259+586 is unusually soft compared with typical X-ray pulsar. The *Ginga* observation in 1989 revealed a complex structure at 5-7 keV, which changes shape with pulse phase (see Fig. 2). The latest *Ginga* observation in 1990 (about 8 month after the observation in 1989) shows no significant structure at 5-7 keV. On the other hand, we found a structure around 10 keV, about twice the energy of the former. We interpret that these structures are due to first and second harmonics of cyclotron absorption. Then the strength of magnetic field is estimated to be $4\text{-}6 \times 10^{11}$ G.

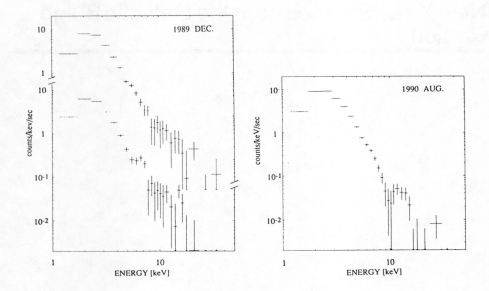

FIGURE 2 Spectra of 1E2259+586. (*Left*) Two of four pulse-phase-resolved spectra in 1989 show different shapes from each other. (*Right*) One of the pulse-phase spectra in 1990.

REFERENCES

1. P. C. Gregory and G. G. Fahlman, *Nature*, **287**, 805 (1980).
2. F. Makino and the Astro-C team, *Astrophys. Letters Commun.*, **25**, 223 (1987).
3. K. Koyama et al., *Publ. Astron. Soc. Japan*, **41**, 461 (1989).
4. K. Koyama et al., *Publ. Astron. Soc. Japan*, **39**, 801 (1987).
5. C. G. Hanson et al., *Astron. Astrophys.*, **195**, 114 (1988).
6. S. R. Davis at al., *M. N. R. A. S.*, **245**, 268 (1990).

Shigeo Yamauchi and Katsuji Koyama
Department of Astrophysics, Nagoya University, Japan

New X-ray Sources near the Galactic Bulge Region

Nine new X-ray sources were discovered during survey observations near the Galactic bulge region with *Ginga*.[1] Seven of the sources were found to have a hardness ratio similar to that of low-mass X-ray binary sources (LMXBs). One source has a hard spectrum with large low-energy absorption similar to that of X-ray pulsars. The other source, GS1734−275, exhibited an unusually soft spectrum. We also found that the *EXOSAT* new source, GPS1742−326, has a typical LMXB spectrum.

1. POSITIONS OF NEWLY DISCOVERED X-RAY SOURCES

Ginga scanning observations in the Galactic bulge region and Galactic center region detected nine new X-ray sources. In order to determine the positions, we carried out a scan fitting with the LAC collimator response function. We list the results in Table 1. Except for source E (GS1734−275), we could not determine the position to better than 4 deg in the direction perpendicular to the scan path since the scan was one-dimensional.

TABLE 1 Positions of Newly Discovered X-ray Sources

Source name	Position	
	l	b
Source A	1.6 ± 0.1	12.9 ~ 16.9
Source B	3.0 ± 0.1	10.3 ~ 14.3
Source C	1.7 ± 0.1	10.3 ~ 14.3
Source D	−2.4 ± 0.1	7.5 ~ 11.5
Source E	0.2 ± 0.1	2.3 ± 0.1
Source F	−3.9 ± 0.1	−4.3 ~ −8.3
Source G	−0.2 ± 0.1	−4.9 ~ −8.9
Source H	−1.4 ± 0.1	−7.6 ~ −11.6
Source I	1.5 ± 0.1	−7.9 ~ −11.9

2. X-RAY COLOR-COLOR DIAGRAM

In order to investigate spectral features of the new sources, we made a color–color diagram (a correlation plot between the softness and hardness ratios; see Fig. 1). Source A has a hard spectrum similar to that of an X-ray pulsar. Source A may be a candidate X-ray pulsar. GPS 1742−326 and the new sources, except for source A and source E, are located near the cataloged LMXBs in the figure, although errors are large. These new sources may be LMXBs. The large softness ratios of these sources could be due to smaller interstellar absorption since they exist at a higher latitude than do the cataloged LMXBs.

Only two X-ray pulsars (GX1+4 and GS1722−36) were found in the survey region. With samples of bright binary X-ray sources, the LMXBs show a concentration toward the Galactic bulge region, while X-ray pulsars are located within the Galactic arm region.[2] The present observation extrapolates this phenomenon to the lower limit.

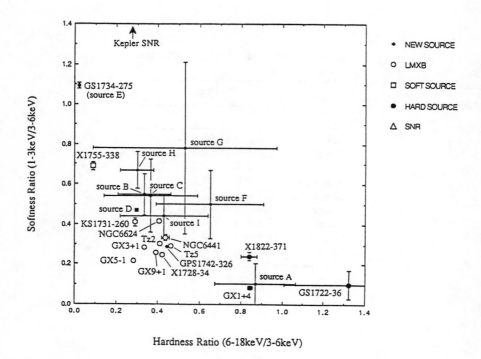

FIGURE 1 A color–color diagram of the X-ray sources observed in *Ginga* scans within 10 deg of the Galactic center

REFFERENCES

1. F. Makino and the Astro-C team, *Astrophys. Letters Commun.*, **25**, 223 (1987).
2. E.g., H. V. D. Bradt and J. E. McClintock, *Ann. Rev. Astr. Ap.*, **21**, 13 (1983).

Andrew M. Abrahams and Stuart L. Shapiro
Center for Radiophysics and Space Research, Cornell University, Ithaca, NY 14850, USA

EOS for Neutron Star Atmospheres: Finite Temperature and Gradient Corrections

Because of the intense magnetic fields ($B \sim 10^{12}$ G) believed to exist on the surfaces of neutron stars, the structure of bulk matter in these extreme conditions is of obvious interest. In a recent study[1] we improved on previous statistical calculations by taking into account two different effects on the magnetic equation of state (EOS): finite temperature and gradient corrections. To investigate the influence of *finite temperature*, we constructed a magnetic Thomas-Fermi (TF) model characterized by four distinct length scales: the Bohr radius, the cyclotron radius, the thermal de Broglie wavelength, and the Wigner-Seitz lattice cell radius. The model is convenient for astrophysical applications; once the EOS is computed for a single element and magnetic field strength it can be easily scaled to other elements and field strengths. Our results smoothly extend from the cold, degenerate limit to the hot, Maxwell-Boltzmann regime. We find that the effects of finite temperature on the magnetic EOS have a significant influence on neutron star atmospheric profiles (see Fig. 1).

Zero-temperature atomic and condensed matter models based on TF-like theories have several major defects. For example, the electron density diverges at the nucleus (as $r \to 0$) and TF atoms have a power-law tail rather than the correct exponential asymptotic falloff at large radii. In Thomas-Fermi-Dirac (TFD) theory, atomic profiles must be truncated at a finite radius where the electron density goes to zero. Also, zero-pressure TFD matter is identical to TFD atoms and so does not

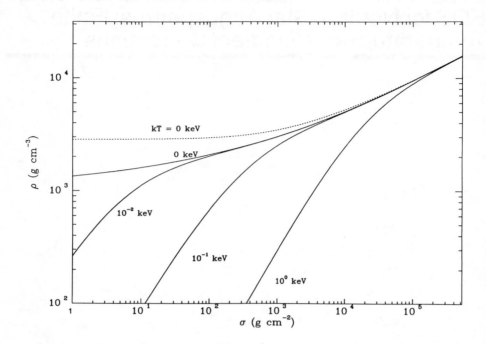

FIGURE 1 The effect of finite temperature on the atmospheric density profile of a cold neutron star. The mass density ρ is plotted as a function of column density σ. Profiles are plotted for pure $^{56}_{26}$Fe in an isothermal atmosphere with surface gravity $g = 10^{14}$ cm s^{-2} and $B_{12} = 1$. Results are shown for TF (*solid*) and zero-temperature TFD-1/9W (*dotted*) models.

represent a condensed state. Most significantly, *no-binding* theorems[2] have shown that molecules and chains are impossible in these theories.

In Thomas-Fermi-Dirac-Weizsäcker (TFD-λW) theory, electron wavefunctions are *modified* plane waves; as a consequence a *gradient* term is introduced into the kinetic energy functional (with coupling coefficient λ). TFD-λW theory gives atomic structure which agrees qualitatively with quantum-mechanical predictions, a zero-pressure condensed matter state distinct from the atom,[3] and binding of atoms into molecules. We formulate the TFD-λW model as a system of elliptic equations amenable to numerical study in one dimension (atoms) and two dimensions (molecules and chains). We computed the cold EOS for $^{56}_{26}$Fe at several magnetic field strengths. Zero-pressure densities from TFD-λW theory are appreciably higher

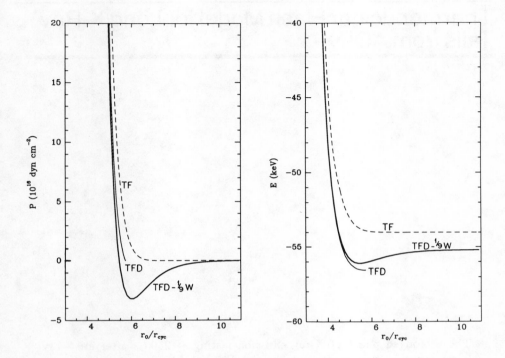

FIGURE 2 In the left-hand plot we show pressure P versus cell radius r_0 (in units of the cyclotron radius r_{cyc}) for $^{56}_{26}$Fe in a magnetic field of 10^{12} G and at zero temperature. In the right-hand plot we show total cell energy E versus cell radius r_0. Curves are shown for TF (*dashes*), TFD (*solid*), and TFD-1/9W (*heavy solid*) calculations.

than those from TFD theory, so gradient corrections are important for the first centimeter of neutron star atmospheres. We find cohesive energies for $^{56}_{26}$Fe of 0.91 keV at $B = 10^{12}$ G and 4.9 keV at $B = 10^{13}$ G (see Fig. 2).

REFERENCES

1. A. M. Abrahams and S. L. Shapiro, *Astrophys. J.*, <u>374</u>, 652 (1991).
2. I. Fushiki et al., Nordita Preprint 89/55 AS (1989).
3. A. M. Abrahams and S. L. Shapiro, *Phys. Rev.* **A42**, 2530 (1990).

Ikko Fushiki,* Ronald E. Taam,† S. E. Woosley,* and D. Q. Lamb‡
*University of California at Santa Cruz, †Northwestern University, ‡University of Chicago

Thermonuclear Flash Model for Long X-Ray Tails from AQL X-1

The prolonged phase of X-ray emission lasting ~ 2500 s after an X-ray burst event in the recurrent transient Aql X-1 is interpreted as an extended phase of hydrogen burning in the accreted envelope of a neutron star. The nuclear burning is accelerated by electron capture processes at the high densities ($\sim 10^7$ g cm^{-3}) characteristic of the accreted layer. For a neutron star characterized by a mass and radius of 1.4 $M_\odot$ and 9.1 km respectively the occurrence of the long X-ray tail requires that the mass of the accumulated layer be $\sim 10^{23}$ g and the envelope temperatures of the neutron star be between 10^7 and 1.5×10^7 K.

1. INTRODUCTION

It is generally well accepted that X-ray bursts result from a thermonuclear shell flash of the accreted hydrogen-helium rich layers on the surface of neutron stars.[1] Although the gross characteristics of the observed bursts can be understood within the general framework of such a model, a number of discrepancies still remain upon detailed quantitative comparisons between observation and theory. Among them is the observation by Czerny, Czerny, and Grindlay[2] of Type I X-ray bursts from the recurrent soft X-ray transient source Aql X-1. The characteristic which distinguished the first X-ray burst (seen at least 9 days after the maximum peak of the outburst) from typical X-ray bursts observed from other sources was the discovery that it exhibited an extremely long, relatively flat, X-ray tail for more than 2500 s. A second burst was observed 4 days later. The luminosity profile of the latter burst was similar to typical X-ray bursts: however, it, too, exhibited a tail for 500 s, but much less pronounced than in the first burst observed.

2. RESULTS

The construction of the initial model and the evolution of the accreted layers on the neutron star surface were followed with the KEPLER computer code modified to study transient sources.[3,4] Of primary importance in the present study is the inclusion of the energy generation and compositional variations associated with the rapid proton (rp) processd.[5] and electron captures in the reaction $H(e^-,\nu)n(p,\gamma)D$. We adopt a neutron star mass and radius of 1.4 $M_\odot$ and 9.1 km respectively. Based upon the observations of Aql X-1 we choose a mass accretion rate of 10^{-9} $M_\odot$ yr^{-1}. From the nonsteady models of Fushiki and Lamb[6] the initial temperature, T_b, at the base of the accreted layer of the neutron star is chosen to be 10^7 K. The abundance of the accreted matter is taken as X = 0.7, Y = 0.299, and Z = 0.001.

Within 18 days from the onset of accretion a combined hydrogen-helium flash instability developed in a layer of $\sim 9 \times 10^{22}$ g. The decay time of the first burst is $\sim$ 6000 s. Since the densities and temperatures at the base of the accreted layer during this phase exceed 5×10^6 g cm^{-3} and 10^9 K respectively, the first step in the proton burning chain is accelerated by the electron capture process. In comparison with the first burst it is seen that the time scale to accrete a critical amount of mass is significantly decreased to 6 hours. The decay time is reduced to $\sim$ 240 s. This is a consequence of the much smaller amount of mass accumulated ($\sim 1.5 \times 10^{21}$ g). The electron capture processes are unimportant in the second and subsequent bursts. To determine the sensitivity of our results to the thermal state of the neutron star, T_b was increased from 10^7 K to 1.5×10^7 K. The decay times of the first and second bursts in this model are 1300 s and 870 s, respectively. We conclude from our calculations of two models that Aql X-1 has an initial temperature, T_b, between 10^7 K and 1.5×10^7 K.

In order to determine the consistency of the initial thermal state of the neutron star core prior to the recurrent transient outburst, the accretion rate was artificially decreased to 2×10^{-12} $M_\odot$ yr^{-1} after the third burst, and it was found that the temperature of the envelope relaxed to its initial thermal state after a period of 1 month.

ACKNOWLEDGMENT

This research has been supported in part by NASA under grants NAGW-768 and NAGW-1284, and by NSF under grant AST 88-13649.

REFERENCES

1. R. E. Taam, *Ann. Rev, Nuc. Part. Sci.* **35**, 1 (1985).
2. M. Czerny, B. Czerny, and J. E. Grindlay, *Ap. J.* **312**, 122 (1987).
3. R. K. Wallace, S. E. Woosley, and T. A. Weaver, *Ap. J.* **258**, 696 (1982).
4. S. E. Woosley and T. A. Weaver, *High Energy Transients in Astrophysics*, AIP Conf. Proc. No. 115, S. E. Woosley (ed.), (AIP, New York, 1984), p. 273.
5. R. K. Wallace and S. E. Woosley, *Ap. J.* **43**, 389 (1981).
6. I. Fushiki and D. Q. Lamb, *Ap. J. Lett.* **323**, L55 (1987)

T. Murakami
Institute of Space and Astronautical Science (ISAS), 3-1-1, Yoshinodai, Sagamihara, Kanagawa 229, Japan

Gamma-Ray Bursts and Dead Pulsars

1. CYCLOTRON HARMONIC

Since the launch of *Ginga* in 1987,[1] the GBD detector has detected at least three GRBs with spectral feature.[2,3] The features show energies equally spaced at 20 and 40 keV. In Table 1 characteristics of the three events are summarized. These are naturally interpreted as cyclotron harmonic in strong magnetic fields of 10^{12} gauss, which is realized only on neutron stars. The fact that magnetic fields in GRBs and in X-ray pulsars are clustering in a very narrow range of about $(1-3) \times 10^{12}$ gauss is interesting.[4]

TABLE 1 Characteristics of the three GRBs in unit of keV. E_1 and E_2 are center energies of the 1st and 2nd harmonic. W_1 and W_2 are the width of each line in unit of keV(FWHM)

Event ID	E_1	W_1	E_2	W_2	E_2/E_1
GB870303	20.4±0.7	3.5±2.7	40.6±2.6	12.3±6.3	1.99±0.3
GB880205	19.3±0.7	4.1±2.2	38.6±1.6	14.4±4.6	2.0±0.15
GB890929	26.3±1.5	4.2	46.6±1.7	7.7	1.8±0.2

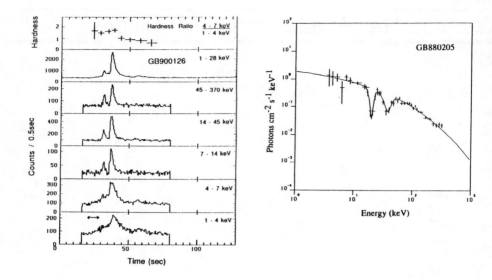

FIGURE 1 (*Left*) Typical example of the cyclotron feature in a GRB spectrum observed with *Ginga*. (*Right*) This is the case of GB880205. The GB900126 detected with *Ginga* showed an X-ray initial phase (precursor) before the first flash of γ-photons. The spectrum during the initial phase is best represented by a blackbody with temperature of about 1.5 keV.

2. X-RAY PRECURSOR AND X-RAY TAIL

About 40% of GRBs among those observed with *Ginga* have shown soft X-ray emission with a time scale of several tens to a hundred seconds after the harder γ-photons have terminated. We called this type of emission a soft X-ray tail in our first papers.[5] A GRB detected on January 26, 1990, also has shown the soft X-ray tail. Besides the tail this GRB is noted for its peculiar behavior in soft X-ray initial emission (precursor) before the first onset of γ-photons. Both spectra in the soft X-ray tail and the initial phase are well represented by a thermal spectrum as blackbody with temperature of about 1.5 keV. The thermal spectra, long cooling time of more than several tens of seconds in the X-ray tail, and 8-s slow rise with thermal spectrum in the initial phase strongly suggest that the energy source of this GRB resides deep in a neutron star.[6] Although there are many proposed models of GRBs, thermonuclear flash and/or neutron star glitch whose energy is released deep in the atmosphere are the most plausible models.[7,8]

3. DISCUSSION

Assuming GRBs originate from strongly magnetized neutron stars, the number of GRBs detected per year enables us to estimate neutron star density. A typical detection rate of GRBs[8] is about several hundred a year at a flux level of 10^{-6} ergs cm^{-2}. We require two assumptions to calculate a number density of neutron stars. One is their unknown recurrence period and the other is their typical distance. Based on the high isotropy of the detected GRBs,[9] we safely set their distances at typically not more than 1 kpc. Thus the required number density of neutron stars is estimated to be more than $10^{-4}(T_{100})(R_{kpc})^{-3}$ pc^{-3}, where T is a recurrence time scale in units of 100 years and R is the distance in units of 1 kpc. This relation requires old neutron stars and short recurrence periods for GRBs because density of neutron stars derived from the birthrate of SNRs is about 10^{-3} pc^{-3}. We think that neutron stars are born as radio pulsars and after their death they might contribute to GRB activity. If this were true, the strong magnetic fields of order 10^{12} gauss of old neutron stars conflict with the current decay time scale of 10^7 year derived from a statistical analysis of radio pulsars.[10] New ideas such as mass accretion-induced decay for magnetic fields and starquake-induced nuclear flash for an energy source are possible explanations.[7,11]

REFERENCES

1. T. Murakami et al., *Publ. Astron. Soc. Japan,* **41**, 405 (1989).
2. T. Murakami et al., *Nature,* **335**, 234 (1988).
3. 3. E. E. Fenimore, et al., *Ap. J (Letters),* **335**, L71 (1988).
4. F. Nagase, *23rd ESLAB symp. Bologna* (ESA, 1989), p. 45.
5. A. Yoshida et al., *Publ. Astron. Soc. Japan,* **41**, 509 (1988).
6. T. Murakami et al., *Nature,* **350**, 592 (1991).
7. O.Blaes et al., *Ap. J.* **363**, 612 (1990).
8. E. P. Liang and V. Petrosian *Gamma-Ray Bursts* (AIP #141, 1984).
9. J.-L. Atteia et al., *Ap.J(Suppl.),* **64**, 305 (1987).
10. A. G. Lyne, R. N. Manchester, J. H. Tayler, *MNRAS,* **213**, 613 (1985).
11. N. Shibazaki, T. Murakami, et al., *Nature,* **342**, 656 (1990).

T. Kunihiro
Faculty of Science and Technology, Ryukoku University, Seta, Otsu-city 520-21, Japan

Strangeness in the Proton and Kaon Condensation in High-Density Nuclear Matter

Recently much attention has been paid to flavor-mixing effects seen in the baryon sector. The relevant quantities include the πN and KN sigma terms[1] $\Sigma_{\pi N} = (m_u+m_d)/2 <P|\bar{u}u+\bar{d}d|P>$ and $\Sigma_{KN} = (m_u+m_d)/2 + m_s/2 <P|(\bar{u}u+\bar{d}d)/2+\bar{s}s|P>$. $\Sigma_{\pi N}$ is quoted as 51–53 MeV with an estimated error 6–8 MeV, while the empirical value of Σ_{KN}, which gives the strength of the driving force of the kaon condensation,[2] is still uncertain.

It is known that the first-order chiral perturbation with the current quark mass ratio $2m_s/(m_u+m_d) \simeq 25$ requires a huge s-quark content of the proton to reproduce the empirical value $y \equiv 2<\bar{s}s>_N / <\bar{u}u+\bar{d}d>_N = 0.48$. With such a large value of y, Σ_{KN} becomes as large as 500–600 MeV, so that the onset density ρ_c^K of the kaon condensation becomes as low as about $4\rho_0$ ($\rho_0 = 0.17$ fm^{-3} being the normal nuclear matter density hence could be reachable in neutron stars. In fact, the ρ_c^K can be approximated by the formula[2]

$$\rho_c^K \sim f_K^2 m_K^2 / \Sigma_{KN}, \qquad (1)$$

where f_K and m_K are the kaon decay constant ($\sim$ 115 MeV) and the kaon mass, respectively.

In this short report, we will report a realistic calculation of the strangeness content of the proton and hence Σ_{KN},[3] and discuss its implications for the kaon condensation in neutron stars.

As first noted by Gasser,[4] the Feynman-Hellman (FH) theorem tells us that once the proton mass M_P is given in terms of the current quark mass m_i ($i = u, d, s$), the quark contents of the proton are given by $<\bar{q}_i q_i>_P = \partial M_P/\partial m_i$. The problem is how to get a formula for M_P in terms of m_i. As a first approximation, the additive quark model was adopted;[5,6] $M_P = 2M_u + M_d$, where $M_u(M_d)$ is the constituent quark mass. As a natural extension of the previous works,[5,6] we have incorporated the short-range interaction via gluons using the nonrelativistic quark model.[9] Note that the crucial difference from the conventional constituent quark model lies in the fact that the constituent quark mass is identified with the dynamical mass generated by the spontaneous breaking of chiral symmetry (SBCS). We take the Nambu-Jona-Lasinio (NJL) model[7] with the 't Hooft term[8] incorporated as an effective Lagrangian of QCD to describe SBCS. The model well reproduces the fundamental physical quantities with accuracy of O(10%-15%).[5,3]

Applying the FH theorem with the mass formula obtained thus, we get for the quark content of the proton

$$<\bar{u}u>_P = 5.0, \quad <\bar{d}d>_P = 4.0, \quad <\bar{s}s>_P = 0.53, \text{ hence } y = 0.12. \qquad (2)$$

Thus

$$\Sigma_{\pi N} = 49 \text{MeV} \quad \text{and} \quad \Sigma_{KN} = 350 \text{MeV}. \qquad (3)$$

$\Sigma_{\pi N}$ is consistent with the empirical value within the error, while the ratio y remains as small as 0.12.

Now one sees that Σ_{KN} is still as small as 350 MeV even with the gluon-exchange interaction. Because of the formula $\rho_c^K \sim f_K^2 m_K^2/\Sigma_{KN}$, this implies the onset density ρ_c^K is as high as $\sim 7\rho_0$, which is higher than the central density of neutron stars obtained without anomalous phases such as the pion-condensed phase. Thus kaon condensation might be unrealistic even in neutron stars, except in the deep interior of a neutron star with pion condensation.

In the present calculation, we have used $m_u = m_d = 5.5$ MeV for the nonstrange current quark mass. Current quark mass dependence of the results discussed here is examined in Ref. 10.

ACKNOWLEDGMENT

We acknowledge T. Hatsuda for a collaboration for the work on which this report is based.

REFERENCES

1. For a review, see R. L. Jaffe and C. L. Korpa, *Comm. Nucl. Part. Phys.* **17** (1987) 163.
2. D. B. Kaplan and A. E. Nelson, *Phys, Lett.* **B175** (1986) 57; G. E. Brown et al., *Phys, Lett.* **B192** (1987) 273; G. E. Brown et al., *Phys. Rev.* **D37** (1988) 2042; T. Tatsumi, *Progr. Theor. Phys.* **80** (1988) 22.
3. T. Kunihiro and T. Hatsuda, *Phys, Lett.* **B240** (1990) 209; Hatsuda and T. Kunihiro, *Z. Phys.* **C51** (1991) 49.
4. J. Gasser, *Ann. of Phys.* **136** (1981) 62.
5. T. Kunihiro and T. Hatsuda, *Phys, Lett.* **B206** (1988) 385; T. Kunihiro, *Progr. Theor. Phys.* **80** (1988) 34; T. Kunihiro, *Soryushiron-Kenkyu* (Kyoto) **77** (1988) D16. No. 4.
6. V. Bernard, R. L. Jaffe and U. -G. Meissner, *Nucl. Phys.* **B308** (1988) 753; Y. Kohyama, K. Kubodera and N. Takizawa, *Phys, Lett.* **B208** (1988) 165.
7. Y. Nambu and G. Jona-Lasinio, *Phys. Rev.* **122** (1961) 345; ibid. **124** (1961) 246.
8. G. 't Hooft, *Phys. Lett.* **D14** (1976) 3432; see also M. Kobayashi, H. Kondo, and T. Maskawa, *Progr. Theor. Phys.* **45** (1971) 1955.
9. N. Isgur and G. Karl, *Phys. Rev.* **D18** (1978) 4187; ibid., **D19** (1979) 2653.
10. T. Hatsuda and T. Kunihiro, submitted to *Nucl. Phys.* B.

T. Hanawa
Department of Astrophysics, Nagoya University, Nagoya 464-01, Japan

High-Energy X-ray Production in Accreting Neutron Star

1. INTRODUCTION

Accreting neutron stars in binary systems are strong X-ray sources and their X-ray spectra give us plenty of information on the neutron stars and their environment. Recent observations[1,2] have revealed that some accreting neutron stars have hard X-ray spectra approximated by a power law. We propose two mechanisms to produce the hard X-ray component. Gas is accreted onto a neutron star either through an accretion disk or nearly radially. See our recent papers[3,4,5] and papers[3,4,5,6] for further details.

2. CASE OF RADIAL ACCRETION

When the flow is nearly radial, the accreting gas is accelerated by gravity up to half the light speed ($v \sim \sqrt{2GM/r} \sim 0.5c$) near the neutron star surface. It is decelerated through a shock wave to a very low speed on the neutron star surface. Soft X-rays emitted from the neutron star are scattered in part in the accreting plasma falling at $v \sim 0.5c$. The scatterd X-rays change their energy by Compton effect ($\Delta E/E \sim v/c \sim 0.5$). Some of the scattered X-rays reenter the neutron star

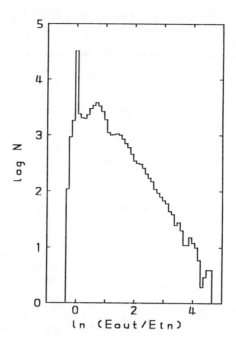

FIGURE 1 The histogram as a function of the energy gain, $\ln(E_{\text{out}}/E_{\text{in}})$, for the case of radial accretion. The optical depth of the accreting plasma is $\tau = 0.4$ and the velocity is $v = 0.5c$. This figure is reproduced from Ref. 5.

and scatter repeatedly between the neutron star and the accreting plasma. X-rays gain energy by scattering and become high-energy ones. Fig. 1 shows an example of the Monte Carlo simulation that computes the probability of energy gain by multiple Compton scatterings. The abscissa is the logarithm of the relative energy gain, $\ln(E_{\text{out}}/E_{\text{in}})$, and the ordinate is the logarithm of the event number. The total number of trials is 10^5. It is shown that the probability of high energy gain is proportional to the power of $(E_{\text{out}}/E_{\text{in}})$. Thus, the resultant X-ray spectrum after multiple scatterings has a power-law component in the high-energy range.

3. CASE OF DISK ACCRETION

When an accreting gas has angular momentum, the accreting neutron star has a boundary layer rotating rapidly on the surface. The rotation velocity and the optical depth of the boundary layer are estimated to be $v \sim 0.5c$ and $\tau \sim 1$. As in the case of the radial accretion, some X-rays from the neutron star are scattered between the boundary layer and the neutron star main body. Scattered X-rays gain or lose their energy depending on the scattering angle. As total X-rays diffuse in energy space by scattering. Statistically some of the X-rays become very high energy ones. It is shown by the Monte Carlo simulation that high energy X-rays are produced efficiently when the boundary layer is rotating at a semirelativistic speed ($v \sim 0.5c$) and is semitransparent ($\tau \sim 1$).

REFERENCES

1. K. Mitsuda et al., *Pub. Astr. Soc. Japan*, **41**, 97 (1989).
2. N. Kawai et al., *Pub. Astr. Soc. Japan*, **42**, 115 (1990).
3. K. Hirotani, T. Hanawa, and N. Kawai, *Ap. J.*, **355**, 577 (1990).
4. T. Hanawa, *Ap. J.*, **355**, 585 (1990).
5. T. Hanawa, *Ap. J.*, **366**, 495 (1991).
6. T. Hanawa, *Ap. J.*, **373**, 222 (1991).

Shinpei Shibata
Department of Physics, Yamagata University, Yamagata 990, Japan

The Current System in the Magnetosphere of Neutron Stars

The loss rate of the rotational energy of pulsars is estimated to be

$$\dot{E}_{\rm rot} = f \frac{\mu\lambda 2\Omega\lambda 4}{c\lambda 3},\qquad(1)$$

where μ and Ω are the magnetic moment and angular velocity of the neutron star, c is the speed of light, and f is a factor of order unity. Although most authors agree with this estimate, there is no reliable theory giving the factor f and predicting what fraction in $\dot{E}_{\rm rot}$ is radiated by each part of the magnetosphere. Local models do not overcome this difficulty; local models give, for example, the energy loss rate from a part of the magnetosphere as a *function* of the local field strength, current density, and other parameters but do not give the value itself. A global view should be invoked to solve this problem.

The pulsar wind, inner gap, and outer gap are know as dominant modes of energy loss. These parts are linked to each other. For example, owing to the voltage drop in the inner and outer gaps, the voltage applied to the wind is reduced, and as a result the base of the wind rotates more slowly than the star.

In this paper we consider the magnetospheric structure in a global sense assuming a large-scale magnetospheric current system as shown in Fig. 1. The current system forms a DC circuit which includes the electromotive source of the rotating magnetic neutron star and three loads: the outer gap, the pair-plasma wind, and

The Current System in the Magnetosphere of Neutron Stars

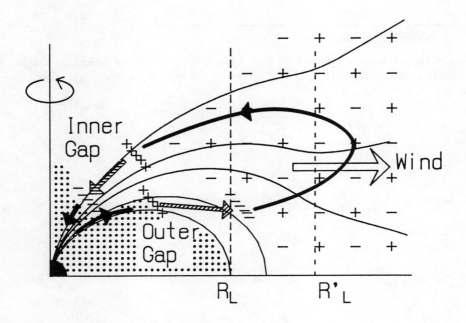

FIGURE 1 Schematic of a global current system in the pulsar magnetosphere

the inner gap. First we construct local models for the three parts of the magnetosphere *with adjustable free parameters,* and next we link them to a global model by imposing global link conditions. The global conditions are that (1) the sum of voltage gaps applied to the magnetic current circuit, (2) the total energy loss rate is the angular velocity of the star times the total angular momentum loss rate, and (3) the magnetic flux in the circuit is composed of an open flux in the pulsar wind and a closed flux in the outer gap. The procedure of our global modeling is analogous to obtaining operating points of an electric circuit when characteristic of circuit elements are given.

We find the following results: (1) the outer gap is an indispensable element for the magnetospheric current circuit. (2) The plasma is strongly accelerated in the pulsar wind so that its energy flux becomes comparable with the electromagnetic energy flux (Poynting flux). (3) As the rotation slows down, operating points disappear, and this is thought to be the turnoff of pulsars. Before the pulsar turns off, the potential drop above the polar cap increases by up to 10%–25% of the total electromotive force. (4) The operating points for the Crab pulsar show that about 99.4% of the total pulsar power is carried away by the pulsar wind and that the remaining power, ~0.6%, is radiated from the inner and outer gaps.

ACKNOWLEDGMENT

This work has been supported in part by a Grand-in-Aid for Scientific Research (02234101 and 02740128) of the Ministry of Education, Science, and Culture in Japan.

Lee Lindblom and Gregory Mendell
Department of Physics, Montana State University, Bozeman, MT 59717, USA

Superfluid Effects on the Stability of Rotating Neutron Stars

Gravitational radiation tends to make all rotating stars unstable, while internal dissipation (e.g., viscosity) tends to counteract this instability. The purpose of this paper is to investigate how the internal dissipation mechanisms that exist within superfluid neutron-star matter influence the gravitational-radiation instability. We argue here that mutual friction between the electrons and the quantized neutron vortices completely suppresses the gravitational-radiation instability in rotating neutron stars cooler than the superfluid-transition temperature.

1. DISSIPATION IN ROTATING SUPERFLUIDS

Below the superfluid-transition temperature ($T \approx 10^9$ K) the dissipation mechanisms are significantly different from those in warmer neutron-star matter. The neutrons and protons form Cooper pairs and condense into superfluid states. Scattering among these particles is prevented by the energy gap that separates these states from the normal-particle states. Thus scattering is confined primarily to the electrons in superfluid neutron stars. These may scatter with each other to dissipate energy via ordinary shear viscosity. In rotating neutron stars another dissipation mechanism exists: mutual friction—the scattering of normal particles off the quantized vortices of the superfluid condensates.[2] In the case of a neutron-star superfluid the scattering of electrons with the vortices of the neutron superfluid dominates this dissipation. Due to Fermi-liquid "drag" effects the neutron vortices carry a substantial magnetic flux with which the electrons may scatter.[1] The hydrodynamic equations (including dissipation) needed to analyze in detail the large-scale dynamics of rotating superfluid neutron-star matter have been developed recently by Mendell.[4] We use some of the results of that work to estimate here the effectiveness of mutual friction in suppressing the gravitational-radiation instability in rotating neutron stars.

2. INSTABILITIES IN SUPERFLUID NEUTRON STARS

Consider the perturbations of a neutron star having time dependence $e^{i\omega t - t/\tau}$, where ω and τ are real. Such a perturbation is stable whenever $1/\tau \geq 0$. Thus in a sequence of rotating stars parameterized by Ω (the angular velocity), those stars rotating slower than the smallest root of $1/\tau(\Omega_c) = 0$ are stable. (Assuming of course that the star having $\Omega = 0$ is stable.)

In superfluid neutron stars the important dissipative damping times are τ_{η_e} due to shear viscosity, τ_{MF} due to mutual friction, and τ_{GR} due to gravitational radiation. These determine the total damping time by $1/\tau = 1/\tau_{\eta_e} + 1/\tau_{MF} + 1/\tau_{GR}$. The viscous and gravitational-radiation damping times are discussed in some detail in Lindblom's review in this volume. That discussion will not be repeated here. The mutual-friction damping time τ_{MF} may be expressed as an integral involving the perturbations:[4]

$$\frac{1}{\tau_{MF}} = \frac{\Omega}{2E} \int \beta_{MF} \left[\Delta \vec{v} \cdot \Delta \vec{v}^* - |\Delta \vec{v} \cdot \vec{z}|^2 \right] d^3x, \qquad (1)$$

where E is an energy of the perturbations, β_{MF} is the strength of the mutual-friction interaction, $\vec{z}$ is the unit vector parallel to the rotation axis, and $\Delta \vec{v}$ is the perturbation of the relative velocity between the neutron superfluid and the charged fluids. (The charged fluids are coupled electromagnetically on timescales much shorter than τ_{MF}.[1]) An estimate for β_{MF} has been given by Mendell[4] based on the calculation of electron-vortex scattering by Alpar, Langer, and Sauls:[1]

$$\beta_{MF} = 1.0 \times 10^{-4} \rho_p^{7/6} \left(1 - \frac{m_n}{m_n^*} - \frac{m_p}{m_p^*}\right)^2 \left(1 - \frac{m_p^*}{m_p}\right)^2 \left(\frac{m_p}{m_p^*}\right)^{1/2}, \qquad (2)$$

where m_n/m_n^* and m_p/m_p^* are the ratios of the masses to the effective masses of the neutrons and protons respectively, and ρ_p is the proton mass density.

The mutual-friction damping times have yet to be evaluated for the pulsations of rotating neutron stars. The magnitude of τ_{MF} may be estimated, however, from Eq. (1) for the plane-wave perturbations of uniform-density neutron-star matter.[4] For perturbations with wavelength $\lambda = 2\pi R/l$, the ratio of the mutual friction to the viscous damping time is found to be

$$\frac{\tau_{MF}}{\tau_{\eta_e}} < 10^{-6} \, l^2 \left(\frac{10^6 \text{ cm}}{R}\right)^2 \left(\frac{10^3 \text{ s}^{-1}}{\Omega}\right) \left(\frac{10^9 \text{ K}}{T}\right)^2, \tag{3}$$

This limit on the mutual-friction damping time can be used to place limits on the roots of $1/\tau(\Omega_c) = 0$. We use the angular-velocity dependence of τ_{η_e} and τ_{GR} computed by Ipser and Lindblom[3] for the $2 \leq l = m \leq 6$ f-modes of rotating neutron stars. We find *no* roots of $1/\tau(\Omega_c) = 1/\tau_{\eta_e}(\Omega_c) + 1/\tau_{MF}(\Omega_c) + 1/\tau_{GR}(\Omega_c) = 0$ for these modes if the temperature is less than 10^9 K. Thus, we conclude that mutual friction eliminates the gravitational-radiation instability in all rotating neutron stars cooler than the superfluid transition temperature.

ACKNOWLEDGMENT

This research was supported by NSF grant PHY-9019753.

REFERENCES

1. M. A. Alpar, S. A. Langer, and J. A. Sauls, *Ap. J.* **282**, 533 (1984).
2. H. E. Hall and W. F. Vinen, *Proc. R. Soc.* **238A**, 215 (1956).
3. J. R. Ipser and L. Lindblom, *Ap. J.* **373**, in press (1991).
4. G. Mendell, *Ap. J.* submitted (1991).

Takumi Muto† and Toshitaka Tatsumi‡
†Yukawa Institute for Theoretical Physics, Kyoto University, Kyoto 606, Japan; and
‡Department of Physics, Kyoto University, Kyoto 606, Japan

Dissipation of Vibrational Energy of Neutron Stars in the Pion-Condensed Phase

Pulsation and accompanying dissipation of the vibrational energy of a neutron star reflect characteristic features of high-density nuclear matter. The dissipation mechanism of vibrational energy in the normal phase has been considered to be a nonequilibrium weak process such as non-leptonic weak decay (Σ^- process).[1] We consider here the dissipation mechanism in the charged-pion (π^c) condensed phase. Our discussion is closely related to our previous treatment of the dynamical behavior of a pion-condensed neutron star;[2,3] vibration is driven by the minicollapse of a neutron star with a metastable (supercompressed) π^c condensed core to one with a stable π^0-π^c condensed (CPC) core (Fig. 1).

In this case, dissipation is mainly caused by the nonequilibrium quasi-particle (η) β decay,

$$\eta \rightarrow \eta' + e^- + \bar{\nu}_e \quad (a), \qquad \eta + e^- \rightarrow \eta' + \nu_e \quad (b). \tag{1}$$

The π^c condensed phase is composed of quasi-particles (η-particles), condensed pions, and electrons. They are specified as their density ρ, ρ_π, and ρ_e, respectively. Due to vibration of a neutron star in a radial mode with period τ, baryon number density ρ changes from the initial equilibrium value (referred to as "eq") to $\rho(t) = \rho_{eq} + \delta\rho \cos(2\pi/\tau)t$. The time scale for Eq. (1) at the typical density $\rho = 3.5\rho_0$ and temperature $T = 10^{10}$ K is calculated to be $\sim 10^3$ s, which is much larger than the vibrational time scale of a neutron star, $\tau = 10^{-4}$–10^{-3} s. Then the density change

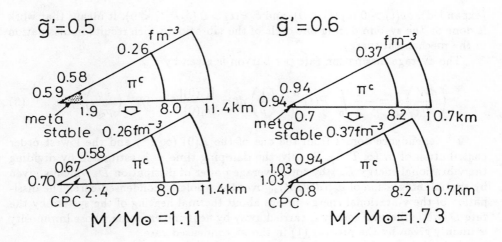

FIGURE 1 Change in internal structure of a neutron star induced by a minicollapse

of condensed pions (electrons) from the initial equilibrium value $\delta\rho_{\pi(e)}(t)$ can be expressed as a sum of the contribution caused purely by the volume change with the weak process (1) kept frozen, $\delta_v\rho_{\pi(e)}(t)$, and the one caused purely by the deviation of the process (1) from chemical equilibrium with the volume fixed, $\delta_c\rho_{\pi(e)}(t)$.[3]

The deviation from chemical equilibrium with respect to the process (1) is measured by the function $u_t \equiv (\mu_e(t) - \mu_\pi(t))/T$, where μ_π (μ_e) is the pion (electron) chemical potential and T the temperature. It is written up to first order in $\delta\rho_\pi(t)$ and $\delta\rho_e(t)$ as $u_t = \tilde{\mu}'/T \cdot \delta\rho(t)/\rho^0 + \tilde{\mu}/T \cdot \delta_c\rho_e(t)$, with $\tilde{\mu}' \equiv \rho_e^{eq}(\partial\mu_e/\partial\rho_e)_{eq}^v - \rho_\pi^{eq}(\partial\mu_\pi/\partial\rho_\pi)_{eq}^v$ and $\tilde{\mu} \equiv \pi^2/(\mu_e^{eq})^2 + \mu_\pi^{eq}/\rho_\pi^{eq}$. Then $\delta_c\rho_e(t)$ is related to the reaction rate $R(u_t)$ for Eq. (1a) and $R(-u_t)$ for Eq. (1b) by

$$\frac{d}{dt}\delta_c\rho_e(t) = R(u_t) - R(-u_t) = 2\frac{I_2'(0)}{I_3(0)}\tilde{\epsilon}(0)T^5 u_t + O(u_t^2), \qquad (2)$$

with

$$I_n(u) \equiv \int_0^\infty dx\, x^n \frac{\pi^2 + (x+u)^2}{1 + \exp(x+u)},$$

where $\tilde{\epsilon}(0)$ is the emissivity factor except for the T dependence in the weak process (1).

The change of the pressure $P(t)$ is also described as the sum of $\delta_v P(t)$ and $\delta_c P(t)$. The latter is responsible for dissipation of the vibrational energy and written as $\delta_c P(t) = \tilde{\mu} \cdot \delta_c\rho_e(t)$ with $\tilde{\mu} \equiv (\partial P/\partial\rho_e)_0^\epsilon - (\partial P/\partial\rho_\pi)_0^\epsilon$. It can be shown that $\tilde{\mu} = \tilde{\mu}'$ with the help of thermodynamical relations. When the system is compressed (expanded), $\delta\rho(t) > 0$ ($\delta\rho(t) < 0$), and $\delta_c P(t) > 0$ ($\delta_c P(t) < 0$). It means that work is done on the system during one cycle of the vibration, which results in dissipation of the mechanical energy.

The average dissipation rate per baryon is given by

$$\left(\frac{dW}{dt}\right)_{av} = -\frac{1}{\tau}\int_0^\tau P(t)\frac{d}{dt}\delta\left(\frac{1}{\rho}\right)dt = \frac{|I_2'(0)|}{I_3(0)}\tilde{\mu}^2\tilde{\epsilon}(0)T^4\left(\frac{\delta\rho}{\rho_{eq}}\right)^2/\rho_{eq}. \quad (3)$$

T^4 dependence comes from the one of the $R(0)$ ($\propto T^5$) and the lowest order contribution of u_t ($\propto T^{-1}$). Finally, the damping time τ_d is estimated by dividing the vibrational energy E_{vib} by total average power of dissipation D, which is given by spatial integration of $\rho_{eq}(dW/dt)_{av}$ over the whole π^c condensed core. The dissipation of the vibrational energy brings about thermal heating of the system by the rate D, while thermal energy is carried away by neutrino emission whose luminosity is mainly given by the process (1) in the π^c condensed case.

Pulsation may be realized through dynamical phenomena in neutron stars. As an example, application of the above dissipation mechanism in the π^c condensed phase to the damping behavior of the observed time profile of the anomalous γ-ray burst on 5 March 1979 is discussed in Ref. 3.

REFERENCES

1. W. D. Langer and A. G. W. Cameron, *Astrophys. Space Sci.* **5**, 213 (1969).
2. T. Muto and T. Tatsumi, *Prog. Theor. Phys.* **83**, 499 (1990).
3. T. Tatsumi and T. Muto, this volume (1992).

Formation

S. E. Woosley† and T. A. Weaver‡
†Board of Studies in Astronomy and Astrophysics, UCO/Lick Observatory, University of California, Santa Cruz, CA 95064, USA; and ‡General Studies Group, Lawrence Livermore National Laboratory Livermore, CA 94550, USA

Theory of Neutron Star Formation

After a discussion of the physics that determines the mass of the iron core that collapses in a massive star, results from a grid of recent stellar evolution models are presented. In some cases these masses will be substantially altered as the core collapses and the star explodes, so we are led to consider the Type II supernova explosion mechanism. When all known effects are included, but still with substantial uncertainty, we "predict" neutron star masses in the range 1.15–2.0 $M_\odot$ with a preponderance between 1.3 and 1.6 $M_\odot$. Our paper concludes with a discussion of some recent developments in the study of accretion-induced gravitational collapse and a short summary of open issues.

1. INTRODUCTION

In the spirit of a workshop, this paper reports recent, mostly unpublished work on several problems relevant to neutron star formation. Section 2 discusses the physics which determines the mass of the iron core of a massive star and reports the results of a grid of stellar evolution calculations which span the expected range of masses, convective diffusion coefficients, and critical nuclear reaction rates. Next comes a brief review of the explosion mechanism, that dark veil that stands between our understanding of presupernova models, which are probably reasonably accurate, and the actual properties—mass, rotation rate, kick velocity, and magnetic field strength—of the young neutron star. Section 4 discusses a variant of core collapse which occurs when an accreting white dwarf collapses directly to a neutron star. Emphasis is given to the observational manifestation of such an event (gamma-ray burster or what?) and limits that can be placed on the occurrence of such events by their nucleosynthesis and gamma-ray line emission. The final section briefly summarizes, without attempting to answer, some critical remaining questions regarding neutron star birth.

2. NEUTRON STAR MASSES AT BIRTH

One frequently reads the simplistic statement in the literature that the iron core of a massive star collapses because it (a) exceeds the Chandrasekhar mass and (b) can no longer generate nuclear energy. Some people even go so far as to claim that this explains why neutron stars should have gravitational masses near 1.4 $M_\odot$. While all these statements are true in an approximate, general sense, they obscure some very interesting physics. Consider the traditional Chandrasekhar mass:

$$M_{ch} = 5.83\, Y_e^2, \tag{1}$$

which is 1.457 $M_\odot$ for $Y_e = 0.50$. There are numerous corrections that must be added (and subtracted) to this when considering a real stellar core, which after all is not an isolated white dwarf.

First, the central density is not infinite when the white dwarf becomes unstable to general relativistic gravity.[1] A structural adiabatic index somewhat greater than 4/3 is unstable when the post-Newtonian corrections are added, so really instability sets in at a central density of $\rho = 2.65 \times 10^{10}$ g cm^{-3}. Also, at this point, not all of the electrons are (special) relativistic, especially in outer layers. Taken together these corrections imply a 2.7% reduction in M_{ch} so that 1.457 $M_\odot$ becomes 1.418 $M_\odot$.

Second, the gas is not ideal. The pressure is reduced by the Coulomb corrections. To rough approximation

$$M_{ch} \approx M_{ch0}\left[1 - 0.0226\left(\frac{Z}{6}\right)^{2/3}\right] \tag{2}$$

which reduces 1.42 $M_\odot$ to 1.39 $M_\odot$ for a carbon white dwarf. For ^{56}Fe the Coulomb correction is larger and, moreover, $Y_e = 0.464$, so $M_{ch}(^{56}\text{Fe}) = 1.15\, M_\odot$, though, as we shall see later, no real iron cores have masses this small.

Third, the surface boundary pressure plays a very important role. This decreases the effective Chandrasekhar mass, but the effect varies as overlaying silicon- and oxygen-burning shells are ignited. Frequently a core may begin to collapse, but as it contracts, an overlaying shell of unburned fuel will be ignited. This generates energy, which raises the overlaying material to larger radii, relieving the boundary pressure on the core. At this point the core ceases contracting and may even expand. Additional burning shells may ignite or convective shells whose base is interior to the Chandrasekhar mass may extend outward beyond that mass. This extension of convective shells is very critical and can add a degree of nonmonotonic behavior to the iron core mass. That is, the size of the collapsing core may be as sensitive to the number of episodes of silicon shell burning as to the mass of the presupernova star.

Fourth, and very important for the core of a massive star, is the fact that such cores have finite entropy, that is, they are not completely degenerate. To rough approximation[2-4]

$$M_{ch} \approx M_{ch0}\left(1 + \frac{\pi^2 k^2 T^2}{\epsilon_F^2}\right), \quad (3)$$

where ϵ_F is the Fermi energy for relativistic degenerate electrons,

$$\epsilon_F = 1.11\,(\rho_7 Y_e)^{1/3}\ \text{MeV}. \quad (4)$$

The distinction between ϵ_F and the chemical potential, μ_e, is important in deriving the above equation. To a first approximation,[4]

$$\mu_e \approx \epsilon_F\left[1 - \frac{1}{3}\left(\frac{\pi k_B T}{\epsilon_F}\right)^2\right]. \quad (5)$$

The Chandrasekhar mass may also be expressed in terms of the electronic entropy per baryon

$$M_{ch} \approx M_{ch0}\left[1 + \left(\frac{s_e}{\pi Y_e}\right)^2\right] \quad (6)$$

where s_e in units of Boltzmann's constant, k, is[5]

$$s_e = \frac{S_e}{N_A k} = \frac{\pi^2 T Y_e}{\epsilon_F} \\ \approx 0.50 \rho_{10}^{-1/3}\left(\frac{Y_e}{0.42}\right)^{2/3} T_{\text{MeV}}, \quad (7)$$

when near complete relativistic degeneracy is assumed. For typical conditions during silicon burning in a massive star ($\rho_{10} = 0.01, Y_e = 0.46, T_{\text{Mev}} = 0.3$) this correction can increase the "Chandrasekhar mass" by 25%. At collapse the correction is smaller, typically 10%. The core must lose entropy to collapse.

Finally, there are minor corrections for the ion and radiation pressure. The star is not entirely supported by electrons. These corrections are small and can be neglected. Perhaps more important, but totally uncertain, is whether a correction should be made for rotation.

Note that in all these correction terms as well as in the fundamental equation for the Chandrasekhar mass, composition, Y_e, and entropy must be properly averaged over the whole stellar core. Typically in the presupernova models Y_e is 0.42 in the center and 0.48 at the edge, thus an "average" value might be 0.45. The composition is mostly iron (Z = 26). Thus the *minimum* core mass capable of collapse, assuming unrealistically that the entropy is zero, is $M_{min} \approx 1.09$ $M_\odot$. When comparing with the neutron star mass that is measured gravitationally, the mass that collapses must be reduced by the neutrinos that are lost; that is, the binding energy of the neutron star must be subtracted. This mass decrement is reasonably independent of the equation of state so long as a black hole is not formed and is, roughly independent of equation of state[6] (also, J. M. Lattimer, private communication),

$$\Delta M \approx 0.084 \left(\frac{M_{grav}}{M_\odot}\right)^2 M_\odot, \qquad (8)$$

or about a 15% reduction depending upon the mass. Thus one could, in principle, have neutron stars with gravitational masses as light as 1.0 $M_\odot$. In nature this never happens.

Including realistic entropy and surface boundary pressure corrections, that is, using the stellar models themselves, we shall see later that the iron core that collapses is typically between 1.3 and 2.0 $M_\odot$ with 1.3–1.6 being most typical of all but the rarest (most massive) stars. *However, even corrected for neutrino losses, the iron core mass is not the neutron star mass.* The prompt explosion mechanism does not work. The delayed explosion mechanism (§3) takes roughly a second to develop, during which mass accretes. Thus the final mass can only be determined by an accurate modeling of the explosion mechanism, which, as we all know, is still elusive. Successful delayed explosions, when they develop, typically have "mass cuts," in the few cases that can be sampled, that correspond to a radius of about 2500 km to 3500 km in the presupernova star. So the iron core itself may not be of as great an interest in determining the neutron star properties as the density gradient just outside that core. Only if the density falls off very abruptly at the edge of the core, as is more typical of supernovae on the lower end of the mass range, is the iron core mass itself representative.

And the story may not even end there. For Type II supernovae, there is a "reverse shock" that occurs when the exploding helium core runs into the hydrogen envelope. The sudden increase in ρr^3 leads to a reflection[7,8] that communicates the signal to the underlying material to slow down. In the moving material this appears as a shock. For SN 1987A and other Type II's, this shock arrives back at the center of the star after about an hour (blue supergiant[9]) to a day (red supergiant[7]) and leads to an unknown amount of material being decelerated below the escape velocity. Chevalier[10] estimates the mass in SN 1987A to be ~ 0.1 $M_\odot$, but the actual value is very uncertain. Woosley[9] even finds that, if the explosion energy is less than about 3×10^{50} ergs, solar mass quantities of material may fall back onto the core. This did not happen in SN 1987A, because we know the explosion energy there to have been much greater than 3×10^{50} ergs, but it could, depending on details of the delayed explosion mechanism, happen in other supernovae. Thus it is possible to make a substantial black hole in the middle of an optically brilliant supernova. All of the radioactivity would be lost, however (which is another reason we know that this did not happen in SN 1987A).

For the accretion-induced collapse of a white dwarf (§4) one obtains a nearly unique value of neutron star mass. The baryon mass at collapse is that of a white dwarf[11-13] having a central density of about 10^{10} g cm^{-3}, that is, about 1.39 $M_\odot$, implying a final gravitational mass near 1.23 $M_\odot$. This would also presumably be typical of neutron stars coming from the 8–12 $M_\odot$ mass range. The core that collapses there resembles closely a bare white dwarf surrounded by a low-density hydrogen envelope.

With all this in mind, it is now educational to examine some actual presupernova models of more massive stars. Recently we have begun a systematic exploration of the evolution of stars in the 10–80 $M_\odot$ range. In addition to the mass, various other parameters—the efficiency of semiconvection, the $^{12}C(\alpha,\gamma)^{16}O$ reaction rate, the metallicity, and mass loss rate—have all been varied to test the sensitivity of the results to uncertain physics. Some of the results of these calculations have been published,[14,15] but most of the detail is still in preparation for the refereed journals.[16] Table 1 gives some preliminary results.

In the table, the first number is the mass of the star, the letter specifies the prescription employed for convection. The letter "N" means that a very small amount of semiconvection was included in the calculation. The treatment of convection was almost strictly LeDoux, though a finite diffusion coefficient, roughly 10^{-4} that of radiation, was used.[14,16,17] The letter "S" means much more semiconvection was employed, a diffusion coefficient roughly 1000 times larger. This is more like, but not equivalent to, Schwarzschild convection. The last number is the value by which the critical rate for the nuclear reaction $^{12}C(\alpha,\gamma)^{16}O$ was multiplied (based on Caughlan and Fowler[18] as standard). The letter "T" stands for three halves. It turns out that a value near 1.7 for this number gives best agreement with the solar abundances,[15,16] though values in the range $\sim$0–5 are experimentally allowed.

Table 1 shows a number of interesting dependencies of massive stellar evolution. First, we note the dependence upon mass. Not too surprisingly, bigger stars make bigger iron cores and presumably more massive neutron stars as well. However, the behavior is not always monotonic (e.g., Models 13S2, 15S2, and 20S2). As remarked previously the number of convective shell burning episodes is critical (see also Ref. 19). A smaller rate for $^{12}C(\alpha,\gamma)^{16}O$ also makes smaller cores; though again this is a general trend and not a rule. Note Models 25S1 and 25S2 and Models 35S1 and 35S2, but also note Models 15S1 and 15S2 and Models 25S2 and 25S3. The trend toward smaller masses for smaller values of the rate arises from a lower entropy in stars that experience well-defined episodes of carbon and neon burning

TABLE 1 Fe Core Masses

Model	Fe Core	2500 km	3500 km	Model	Fe Core	2500 km	3500 km
12S2	1.37	1.47	1.56	15S1	1.41	1.43	1.50
13S2	1.54	1.55	1.66	25S1	1.45	1.63	1.73
15S2	1.30	1.50	1.58	35S1	1.40	1.61	1.71
20S2	1.66	1.85	2.06	15S3	1.30	1.53	1.60
25S2	1.95	1.90	2.24	25S3	1.52	1.52	1.58
30S2	2.04	1.88	2.23	35S3	1.91	1.90	2.12
35S2	2.03	1.85	2.18	15ST	1.63[a]	1.55	1.68
15N2	1.20[a]	1.34	1.38	25ST	1.61	1.72	1.83
20N2	1.69	1.53	1.69	35ST	1.63	1.84	2.06
25N2	1.49	1.71	1.80	25N1	1.37	1.51	1.59
30N2	1.68	1.84	2.04	35N1	1.62	1.56	1.69
35N2	1.63	1.85	2.06	15N3	1.25[a]	1.44	1.50
				25N3	1.53	1.76	1.90

[a] Implosive burning in progress. Iron core poorly resolved.

(which are bypassed when the carbon abundance is too small). But the effect is clearly obscured by the number of shell-burning episodes that occur.

The dependence upon convective algorithm is also striking. Most of the differences can be traced to the smaller CO core obtained when semiconvection is restricted (Fig. 1). This gives less heavy element production and smaller iron cores.

TABLE 2 Massive Stars with Mass Loss

Model	35WR	40WR	60WRA	60WRB	85WRA	85WRB
Initial M	35	40	60	60	85	85
Final M	15.2	11.1	4.25	6.65	8.34	9.71
$^{12}C(\alpha,\gamma)^{16}O$	2.7	3	2.7	1	2.7	1
M_{Fe}	1.45	1.42	1.40	1.46	1.90	1.59
M_{3500}	1.77	1.67	1.66	1.77	2.18	1.96

Finally, we note (Table 2 from Ref. 20) the possibility that very massive stars endowed with large mass loss rates may lose, not only all of their hydrogen envelope, but a portion of their helium and carbon-oxygen core as well. The final structure of stars such as 60WRA does not closely resemble what one would obtain from the evolution of an isolated helium core of any mass evolved without mass loss. That is, the 4.25 $M_\odot$ residual of Model 60WRA is very different from the end of the evolution of a 4.25 $M_\odot$ bare helium core. Most of the difference occurs because the carbon-oxygen core in the mass-losing calculation is much larger, roughly all of the star in Model 60WRA, though the surface abundance of helium remains nontrivial.

3. THE DELAYED SUPERNOVA MECHANISM

For many years it was hoped that a simple, purely hydrodynamical mechanism could be found to explain the explosion of Type II supernovae. This would reduce reliance upon more intricate calculations of neutrino transport and explain a common natural phenomenon in a simple, physically appealing fashion. Unfortunately, nature appears not to have taken this simple path. For the realistic equations of state of the day and for presupernova models calculated carefully by at least four independent groups, it appears that the so-called prompt mechanism does not work. Nevertheless, the events during the first 20 ms (sound crossing time for the core) following maximum compression remain exceedingly important because, just as the presupernova model set the conditions for the collapse, the propagation and initial death of the shock set the stage for what follows. We shall be skimpy on detail here and refer the interested reader to excellent reviews of the subject.[5,8]

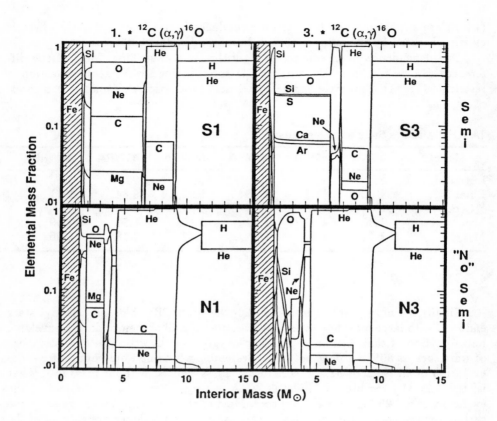

FIGURE 1 Four 25 $M_\odot$ presupernova models calculated with different values of the semiconvective diffusion parameter and rate for the $^{12}C(\alpha,\gamma)^{16}O$ reaction rate. Only the inner 15 $M_\odot$ of the 25 $M_\odot$ models are shown.

We begin our discussion after the prompt shock has already failed, stalling typically at a radius of a few hundred kilometers, where it has lost all outward velocity and become a standing accretion shock. A nearly stationary "neutrinosphere" (surface of near unit optical depth for neutrinos) initially develops at about 50 km, where the density $\rho \sim 10^{12}$ g cm^{-3} and the effective neutrino emission temperature is ~ 5 MeV. As time passes, the neutron star and its neutrinosphere shrink until the radius is only 10-20 km. It is the interaction of the neutrinos from this shrinking core, typically 3×10^{53} ergs of them, with the almost optically thin material above the neutrinosphere that ultimately, in a successful model, slows the accretion and turns material around again.[8,21-27]

The explosion of the infalling mantle is essentially a study in accretion physics. Matter passes through an accretion shock, is abruptly slowed, and "settles" onto the neutron star. This material behind the shock is in near hydrostatic equilibrium but still moving at thousands of kilometers per second. The rate at which material arrives at the shock can be estimated[28] by assuming that material in the mantle has a density distribution

$$\rho(r > R_s) = \frac{H}{r_o^3} \qquad (9)$$

which experiences an acceleration $\alpha^2 g = \alpha^2 \frac{GM(r)}{r^2}$. Neglecting pressure, which has the effect of making $\alpha < 1$, the velocity at radius r will, after some time, be a fraction of the local escape velocity, that is, $v(r) = -\alpha v_{esc}(r)$. This implies a density structure

$$\rho(r,t) = \frac{2}{3} \frac{H}{\alpha[2GM(r)]^{1/2}} t^{-1} r^{-3/2}. \qquad (10)$$

Hence the accretion rate, $4\pi r^2 \rho v$, is

$$\dot{M}(r) = \frac{8\pi}{3} H t^{-1}, \qquad (11)$$

which, interestingly, is independent of α. For $H \sim 10^{32}$ g, typical of massive stars, the accretion rate at the shock is a few tenths of a solar mass per second at a time of 1 s. As time passes, H increases slightly so that the accretion rate, including the t^{-1} dependence, stays near this value, at least until a black hole is formed. Numerically this is consistent with earlier calculations in which we[29] followed the failed explosion of a 25 $M_\odot$ star for 6 s. During that period, 1.8 $M_\odot$ accreted through the shock.

Most interesting for the delayed mechanism is what happens to this material after it passes through the shock and, in particular, the competition between energy absorption from neutrinos flowing out from the neutron star and emission of neutrinos by electron (and positron) capture. In order to smoothly merge with the neutron star, net dissipation (corrected for energy absorption) must carry away energy at a rate equal to the gravitational binding energy of the accreted material; that is $L_{dis} = \dot{M} \frac{GM_n}{R_n} \approx 2.5 \times 10^{52} (50 \text{ km}/R_n)$ ergs s^{-1} where R_n here is the radius of the neutron star, here assumed to be equivalent to the neutrinosphere radius, and $M_n = 1.5$ $M_\odot$. Here we have also assumed $\dot{M} \approx 0.3$ $M_\odot$ s^{-1}, which is characteristic only of high-mass stars ($M \gtrsim 15$ $M_\odot$). The dissipation is composed of two parts, photodisintegration and neutrino loss. The photodisintegration loss is $\dot{M}Q$, where $Q = 1.7 \times 10^{52}$ ergs per solar mass (assuming complete disintegration to nucleons; 8.8 MeV/nucleon), so, when $R_n \approx 30$ km for example, the neutrinos must carry off roughly 3×10^{52} erg s^{-1} in steady state. If they do not, then the neutron star will begin to "resist" the accretion. Loosely speaking, the sink will back up and the shock consequently will move outward in radius. This does not necessarily imply that any *mass* moves outward, though it certainly may do so *with or without any additional energy input*. An interesting case was considered in our 1982 paper.[29]

When all dissipation was turned off, the accretion shock moved rapidly out in radius to several thousand kilometers in 1.3 seconds (note an error in the labeling of the axes in Ref. 29; in our Fig. 11 it should be "R*10," not "R/10"). Positive velocities of roughly 1000 km s^{-1} were also observed behind the shock. While it is certainly not physical to set $L_{dis} = 0$, this calculation demonstrates the important role of dissipation in the accretion process.

Electron and positron capture on nucleons occurs in a narrow region above the neutrinosphere, so another condition for the sink not to back up is that the accretion rate into that loss region must equal the rate at the shock. If sufficient power is absorbed between the shock and the loss region to slow $\dot{M}$, material may also begin to move out. Bethe[8] suggests that only energy deposited outside the loss region can be effectively utilized in the explosion, but we do not agree. Decreasing the losses, even in a region where the net energy generation retains a negative sign, must be useful to explosion. However, it is very important that not too much of the energy absorbed in the "gain region" be advected into the loss region. Slowing this matter, however, takes far less than the 3×10^{52} ergs s^{-1} being lost in steady state to electron capture. The kinetic energy of the inward-moving material is very small compared with gravity, though the thermal energy is comparable. This is what the shock and gravitational compression after the shock have accomplished. Thus the farther out the energy is deposited beyond the loss region, the less work is required to reduce $\dot{M}$. Typically the internal energy of the matter is a few times 10^{18} ergs g^{-1} in the region of interest. Absorption of energy comparable to this *in the time that it takes the matter to move to the loss region* will begin to reverse the flow. Since the matter enters the loss region at a few thousand km s^{-1} and the region where it might gain energy by neutrino capture is only ~ 100 km in extent, the available time is short, a few hundredths of a second. Thus the energy absorbed from neutrinos must be $\sim 10^{20}$ ergs g^{-1}. This implies neutrino luminosities (in each flavor) of several times 10^{52} ergs s^{-1}. Values comparable to this are observed in the Wilson and Mayle models during the first second or so. It is this coincidence that allows the explosion to develop.

Clearly it is to the star's advantage to slow the accretion at the earliest possible moment. The less energy advected into the loss region, the more useful energy will be available for the explosion. The best way to halt the accretion is to increase the neutrino luminosity, and so the most successful models of Wilson and Mayle are those that amplify the early luminosity by considering convection in the neutron star. Convection outside the neutrinosphere also plays an important role, in part by cooling the loss region and reducing L_{dis} (Wilson, private communication), but also by allowing energy that would have been advected into the loss region to stay out of it.[8] Convection is probably an essential aspect of the delayed mechanism and the fact that others have not included it may account for the fact that no one, so far, has replicated the successful explosions of Wilson and Mayle.

Once the accretion is halted, a mass separation begins to develop as material actually begins to move outward. This decreases the efficiency for neutrino capture on nucleons and a second stage of heating commences where the bulk of the final supernova energy is developed. Scattering of neutrinos on pairs and, possibly, neutrino annihilation[30] play an important role here. The density becomes too low, except in a very narrow region near the neutron star, for capture on baryons to be important. Whether or not a full 10^{51} ergs of final kinetic energy is finally developed in the delayed explosion model is presently a point of some contention. Wilson (private communication, 1991) has run a 20 $M_\odot$ model a full 10 s beyond core bounce and obtained an explosion energy of 10^{51} ergs, but other model builders have yet to repeat the calculation or, for that matter, to obtain any explosion.

4. ACCRETION INDUCED COLLAPSE

Since the workshop the work presented on this subject at the meeting has been submitted for publication,[31,32] so the present discussion can be brief. Many groups have studied the possibility that accretion at certain rates onto a white dwarf may lead directly to neutron star formation. The event could occur, for appropriate conditions, either for a carbon-oxygen white dwarf or a neon-oxygen white dwarf and the implosion might be accompanied by very little mass ejection or optical display. Most active in studying these sorts of models have been Nomoto and his colleagues[11] and Canal, Isern, and Labay.[12,13]

What we have recently examined are the details of flame propagation,[31] the hydrodynamics of the collapse, and the first second of the neutron star's life.[32] In the first case a critical central density at nuclear ignition has been determined, about 6×10^9 g cm^{-3} for carbon-oxygen dwarfs and 8×10^9 g cm^{-3} for neon-oxygen ones, above which collapse to a neutron star is likely. For larger densities, electron capture behind the flame leads to its stabilization against Rayleigh-Taylor modes larger than a certain (small) wavelength. This keeps the effective speed of the flame slow enough, typically a few percent of the sound speed at most, that explosion does not occur. At lower densities, the normal conductive speed of the flame actually becomes smaller, but the increased range of length scales susceptible to the Rayleigh-Taylor instability more than compensates. The flame is born slow but rapidly accelerates due to the highly nonlinear increase in its surface area. Ultimately it reaches a substantial fraction of the sound speed and a Type Ia supernova results. For models in the present literature, it appears that neon-oxygen dwarfs will always collapse, whereas carbon-oxygen dwarfs may either collapse or explode depending upon the accretion rate.[11,13]

Once the decision to collapse has been made, negative velocities, whose origin may be traced to the pressure deficit from electron capture, begin to accelerate. Electron capture is augmented by photodisintegration, and the subsequent evolution of the core resembles closely that of the collapsing iron core of a massive star. A steady burning front persists briefly at a radius of several hundred kilometers ultimately to be swept inward as the rest of the white dwarf falls through the front, but the energy release by nuclear reactions is of little consequence. Collapse continues to well beyond nuclear density ($\rho_{max} = 6.1 \times 10^{14}$ g cm^{-3}; $T_c = 19.3$ MeV) and a bounce shock forms at 0.75 $M_\odot$ (13 km). Meanwhile collapse continues in the outer layers, falling though a flame at ~ 0.8 $M_\odot$. About 6 ms after maximum central compression the shock stalls at 1.25 $M_\odot$ (150 km) and becomes an accretion shock. The accretion shock reaches the outer zones 50 ms later while they are still located at about 700 km.

Over the next few hundred milliseconds a dynamic neutrino-energized wind begin to blow from the surface of the collapsing star. Typically, at say 200 ms after maximum central compression, the neutrinosphere is situated at 30 km with a luminosity of about 10^{53} erg s^{-1} divided equally among the six varieties of neutrinos (e, μ, τ, and their antiparticles). Capture of a portion of these neutrinos by nucleons situated above the neutrinosphere (corrected for the inverse process of electron and positron capture on nucleons) plus neutrino-electron scattering deposits about 5×10^{50} erg s^{-1} of internal energy in the optically thin (to neutrinos) material. The atmosphere, which is in near hydrostatic equilibrium at its base, expands in response to the energy deposition. A steady mass loss rate of about 0.005 $M_\odot$ s^{-1} is maintained over the next second or so for a total mass loss near 0.01 $M_\odot$. Terminal velocities of this material are about 5%–10% the speed of light.

Woosley and Baron[32] also calculated that the ejected material would be quite neutron-rich. In fact one of the most abundant isotopes ejected was ^{88}Sr, which composed about one-sixth of the 0.004 $M_\odot$ of ejecta having Y_e near 0.45. Based upon this copious production of a rare nucleus, they estimate that accretion-induced collapse must be a rare phenomenon (compared say with neutron star production by Type II and Ib supernovae), probably occurring no more frequently than once every thousand years in the Galaxy. Note that this limit is about a factor of 10 less restrictive than originally presented at the meeting. The change is due to the incorporation of the ^{4}He$(\alpha n, \gamma)^9$Be reaction in the network and a more careful consideration of the distribution of neutron excess in the ejecta.

Interesting synthesis of radioactive nuclei, including perhaps a few thousandths of a solar mass of ^{56}Ni, also occurred, which might lead to a detectable (in our Galaxy) electromagnetic display. However, no bright γ-ray burst was produced, either by shock breakout or by neutrino annihilation[30] above the photosphere, contrary suggestions by Paczynski[33] and Ramaty and Dar[34] not withstanding.

5. SOME OPEN ISSUES

We have treated in this paper chiefly those aspects of supernovae and presupernova stars that determine the *mass* of the neutron star and have dwelt upon the basic mechanism that might be responsible for the explosion. Several other interesting issues deserve much greater attention than they have received in the past. These are the evolution of the magnetic field, the role of rotation, and the possibility of "kick" velocities.

On the subject of magnetic fields, it is worth noting that the core that collapses has, within the week, experienced convective silicon burning and, a few months earlier, oxygen burning. One would expect this convection to influence the nature of the magnetic field, perhaps leading to a tangled field in the core. Even more relevant may be the accretion of up to several tenths of a solar mass of material as the delayed explosion develops. This material is also thought to be convective during the accretion process.[8] Does the same field permeate the core and the accreted material? Does the field emerge instantly following the explosion? And is all this affected by differential rotation during the collapse? Less mass is accreted in the explosion of 8–12 $M_\odot$ stars and none in the case of accretion induced collapse. Should the surface magnetic fields differ in the neutron stars derived from such events?

Continuing on the subject of rotation, it is a total mystery how much angular momentum is contained within the collapsing core. The specific angular momentum on the main sequence, even if the star rotates rigidly, would imply a neutron star born rotating at breakup, but the core goes through six major convective burning episodes which transport angular momentum out at an unknown rate. Perhaps it is reasonable to assume that the core rigidly rotates out to the extremity of the oxygen convective shell or beyond, in which case the neutron star in a massive star may *not* be born rotating at breakup. If, on the other hand, most neutron stars are born as millisecond rotators, how and when do they get slowed down?

Finally, with regard to kicks, the delayed explosion mechanism offers the possibility that the neutrino luminosity from one side of the neutron star to the other may vary. Convection[26] and even macroscopic core overturn[35] may contribute to this asymmetry.[36] No sufficiently accurate calculation presently exists to determine this effect accurately, but for now we point out that a 0.1% asymmetry in neutrino emission from one hemisphere to another could lead to a kick velocity of 50 km s^{-1}. This results because about 15% of the mass is emitted at the speed of light in the form of neutrinos.

ACKNOWLEDGMENT

This work has been supported by the NSF (AST 88-13649), NASA (NAGW-1273 and 2525) and the DOE (W-7405-Eng-48).

REFERENCES

1. S. L. Shapiro and S. A. Teukolsky, *Black Holes, White Dwarfs, and Neutron Stars* (John Wiley: New York), p. 156 (1983).

2. S. Chandrasekhar, *Stellar Structure* (Dover: New York), p. 392 (1938).
3. F. Hoyle and W. A. Fowler, *Ap. J.*, **132**, 565 (1960).
4. E. Baron and J. Cooperstein, *Ap. J.*, **353**, 597 (1990).
5. J. Cooperstein and E. Baron, in *Supernovae*, ed. A. G. Petschek (Springer Verlag: New York), p. 213 (1990).
6. A. Burrows, in *Supernovae*, ed. A. Petschek (Springer Verlag: New York), p. 159 (1990).
7. T. A. Weaver and S. E. Woosley, *Ninth Texas Symp. on Rel. Ap.*, published in *Ann. N. Y. Acad. Sci.*, **336**, p. 335 (1980).
8. H. A. Bethe, *Rev. Mod. Phys.*, **62**, No. 4, 801 (1990).
9. S. E. Woosley, *Ap. J.*, **330**, 218 (1988).
10. R. A. Chevalier, *Ap. J.*, **346**, 847 (1989).
11. K. Nomoto and Y. Kondo, *Ap. J. Lettr.*, **367**, L19 (1991).
12. R. Canal, J. Isern, and J. Labay, *Ann. Rev. Astron. Ap.*, **28**, 183 (1990).
13. J. Isern, R. Canal, and J. Labay, *Ap. J.*, in press (1991).
14. S. E. Woosley and T. A. Weaver, in *Les Houches, Session LIV: Supernovae*, ed. J. Audouze et al. (Elsevier Science Publishers: North-Holland), in press (1991).
15. T. A. Weaver and S. E. Woosley, *BAAS*, **23**, No. 2, 975 (1991).
16. T. A. Weaver and S. E. Woosley, in preparation for *Ap. J.* and *Rev. Mod. Phys.* (1992).
17. S. E. Woosley and T. A. Weaver, *Physics Reports*, **163**, 79 (1988).
18. G. A. Caughlan and W. A. Fowler, *Atomic Data and Nuclear Data Tables*, **40**, 238 (1988).
19. Z. Barkat and A. Marom, in *Supernovae*, Volume 6 of the Jerusalem Winter School for Theoretical Physics, ed. J. C. Wheeler, T. Piran, and S. Weinberg (World Scientific: Singapore), p. 95 (1990).
20. S. E. Woosley, N. Langer, and T. A. Weaver, in preparation for *Ap. J.* (1991).
21. J. R. Wilson, in *Numerical Astrophysics*, ed. J. M. Centrella, J. M. LeBlanc, and R. L. Bowers (Jones and Bartlett: Boston), p. 422 (1985).
22. H. A. Bethe and J. R. Wilson, *Ap. J.*, **295**, 14 (1985).
23. R. W. Mayle, Ph. D. thesis, Lawrence Livermore National Laboratory reprint, UCRL−53713 (1985).
24. J. R. Wilson, R. W. Mayle, S. E. Woosley, and T. A. Weaver, *Ann. N.Y. Acad. Sci.* **470**, 267 (1986).
25. R. W. Mayle and J. R. Wilson, *Ap. J.*, **334**, 909 (1988).
26. R. W. Mayle, in *Supernovae*, ed. A. G. Petsheck (Springer Verlag: New York), p. 267 (1990).
27. R. W. Mayle and J. R. Wilson, in *Supernovae*, Proceedings of the Tenth Santa Cruz Summer Workshop, ed. S. E. Woosley (Springer Verlag: New York), p. 333 (1991).
28. J. H. Cooperstein, H. A. Bethe, and G. E. Brown, 1984, *Nuc. Phys. A*, **429**, 527 (1984).

29. S. E. Woosley and T. A. Weaver, in *Supernovae: A Survey of Current Research*, ed. M. J. Rees and R. J. Stoneham (D. Reidel: Dordrecht), p. 79 (1982).
30. J. Goodman, A. Dar, and S. Nussinov, *Ap. J. Lettr.*, **314**, L7 (1987).
31. F. X. Timmes and S. E. Woosley, in press *Ap. J.* (1992).
32. S. E. Woosley and E. Baron, in press *Ap. J.* (1992).
33. B. Paczynski, *Ap. J.*, **363**, 218 (1990).
34. R. Ramaty and A. Dar, preprint, also published in proceedings of COSPAR meeting (1990).
35. L. Smarr, J. R. Wilson, R. T. Barton, and R. L. Bowers, *Ap. J.*, **246**, 515 (1981).
36. S. E. Woosley, in *The Origin and Evolution of Neutron Stars*, IAU Symp. 125, eds. D. Helfand and J.-H. Huang (D. Reidel: Dordrecht), p. 255 (1987).

K. Koyama
Department of Astrophysics, Nagoya University, Japan

X-Ray Observations of SNRs: Birth of Neutron Stars and Their Evolution

The *Ginga* satellite discovered several important facts related to neutron stars. The following topics are presented here: (1) most of the Crab-like SNRs so far observed have power-law spectra with photon index of about 2, (2) the search for low-mass X-ray binaries (LMXBs) near the Galactic center revealed that the total number of LMXBs does not increase drastically, (3) observations of newly discovered X-ray pulsars in the 5-kpc Arm suggest that they have smaller magnetic fields than those of usual X-ray pulsars, and (4) a peculiar pulsar in the SNR G109-1.0 was found to have a relatively small magnetic field in spite of its young age.

1. INTRODUCTION

The X-ray emission from supernova remnants (SNRs) has mainly two origins: One is thermal emission from a thin hot plasma heated by a shock wave, and the other is nonthermal synchrotron emission powered by rotational energy of a central neutron star. In general, the former emission shows shell structure, while the latter shows filled center structure. From the morphology, SNRs are often classified into shell-like and Crab-like SNRs.

Since neutron star are born in a supernova explosions, the study of Crab-like SNRs, which may include neutron stars is of particular interest for the birth and evolution of neutron stars.

Comprehensive work on X-ray emitting SNRs has been carried out with the *Einstein Observatory* by Seward[1] and others. They observed about 80 Galactic SNRs, and 47 SNRs were found to be X-ray emitters. In the X-ray catalogue of SNRs, about 20 SNRs belong to the Crab-like class.[1] However, only 7 SNRs have been found to exhibit pulsations either in X-ray or radio bands. Therefore we need to search for pulsations from other Crab-like SNRs. In section 2, we report timing and spectral study of Crab-like SNRs with the *Ginga* satellite.

Recently progress has been made in neutron star physics by the discovery of rapidly rotating neutron stars (the millisecond pulsars) in globular clusters. Since most of these millisec pulsars are in binary systems with relatively weakly magnetized neutron stars, several authors have proposed the model that low mass X-ray binaries (LMXBs) will evolve to milli-sec pulsars after exhausting their accreting gas from the companion stars. This model, however, contains a serious problem in that the birthrate of millisec pulsars is estimated to be one or two orders of magnitude larger than that of the progenitor LMXBs. In order to reestimate the total number of LMXBs, we will report the results of a search for low-luminosity LMXBs near the Galactic center. One of the remarkable discoveries of the *Ginga* satellite is the detection of cyclotron absorption features from many of the binary X-ray pulsars. This established the picture that the massive X-ray binary sources (X-ray pulsars) have magnetic fields of more than 10^{12} gauss. This issue is discussed in detail by Makishima (this volume). In sections 4 and 5, we will discuss a class of binary X-ray pulsars with smaller magnetic fields than those of typical X-ray pulsars.

2. GINGA OBSERVATIONS OF CRAB-LIKE SNRS
2.1 PULSAR SEARCH

Aoki et al. (private communication, 1990) have carried out folding and Fourier analyses of the *Ginga* data from four Crab-like SNRs: Kes 73, CTB 80, Vela-X, and 3C 58. No positive result was obtained from these SNRs. They set an upperlimit for the pulse amplitude of less than 5%. The flux ratios of pulse amplitude to the total synchrotron nebula in the 2–10 keV range are all less than 10%. We note that the ratios of the Crab nebula, MSH 15-52, and PSR 0540-69 are about 10%, 10%, and 40%, respectively. Therefore these upper-limits are significantly smaller than those of known pulsars. One possibility is that the X-ray beams from the central pulsars in these SNRs are almost out of the line of sight.

2.2 X-RAY SPECTRUM

The X-ray spectrum from the Crab nebula shows a nonthermal power-law spectrum with a photon index of about 2.0. Another Crab-like SNR, Vela-X, shows central X-ray emission which is from a synchrotron nebula. The X-ray spectrum of this synchrotron nebula has been reported as a power-law shape with a photon index of about 2. For MSH 15-52, Kawai et al. (private communication, 1990) observed the X-ray spectrum of the pulsed component and nonpulsed component separately. The photon index of the pulsed component is smaller than 2.0, while that of the nonpulsed component is about 2.0. Asaoka and Koyama[2] have observed other Crab-like SNRs (3C 58, G21.5-0.9, and CTB 87), and found that they all have a power-law spectrum with photon index of about 2 (Fig. 1).

A theoretical question is why all or most of the Crab-like SNRs have nearly the same photon index although their ages and luminosities are differ. Apart from this question, our results may provide another empirical tool to search for Crab-like SNRs. Using the X-ray spectrum, we can identify the synchrotron nebula even if its central pulsar beam is out of the line of sight, because the X-ray flux from the synchrotron nebula would be independent of the beaming angle of the central pulsar. Therefore this method should be very powerful for the statistical study of the birthrate of neutron stars in SNRs.

3. LOW MASS X-RAY BINARIES AND MILLISECONDS PULSARS

The progenitors of millisec pulsars have been suggested to be LMXBs. However, several authors have pointed out that the birthrate of millisec pulsars could be one or two orders of magnitude larger than that of the progenitor LMXBs. In order to estimate the total number of LMXBs, Yamauchi and Koyama [3] have carried out survey observations near the Galactic bulge region and have found nine new X-ray sources above the detection limit of about 10^{35} ergs s^{-1}. They plotted a color-color diagram of these new X-ray sources together with the catalogued X-ray sources. We see seven new sources are located near the region of catalogued LMXBs in the color-color diagram. Thus we can assume that seven out of the nine new X-ray sources are likely to be LMXBs. In the same region of the Galactic center survey, we found catalogued LMXBs which are brighter than 10^{36} erg s^{-1}. Therefore, we suggest that the total number of LMXBs does not increase drastically with a more sensitive survey of the Galaxy. This conclusion is consistent with the result by Grindlay and Herz[4] in which they have summarized and classified serendipitous sources with the *Einstein Observatory* along the Galactic plane. Thus a discrepancy between the birthrates of millisec pulsars and LMXBs remains.

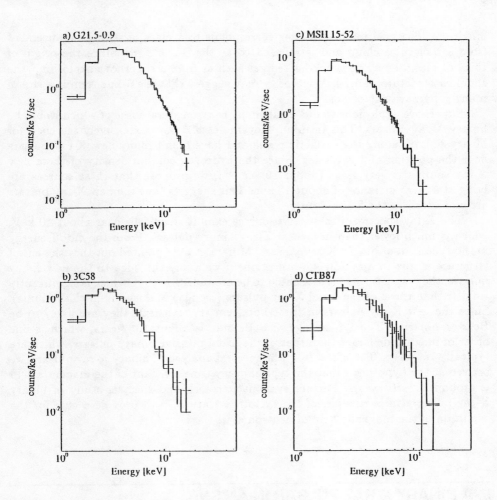

FIGURE 1 The X-ray spectra of four Crab-like SNRs[2]

4. A COLONY OF X-RAY PULSAR AND SNRs

The Tenma satellites have discovered intense 6.7-keV line emission along the inner Galactic plane.[5,6] Since the 6.7-keV line is likely due to highly ionized iron atoms, the origin is optically thin hot plasma. The temperature of the plasma was determined to be about 5–10 keV. From this discovery, Koyama et al.[7] have found that the 6.7-keV line emission could be a powerful tool to search for the Galactic thin hot plasma and have carried out a Galactic plane survey using the 6.7-keV iron line for the first time. From the scan profile, we find a shoulder around 30 deg of Galactic longitude. The position near 30 deg corresponds to a place of strong enhancement from a molecular cloud and is referred to as the 5-kpc Arm. Since the origin of this hot plasma may be shock-heated gas such as from a supernova explosion,[8] the shoulder structure of 6.7-keV intensity may suggest that the 5-kpc Arm is a site of many supernova explosions.

Supernova explosions would make many neutron stars, some of which would be binary X-ray pulsars. This in mind, Koyama et al.[9] have extensively searched for binary pulsars along the Galactic plane and have found many new X-ray pulsars near the position of $\ell = 30$ deg. Since the hydrogen column density of these new X-ray sources is very large (about $10^{23} cm^{-2}$), we propose that these sources are lying at a large distance of about 10 kpc. This suggests that the new X-ray pulsars are embedded in the 5-kpc Arm.

The X-ray spectra of these new sources exhibit steep falls near about 10 keV, which is much lower than in typical X-ray binary pulsars. From the cutoff energy in the spectrum of binary X-ray pulsars, Mihara et al.[10] pointed out that the cutoff structure is due to a cyclotron absorption. If we apply this idea to the new X-ray pulsars, the magnetic field is estimated to be around 10^{11}–10^{12}, which is significantly smaller than those of the usual X-ray pulsars (see also Makishima in this volume). Since the new X-ray pulsars in the 5-kpc Arm are transient, they are likely to be Be star binaries. The typical age of a Be star is about 10^7 years, which is one order of magnitude larger than that of the usual massive binary pulsars, which are typically 10^6 years. Therefore, from the spectral analysis of binary X-ray pulsars, we arrived at the hypothesis that the typical decay time constant of the magnetic field is around 10^6–10^7 years. Further systematic search and spectral study of binary X-ray pulsars (in particular, of the cyclotron feature), will provide a clue for the evolution of the magnetic field of neutron stars.

5. A BINARY X-RAY PULSAR IN AN SNR

Although most scientists believe that X-ray binary pulsars are born in supernova explosions of massive binary systems, no compelling facts for this scenario have been found so far. The X-ray pulsar 1E2259+586 is the first example of a possible binary pulsar lying in an SNR. This pulsar was found in the center of a young SNR, G109–1.0, by Fahlman and Gregory (Fig. 2).[11] No orbital motion from 1E2259+586 has been found (see, e.g., Koyama et al.[12]). However, since the pulse period change is too small to power the X-ray flux, 1E2259+586 would not be an isolated neutron star but would be a binary X-ray pulsar. The spin period history shows a long-term spin-down trend with a small rate of $\dot{P} = 3 \times 10^{-13}$ yr yr^{-1} (Iwasawa et al., this volume). The slow increase of the pulse period indicates that the accretion torque is small although the X-ray luminosity is moderately large (10^{36} ergs s^{-1}). This may be a case of the Alfvén radius being nearly equal to the corotation radius. Using this relation, Koyama et al.[13] suggested that the magnetic field is about 5×10^{11}

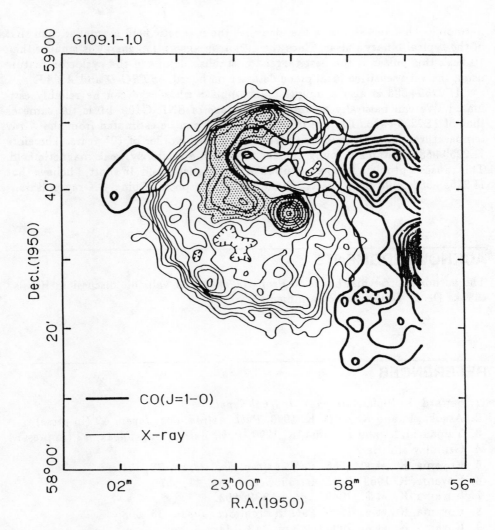

FIGURE 2 The radio and X-ray map of SNR G109–1.0 and X-ray pulsar 1E2259+586 (after Tatematsu et al.[14])

gauss. This value is consistent with the unusually soft X-ray spectrum, if we assume the soft spectrum is due to cutoff structure by the cyclotron absorption. Koyama et al.[12] pointed out a possible cyclotron line at about 7 keV. From the follow-up observations made in 1989 and 1990, Iwasawa et al. (this volume) found a further hint of a cyclotron absorption feature near 6 keV and near 12 keV as a second harmonic. This also supports the idea that the magnetic field is smaller than that of the typical binary pulsar. Since the cyclotron absorption energy is smaller than 10 keV, this pulsar is the best target to establish details of the cyclotron feature using the very sensitive focal plane detector on board *ASTRO-D* and *AXAF*.

1E2259+586 is also a unique binary pulsar whose age can be reliably estimated. We can naturally assume that the age of SNR G109-1.0 is the same as that of 1E2259+586. The age of SNR G109-1.0 can be estimated from the X-ray temperature, diameter, and luminosity and is of the order of 10^4 years. Therefore 1E2259+586 contains a young neutron star with a relatively weak magnetic field. This causes a problem for the evolution of the magnetic field. In short, I believe that 1E2259+586 could be a Rosetta Stone for many aspects of binary X-ray pulsars.

ACKNOWLEDGMENTS

The author express his thanks to Prof. Hayakawa for valuable discussion. He also thanks Dr. K. Leahy for her comments.

REFERENCES

1. Seward, F. 1990, *Astrophys. Journal Suppl.*
2. Asaoka, I., and Koyama, K. 1990, *Publ. Astron. Soc. Japan*, **42** (in press)
3. Yamauchi, S., and Koyama, K. 1990, *Publ. Astron. Soc. Japan*, **42** (in press)
4. Grindlay and Herz
5. Koyama, K., et al. 1986, *Publ. Astron. Soc. Japan*, **38**, 121.
6. Koyama, K. 1989, *Publ. Astron. Soc. Japan*, **41**, 667.
7. Koyama, K., et al. 1989, *Nature*, **339**, 603.
8. Koyama, K., et al. 1986, *Publ. Astron. Soc. Japan*, **38**, 503.
9. Koyama, K., et al. 1990, *Nature*, **343**, 148.
10. Mihara, G., et al. 1990, *Nature*.
11. Fahlman, G. G., and Gregory, P. C. *Nature*, **293**, 202.
12. Koyama, K., et al. 1989, *Publ. Astron. Soc. Japan*, **41**, 461.
13. Koyama, K., et al. 1987, *Publ. Astron. Soc. Japan*, **39**, 799.
14. Tatematsu, K., et al. 1989, *Astron. Astrophys.*

T. Takatsuka
College of Humanities and Social Sciences, Iwate University, Ueda 3-18-34, Morioka 020, Japan

Equation of State of Dense Supernova Matter and Newborn Hot Neutron Stars

The equation of state (EOS) of dense supernova matter is constructed with the finite-temperature Hartree-Fock approach with the effective interaction. The EOS is found to be remarkably stiffer than that of usual cold neutron matter. On the basis of this EOS, characteristic features of neutron stars at their birth are demonstrated. The releasable gravitational energy and the extent of the spin-up, as a result of the evolution to the cold neutron stars with or without pion condensation, are investigated. A comment is given on the consequence of the spin-up for the maximum rotation rate of cold neutron stars, and the releasable energy is discussed in relation to the neutrino events from SN 1987A.

1. INTRODUCTION

It is believed that a prototype of a neutron star with a radius $\sim 10^2$ km is formed in the central part of the collapsing supernova core as a result of the bounce and successful supernova explosion, although the relevant mechanism is not well understood yet. The proto–neutron star thus formed contracts rapidly (from radius $\sim 10^2$ to $\sim 10^1$ km) and very soon goes into quasi-hydrostatic equilibrium. For instance, a simulation study shows that the rapid contraction almost terminates within the first 0.1–1 s after the bounce.[1] Therefore we suitably define the birth of a neutron star at this stage. Afterward, due to the diffusion of neutrinos, this hot neutron star just born cools down and gradually contracts in a time scale of 10–20 s toward a typical cold neutron star.

Study of hot neutron stars at birth is of particular interest not only from the astrophysical viewpoint but also from the viewpoint of a new form of dense matter under extreme conditions. The matter from which neutron stars at the birth era are made, namely, supernova matter, is characterized by almost constant entropy per baryon ($S \simeq 1$–1.5)[2] and also by a high and almost constant lepton fraction ($Y_l \simeq 0.3$–0.4)[2,3] throughout the density ρ. In addition, the maximum density would amount to several times nuclear density ρ_0 (=0.17 nucleons/fm$^3 \simeq 2.8 \times 10^{14}$ g/cc), and at high densities the temperature T could be as high as $T \simeq 10$–50 MeV. These are caused by the important effect that neutrinos are trapped in the supernova core when the ρ exceeds about 10^{11-12} g/cc,[4,5] and the collapse proceeds adiabatically. These characteristics are clearly distinguished from those of ordinary neutron star matter responsible for typical cold neutron stars as well as symmetric nuclear matter associated with finite nuclei; in the former $T \lesssim 10^{-2}$ MeV, $S \simeq 0$, and $Y_l \lesssim 0.05$, and in the latter $\rho \simeq \rho_0$, $T = 0$, $S = 0$, and $Y_l = 0$.

In this report, on the basis of our recent work,[6,7] we discuss the characteristics of dense supernova matter and the consequences for hot neutron stars at birth. The equation of state (EOS) of dense supernova matter is derived by solving the finite-temperature Hartree-Fock equations, and hot neutron star models are obtained by applying the EOS to the TOV equation. We ask how newborn neutron stars are different from usual cold ones, to what extent they are spun up by the contraction in the subsequent cooling stage evolving to cold stars, and how much gravitational energy they can release. In the discussion, we include the interesting case where pion condensation takes place in cold neutron stars as a result of the cooling. The consequence of the spin-up on the maximum rotation rate and the possible relation of the energy release to the neutrino events from SN 1987A are also discussed.

2. OUTLINE OF APPROACH
2.1 EOS OF DENSE SUPERNOVA MATTER

For lower densities $\rho \lesssim \rho_0$, a reliable EOS has been given[8] and here we concentrate on getting the EOS at higher densities $\rho \rho_0$, for which realistic studies do not exist yet. Dense supernova matter with $\rho \rho_0$ is primarily composed of neutrons (n), protons (p), relativistic electrons (e), and degenerate electroneutrinos (ν). Keeping to the essentials of the matter, we neglect neutrinos ($Y_l = Y_e + Y_\nu \to Y_e$) and make the simplification that the matter is composed of n, p and e, with the fractions $1 - Y_p$, $Y_p = Y_e$ (because of charge neutrality) and $Y_e = Y_l$, respectively. This is a reasonable approximation since Y_ν is expected to be small compared with $Y_l (\simeq$

0.3–0.4); $Y_\nu < 0.06$ at $\rho \lesssim \rho_0^{2.8}$ and $Y_\nu/Y_l \sim 0.2$ at $\rho \simeq (1\text{–}5)\rho_0$.[9] In addition, the energy contribution from ν would be absorbed into that of e by taking $Y_e (= Y_l)$ in the range 0.3–0.4. Then the central problem in getting the EOS of dense supernova matter is how to derive the thermodynamic quantities, such as the internal energy E_N, entropy S_N and pressure P_N, for hot and dense asymmetric nuclear matter with the asymmetry parameter $x = (N - Z)/A = 1 - 2Y_p$. The basic idea of our approach[10,11] consists of (a) solving exactly the finite-temperature Hartree-Fock equations under the conservation of total nucleon number and (b) introducing the effective two-nucleon interaction $\tilde{V}$ constructed at $T = 0$. Point (a) is to ensure the thermal equilibrium in variational theory and (b) is to take account of the basic nucleon correlations. The validity of making extended use of $\tilde{V}$ even in the $T > 0$ case has been proved by Hiura and the present author;[11] the temperature dependence of effective interaction has been found to be very weak by performing the G-matrix calculation at $T > 0$. In the sense $\tilde{V} = gV$ with V being the two-nucleon potential, our approach corresponds to the variational method by Schmidt and Pandharipande[12] with introducing the temperature-independent correlation function $g(r)$ of Jastrow type, but is more exact because we are free from the assumption of the single-particle spectra at $T > 0$, which is inevitable in their approach.

We arrive at the EOS by the following steps.

(A) We treat the Hartree-Fock equations both for neutron matter ($x = 1$) and symmetric nuclear matter ($x = 0$) and obtain E_N, S_N, and P_N for these matters as functions of ρ and T. Those for asymmetric nuclear matter ($x = 1 - 2Y_p$) associated with supernova matter are derived by the interpolation between the $x = 1$ and $x = 0$ cases: for example, $E_N(x) = E_N(x = 0) + [E_N(x = 1) - E_N(x = 0)]x^2$.

(B) Similarly, the E_e, S_e, and P_e for electron part are calculated for hot relativistic electron gas with the density $\rho_e = Y_e\rho$ and T, by switching off the Coulomb interaction. Then by the addition, the E, S, and P for the total system are obtained: $E = E_N + E_e$ and so on.

(C) Finally, the isoentropy nature of supernova matter is taken into account. The isothermal EOS in (B) is converted into the isoentropy EOS by using the ρ-T relation under constant entropy: $E(\rho; T, x) \to E(\rho; S = \text{const.}, x)$ and $P(\rho; T, x) \to P(\rho; S = \text{const.}, x)$, which gives the EOS of dense supernova matter.

More details of (A) are as follows. The Hartree-Fock equations are expressed as

$$\varepsilon_\alpha = t_\alpha + \sum_\beta f_\beta \langle \alpha\beta | \tilde{V} | \alpha\beta - \beta\alpha \rangle, \tag{1}$$

$$\rho = \sum_\alpha f_\alpha / \Omega, \tag{2}$$

$$f_\alpha = 1/[1 + \exp[(\varepsilon_\alpha - \mu)/T]], \tag{3}$$

where ε_α is the single-particle energy with α (and β) specifying the nucleon state, t_α the kinetic energy, f_α the occupation probability, μ the chemical potential shifted by the constant energy term appearing in the ρ-dependent Hartree-Fock theory, and Ω the normalization volume. From the solutions, ε_α and f_α, we have

$$\tilde{E}_N = \sum_\alpha f_\alpha(\varepsilon_\alpha + t_\alpha)/2A, \qquad (4)$$

$$\tilde{S}_N = -\sum_\alpha \{f_\alpha \ln f_\alpha + (1-f_\alpha)\ln(1-f_\alpha)\}/A, \qquad (5)$$

$$\tilde{P}_N = \rho^2 \partial \tilde{F}_N/\partial \rho \quad \text{with} \quad \tilde{F}_N = \tilde{E}_N - T\tilde{S}_N, \qquad (6)$$

where $\tilde{F}_N$ is the free energy and A the total nucleon number in Ω. As for $\tilde{V}$, we use the G0-force by Sprung and Banerjee[13] for $\tilde{V}(x=0)$ and the G0-version constructed in neutron matter[14] for $\tilde{V}(x=1)$. These are based on the G-matrix calculation with the RSC potential. Here we pay attention to the observational constraint that the EOS of neutron matter at $T=0$ relevant to usual neutron stars should be stiff to sustain the neutron star mass, $1.44M_\odot$, of PSR 1913+16.[15] Our EOS based on the RSC potential ($\tilde{E}_N$ and $\tilde{P}_N$ for $x=1$ and $T=0$), however, is not stiff enough to satisfy the condition. To cover this situation, we adopt the EOS by Wiringa, Fiks, and Fabrocini[16] ($E_0(\rho; T=0, x=1)$ and $P_0 = \rho^2 \partial E_0/\partial \rho$) as a standard EOS for cold neutron stars, which has been obtained using the Urbana v14 two-nucleon potential plus phenomenological three nucleon interaction. The E_0 is one of the modern and realistic EOSs and gives the maximum mass $\sim 1.82M_\odot$. To be consistent with this choice, our $\tilde{E}_N$ (and also $\tilde{P}_N$) is modified as

$$\tilde{E}_N \to E_N(\rho; T, x) = \tilde{E}_N(\rho; T, x) + [E_0(\rho; T=0, x) - \tilde{E}_N(\rho; T=0, x)], \qquad (7)$$

where $E_N(\rho; T=0, x)$ is obtained by interpolation between the $x=1$ and $x=0$ results given in Ref. 16. The E_N and P_N are actually used in step (A). Note that the entropy part is unaffected ($\tilde{S}_N \to S_N$).

So far we have been concerned about the EOS at $\rho \gtrsim \rho_0$. For $\rho \simeq (0.001\text{–}1)\rho_0$, we construct the isoentropy EOS in the same manner as in (C), by utilizing the results in Ref. 8, where E, P, and S are given at various T and ρ for $Y_l = 0.3$ and 0.4. For still lower densities, we substitute the EOS with the one at $T=0$ given by Baym, Bethe and Pethick.[17] These three EOSs are smoothly matched at the region $\rho \simeq (0.6\text{–}1)\rho_0$ and $(0.001\text{–}0.002)\rho_0$.

2.2 EOS FOR COLD NEUTRON STARS

As mentioned in section 2.1, we use the $E_0(\rho; T=0, x=1)$ (and P_0) for the EOS of cold neutron stars. In later discussion, we will consider the case where pion condensation takes part in cold neutron stars. From the several works so far done, we can expect that pion condensation could be a new phase realized at the important densities relevant to neutron stars. Recently this point has been confirmed by Tatsumi and Muto[18] through the analysis of the Landau-Migdal parameter g' consistent with nuclear experiments.

We construct the pion-condensed EOS by the approximation that the energy gain due to pion condensation $\Delta E_\pi (<0)$ is simply added to the normal EOS (E_0) as $E = E_N = E_0 + \Delta E_\pi$. Concerning ΔE_π, we adopt two cases: the moderate (denoted by π-1) and the strong (π-2) pion condensation. The former is taken from the one[19] obtained by introducing the effective pion-nucleon P-wave interaction and the latter from the one of Muto and Tatsumi ($g' = 0.5$) [20] derived by the σ-model approach. The condensation of charged pions (π^c) occurs at $\rho \simeq 2.1(1.5)\rho_0$ for π-1 (π-2) and that of neutral pions (π^0) joins at $\rho \simeq 4.2(3.4)\rho_0$ (hence their coexistence) for π-1 (π-2). The choice of these two typical cases is expected to cover the present theoretical uncertainties in the problem of pion condensation. The cold neutron stars become more compact by the occurrence of pion condensation to soften the EOS dramatically and the contrast between hot and cold stars would be all the more remarkable.

There is a possibility that pion condensation takes place even in hot neutron stars.[21-23] The effect, however, is not so remarkable as in the cold case, since the high T pushes the onset of π^0 condensation to the higher density side[21-23] and so the realization is restricted to the very central regions of only massive stars. Furthermore, the realization would become less likely if we take account of the macroscopic stability under thermal fluctuations.[21] Also the effect of π^c condensation would be weakened considerably due to the high abundance of protons (Y_p =0.3-0.4). Therefore, for simplicity, we neglect the effect on neutron stars at birth; that is, we assume that pion condensation occurs in neutron stars during cool down.[21]

3. NUMERICAL RESULTS AND DISCUSSION

We consider two typical cases of supernova matter with $(S = 1, Y_l = 0.3)$ and $(S = 1, Y_l = 0.4)$. In the following, we summarize main results and discuss several important points.

3.1 CHARACTERISTICS OF THE EOS OF DENSE SUPERNOVA MATTER

As shown in Fig. 1 (E versus ρ for a and P versus ρ for b), the EOS of dense supernova matter is remarkably stiffer than that of ordinal cold neutron matter.[6,7] The cause is analyzed in Fig. 1a; the stiffening effect due to the high lepton fraction and the high temperature overwhelms the softening effect due to the high abundance of protons. In the stiffening effect, the contribution from lepton energy (in our case, the kinetic energy of relativistic electrons) is larger than that from thermal energy.

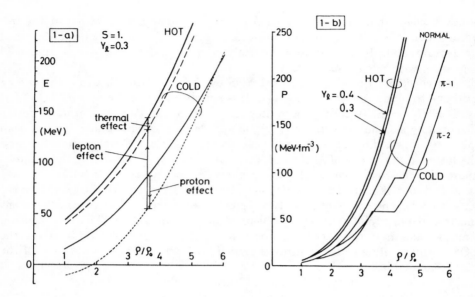

FIGURE 1 The stiffer EOS of supernova matter (bold solid line with HOT) as compared with that of neutron matter (thin solid line with COLD). (a) The internal energy E versus the density ρ. The cause is analyzed as coming from the stiffening effect due to high lepton fraction (lepton effect) and high temperature (thermal effect), which overwhelms the softening effect due to the high abundance of protons (proton effect). (b) The pressure P versus ρ. The inclusion of pion condensation (moderate for π-1 and strong for π-2) is shown to increase the difference between the hot EOS and the cold one.

Fig. 1b shows that the difference between the hot EOS and the cold one becomes all the more remarkable if the latter contains the pion condensation. It is also shown that the hot EOS gets stiffer with increasing Y_l.

3.2 CHARACTERISTIC FEATURES OF HOT NEUTRON STARS

Reflecting the stiff EOS, the neutron star just born has a fat density profile;[6,7] relatively lower central density and larger radius. Also it is hot to the extent that the central temperature amounts to 20–30 MeV. Fig. 2a illustrates the example where a hot star ($S = 1, Y_l = 0.3$) is compared with a cold normal star or a cold pion-condensed star (π-2), with total baryon number $N_B = 2.0 \times 10^{57}$. Note that the comparison should be made for the same N_B stars because only N_B is a conserved quantity in the evolution. The radii R for hot and cold neutron stars are compared

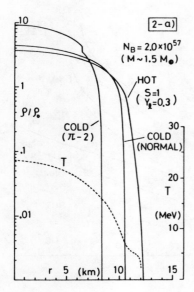 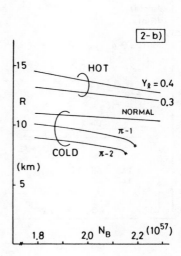

FIGURE 2 Characteristics of hot neutron stars. (a) Density ρ and temperature T profiles of a neutron star at birth (HOT), with r being the distance from the center. Density profiles of cold neutron stars, without (NORMAL) and with (π-2) pion condensation, are also shown in comparison. (b) Radii R of hot and cold neutron stars. Comparison should be made with the same total baryon number N_B. Crosses denote the maximum mass points.

in Fig. 2b. It is known that hot neutron stars contract noticeably by cooling in the ν-diffusion stage. For example, in the case of $N_B = 2.0 \times 10^{57}$, the radius is $R \simeq 13.7$ km for a hot neutron star with $Y_l = 0.4$ and $R \simeq 10.8$ km for a cold normal one or even $R \simeq 8.5$ km for a cold pion-condensed one (π-2).

3.3 SPIN-UP DUE TO CONTRACTION IN THE COOLING STAGE

The contraction of hot neutron stars means their spin-up.[7] By assuming the conservation of total angular momentum, the ratio of angular frequency ω (cold)/ω (hot) is deduced by the ratio of moment of inertia I(hot)/I(cold) from neutron star models. The results are given in Fig. 3a. It is found that noticeable spin-up occurs; for instance, about 20(25)%, 50(60)% and 80(90)% when a hot neutron star with $N_B = 2.0 \times 10^{57}$ and $Y_l = 0.3(0.4)$ is cooled down to a cold

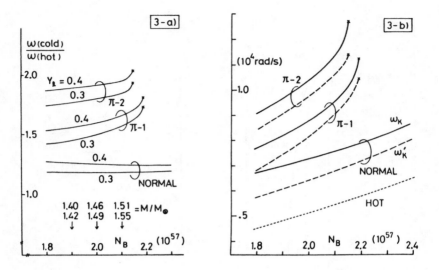

FIGURE 3 (a) Extent of spin-up $\omega(\text{cold})/\omega(\text{hot})$, with ω being the angular frequency, in the evolution of hot neutron stars just born with $Y_l = 0.3$ or 0.4 to the cold ones in normal or pion-condensed states (π-1, π-2). For reference, the gravitational masses M corresponding to N_B are indicated by the numbers attached to arrows; lower (upper) ones are for cold normal (pion-condensed; π-2) neutron stars. (b) Comparison of Keplerian angular frequency $\omega_K(\text{cold})$ (*solid line*) with that including the thermal evolution $\omega'_K(\text{cold}) = \omega_K(\text{hot}) \times$ spin-up ratio (*dashed line*). The dotted line is for $\omega_K(\text{hot})$ with $Y_l = 0.4$. Crosses denote the maximum-mass points.

star in normal, moderately pion-condensed (π-1) and strongly pion-condensed (π-2) states, respectively. It is of interest to see how the spin-up affects the maximum rotation of cold neutron stars currently discussed in terms of the Keplerian angular frequency $\omega_K(\text{cold})$. By using the relativistic Roche model[24] to give $\omega_K \simeq 6.3 \times 10^3 (M_B/M_\odot)^{1/2} (R/10 \text{ km})^{-3/2}$ rad/s, with M_B total baryon mass (N_B times nucleon mass), we calculate $\omega_K(\text{cold})$ and $\omega_K(\text{hot}; Y_l = 0.4)$. In Fig. 3b, the $\omega_K(\text{cold})$ are compared with the $\omega'_K(\text{cold}) = \omega_K(\text{hot}; Y_l = 0.4) \times$ spin-up ratio. The relation $\omega'_K(\text{cold}) < \omega_K(\text{cold})$ is found for all cases considered. The relation holds also for $Y_l = 0.3$ cases. Thus the maximum rotation rate is more severely constrained by $\omega'_K(\text{cold})$ than usual $\omega_K(\text{cold})$.

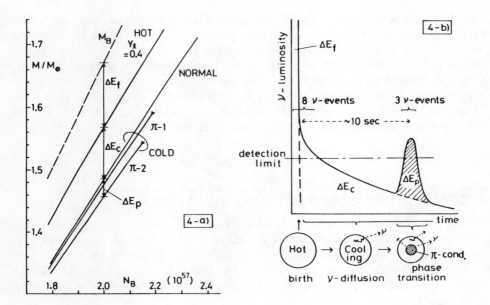

FIGURE 4 (a) The gravitational mass M of neutron stars as a function of N_B. The releasable energy in the stages of formation, cooling, and phase transition to pion condensate is indicated by ΔE_f, ΔE_c and ΔE_p, respectively. Crosses denote the maximum-mass points. (b) Schematic illustration of the ν-luminosity versus time associated with our scenario to explain the last three ν-events from SN I987A in KAMIOKA data.

3.4 GRAVITATIONAL ENERGY RELEASE DUE TO THE CONTRACTION

The contraction also means the release of gravitational energy.[6] In Fig. 4a, the M for hot and cold neutron stars, together with the M_B, are illustrated as a function of the N_B. The amount of releasable energy can be extracted from the comparison of M between hot and cold stars, keeping the N_B constant. For example, in the case of $N_B = 2.0 \times 10^{57}$, $\Delta E_c \simeq 1.5 \times 10^{53}$ ergs is released in the cooling stage due to ν-diffusion, which is comparable to $\Delta E_f \simeq 1.8 \times 10^{53}$ ergs released in the formation stage. The total energy $\Delta E = \Delta E_f + \Delta E_c \simeq (2.6 \to 3.3) \times 10^{53}$ ergs for $N_B = (1.8 \to 2.0) \times 10^{57}$ is consistent with the observational ones; $\Delta E_{obs} \simeq$ 0.9–3.5 $\times 10^{53}$ ergs [25] and 1.6–3.1 $\times 10^{53}$ ergs [26] extracted from the ν-events from SN 1987A,[27,28] assuming the distance to the LMC is 50 kpc. If the pion condensation takes place in the neutron star core during cooling, the late-time energy release $\Delta E_p \sim 5 \times 10^{52}$ ergs is added. This supports our idea[6,21,29] to explain the last three ν-events in KAMIOKA data,[27] unexpectedly observed ~ 10 s after the first eight ν-events. Although the possibility of statistical fluctuations is not completely excluded, we take the phenomenon as indicating the phase transition into pion condensate. It is because ΔE_p can supply the energy 10^{52-53} ergs[30] ($\sim 5 \times 10^{52}$ ergs[26]) corresponding to the last three events, thereby raising the ν-luminosity above the detection limit. The time interval ~ 10 s can be understood as the cooling time necessary for the onset of pion condensation and also the duration time ~ 3 s would be related to the time needed for the core to finish the phase transition. In Fig. 4b, the ν-luminosity of this scenario is illustrated schematically.

4. CONCLUDING SUMMARY AND REMARKS

We have shown that the EOS of supernova matter is remarkably stiffer than that of usual cold neutron matter, which is caused by the particular properties, high temperature and high lepton fraction. Correspondingly, neutron stars just born are characterized by the hot and fat features. The radius is larger by about 20%–30% (30%–60%) compared with cold normal (pion-condensed) stars, for the mass range $(1.4–1.5) M_\odot$.

Neutron stars just born make a gradual contraction in the cooling stage, evolving to cold stars. We have stressed their notable spin-up due to the contraction, roughly speaking, by about 20%–50% (50%–90%) when the cooled stars are in normal (pion-condensed) state. We have also noted that the fastest rotation of a given cold neutron star is constrained by the extent of spin-up from the hot stage, not simply by the Keplerian angular frequency for the cold star currently taken.

The contraction leads to gravitational energy release. We have found that about 1.5×10^{53} ergs is released in the cooling stage, which is comparable to the energy release in the formation stage. The total energy from the sum of these two is consistent with the observational one from SN 1987A. If the pion condensation occurs in a neutron star core during cooling, the late-time energy release of $\sim 5 \times 10^{52}$ ergs is possible, which could explain the time profile of the ν-burst from SN 1987A in KAMIOKA data.

In connection with the spin-up energy, it is worthwhile mentioning that $\sim 10\%$ of the released gravitational energy contributes to the increase of rotational energy, in the maximum rotation case of an $N_B = 2.0 \times 10^{57}$ star. Concerning the maximum mass of a new born neutron star, we remark from Fig. 4a that $M_{\max}(\text{hot}) \simeq (1.64-1.67) M_\odot$ depending on Y_l for the case of evolution to a cold pion-condensed star (π-2), because the hot star with $M > M_{\max}(\text{hot})$ leads to a black hole as a result of the cooling down ($M_{\max}(\text{hot}) \simeq (1.89 \sim 1.92) M_\odot$ in the case of evolution to a cold normal star).

Finally, to be more realistic, our EOS of dense supernova matter should be refined by improving the $\tilde{V}$ used and by including the effect of degenerate neutrinos by getting Y_ν under the constant Y_l and β-equilibrium. Also it is necessary to investigate carefully the effect of pion condensation on hot neutron star models.

REFERENCES

1. A. Burrows and J. M. Lattimer, *Astrophys. J.* **307**, 178 (1986).
2. H. A. Bethe, G. E. Brown, J. Applegate, and J. M. Lattimer, *Nucl. Phys.* **A324**, 847 (1979).
3. R. I. Epstein and C. J. Pethick, *Astrophys. J.* **243**, 1003 (1981).
4. D. Z. Freedman, *Phys. Rev.* **D9**, 1389 (1974).
5. K. Sato, *Prog. Theor. Phys.* **53**, 595 (1975); **54**, 1325 (1975).
6. T. Takatsuka, *Prog. Theor. Phys.* **82**, 475 (1989).
7. T. Takatsuka, *Prog. Theor. Phys.* **85**, No. 3 (1991).
8. J. M. Lattimer, C. J. Pethick, D. G. Ravenhall, and D. Q. Lamb, *Nucl. Phys.* **A432**, 646 (1985).
9. S. Nishizaki and T. Takatsuka, this volume (1992).
10. T. Takatsuka, *Prog. Theor. Phys.* **73**, 1043 (1985); **75**, 201 (1986).
11. T. Takatsuka and J. Hiura, *Prog. Theor. Phys.* **79**, 268 (1988).
12. K. E. Schmidt and V. R. Pandharipande, *Phys. Lett.* **87B**, 11 (1979).
13. D. W. L. Sprung and P. K. Banerjee, *Nucl. Phys.* **A168**, 273 (1971).
14. T. Takatsuka, *Prog. Theor. Phys.* **72**, 252 (1984).
15. J. H. Taylor and J. M. Weisberg, *Astrophys. J.* **345**, 434 (1989).
16. R. B. Wiringa, V. Fiks, and A. Fabrocini, *Phys. Rev.* **C38**, 1010 (1988).
17. G. Baym, H. A. Bethe, and D. J. Pethick, *Nucl. Phys.* **A175**, 225 (1971).
18. T. Tatsumi and T. Muto, Proc. of the International Symposium on Nuclear Astrophysics, *Nuclei in the Cosmos*, 18–22 June 1990, Baden/Vienna, Austria.
19. T. Takatsuka and R. Tamagaki, *Prog. Theor. Phys.* **82**, 945 (1989).
20. T. Muto and T. Tatsumi, *Prog. Theor. Phys.* **78**, 1405 (1987).
21. T. Takatsuka, *Prog. Theor. Phys.* **80**, 361 (1988).
22. M. Takahara, T. Takatsuka, and K. Sato, preprint, UTAP-112/90 (1990).
23. M. Takahara, T. Takatsuka, and K. Sato, this volume (1992).
24. S. L. Shapiro, S. A. Teukolsky, and I. Wasserman, *Astrophys. J.* **272**, 702 (1983).
25. K. Sato and H. Suzuki, *Phys. Lett.* **B196**, 267 (1987).
26. S. H. Kahana, J. Cooperstein, and E. Baron, *Phys. Lett.* **B196**, 259 (1987).
27. K. Hirata et al., *Phys. Rev. Lett.* **58**, 1490 (1987).
28. R. M. Bionta et al., *Phys. Rev. Lett.* **58**, 1494 (1987).
29. T. Takatsuka, *Prog. Theor. Phys.* **78**, 516 (1987).
30. H. Suzuki and K. Sato, *Publ. Astron. Soc. Japan* **39**, 521 (1987).

M. Takahara,* T. Takatsuka,† and K. Sato‡
*Doshisha Women's College of Liberal Arts, Kyoto 610-03, Japan; †College of Humanities and Social Sciences, Iwate University, Morioka 020, Japan; and ‡Department of Physics, University of Tokyo, Tokyo 113, Japan

Supernova Explosions and the Soft Equation of State

We investigate the equation of state of the hot supernova matter with π^0-condensed state on the basis of the alternating-layer-spin model by solving the Hartree-Fock equation at finite temperature. We find that although the phase transition to the π^0-condensed state is suppressed considerably with temperature, this phase transition is realized in the collapsing core. We first calculate the entropy and specific heat of the π^0-condensed state and find that they are much lower than those of the normal state. The application to supernova explosions is discussed in detail, especially the evolution of the collapsing core along the isoentropy line and its effects on the pressure.

1. INTRODUCTION

Since the theoretical suggestion of pion condensation,[1,2] this phase transition has been investigated extensively by many theorists in relation to neutron stars, since it makes the equation of state (EOS) considerably soft and changes the structure of neutron stars. However, the critical density to the pion-condensed state is not well determined yet due to the dependence on theoretical approaches. If the critical density is as low as ~ $(1.5-2)\rho_0$, the pion-condensed state may also be realized in the collapsing core as well as in neutron stars.

If this is true, the soft EOS with pion condensation plays an important role in supernova explosions; the soft EOS enhances the release of gravitational energy and results in the strong shock wave, which is favorable to the prompt explosion mechanism.[3-7] Besides, the soft EOS results in the hotter core and the enhanced neutrino emission, which is favorable to the delayed explosion mechanism.[8] Therefore, the soft EOS with pion condensation will help the supernovae to explode.

Indeed, Takahara and Sato[5] investigated supernova explosions in use of the EOS with phase transitions and found that such EOS is favorable to supernova explosions if the strength of the phase transition is moderate. Since the EOS they used was an idealized one, the next step is to confirm their results using a realistic EOS. This is an important issue, because the explosion mechanism of supernovae has not been determined.

However, there is no EOS with pion condensation for supernova matter, since the pion condensation has been investigated in relation to neutron stars. Supernova matter is quite different from neutron star matter, since it contains many leptons ($Y_e \sim 0.3$–0.5) and is much hotter ($T \geq 10$ MeV) than the neutron star matter. Therefore, the purpose of this investigation is to calculate the EOS with pion condensation for the supernova matter and to simulate supernova explosions. Is the pion condensation realized in a collapsing core and, if so, how much is the increment to the explosion energy?

2. CALCULATION OF THE EOS WITH PION CONDENSATION

We focus our attention on the neutral-pion (π^0) condensation, since this phase transition is considered to be of the first order and the effects on EOS are expected to be much larger than those of the charged-pion (π^c) condensation, which is considered to be of the second order.

We describe the π^0-condensed state by the alternating-layer-spin model.[9-11] We treat the finite-temperature Hartree-Fock equation for both the symmetric nuclear matter and the neutron matter including the normal and the π^0-condensed state. We introduce the effective interaction consisting of the central part $\tilde{V}_C$ and the tensor part $\tilde{V}_T$. The latter is the driving force for the π^0-condensation. As for the $\tilde{V}_C$, we adopt the one based on the RSC potential: G0-force of Sprung-Banerjee[12] for the symmetric nuclear matter and the G0-version suggested by Takatsuka[13] for the neutron matter. As for the $\tilde{V}_T$, we construct the effective two-nucleon tensor potential of the OPEP type, which can simulate the results given by the more fundamental approach in which the isobar Δ degrees of freedom are included explicitly.[14,15]

By solving this Hartree-Fock equation, we obtain the EOS of the normal and π^0-condensed states for both the hot symmetric nuclear matter and the hot neutron matter. The critical density and the coexisting region are calculated by the

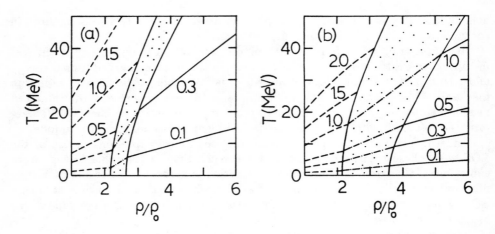

FIGURE 1 Phase diagram of the π^0-condensation: (a) for the neutron matter and (b) for the supernova matter with $Y_e = 0.5$. The density is normalized by the nuclear density $\rho_0 = 0.17 \text{fm}^{-3}$. The dotted region shows the coexisting region of the normal and π^0-condensed states. The dashed and solid lines denote the isoentropy (entropy per baryon in units of the Boltzmann constant) lines for the normal and π^0-condensed state, respectively. The dash-dotted lines connect the corresponding points on the boundaries of the coexisting region.

Maxwell double tangent approximation. Interpolating these EOSs by Y_e and adding the EOS of electrons, we obtain the EOS of supernova matter in normal and π^0-condensed state with various electron fraction, and investigate their thermodynamic properties.

3. COLLAPSE OF THE IRON CORE

It is to be noticed that after the neutrino trapping, the free fall time of the iron core ($\sim$ ms) is much shorter than the diffusion time of neutrinos ($\sim$ 10 s). Therefore, the iron core collapses adiabatically and the entropy of each mass shell is conserved; for the simulation of the core collapse, the EOS for fixed entropy is necessary.

Using the calculated EOS, we performed the simulation of the collapse of the iron core to check whether the pion condensation is realized in the collapsing core. Since this is a preliminary simulation, we construct an idealized presupernova model following Takahara and Sato[4] assuming the density and temperature relation $T \propto$

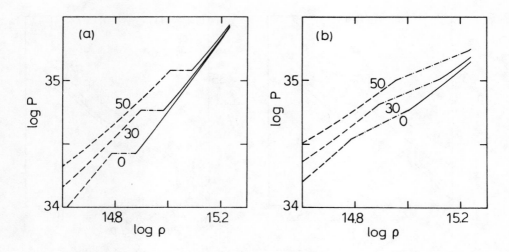

FIGURE 2 Total pressure P versus density ρ: (a) for the neutron matter and (b) for the supernova matter with $Y_e = 0.5$. In these figures, the total pressure and density (in cgs unit) are shown in logarithm. The notation is the same as in Fig. 1.

ρ^α, $\alpha = 4/30$, based on the presupernova model of Woosley and Weaver[16] and Nomoto et al.[17] Using this relation, we solved the structure of an iron core in equilibrium with the electron fraction $Y_e = 0.46$ and obtained an iron core of 1.4 $M_\odot$.

4. RESULTS

The calculated EOS is shown in Figs. 1–5. Figure 1 shows the phase diagram of the pion condensation. It is clearly seen from this figure that the phase transition is suppressed considerably by temperature and the coexistent region changes with Y_e; it is narrow for the neutron matter, and is wide for the supernova matter with $Y_e = 0.5$. It is also seen that the entropy of the pion condensed state is much lower than that of the normal state. Since the entropy of each mass shell is conserved during the collapse, the core becomes considerably hot after the phase transition.

The EOS ($\log P$–$\log \rho$ diagram) is shown in Fig. 2 for the neutron matter and for the supernova matter with $Y_e = 0.5$. It is clearly seen that this phase transition makes the EOS considerably soft. Figure 2a shows that this phase transition is of the first order since the pressure in the coexisting region is constant. However, the pressure in the coexisting region increases in Fig. 2b due to the electron pressure even though this is the first-order phase transition. It is to be noticed that the nucleon pressure of the symmetric nuclear matter is lower than that of the neutron

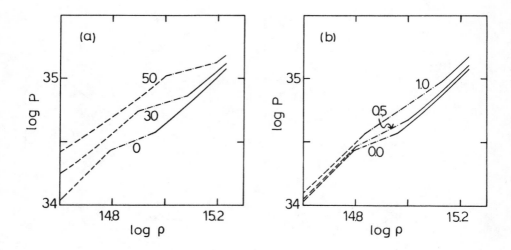

FIGURE 3 Total pressure P versus density for the supernova matter with $Y_e = 0.35$: (a) for fixed temperature and (b) for fixed entropy. The notation is the same as in Fig. 2.

matter, while the total pressure of the supernova matter with $Y_e = 0.5$ is higher than that of the neutron matter due to the electron pressure. The effect of temperature is larger in the lower density region.

The EOS for the supernova matter with $Y_e = 0.35$ is shown in Fig. 3 for fixed temperature and for fixed entropy. The effects of high temperature of the pion-condensed state on the pressure is not so large, since the density of the condensed state is much higher than that of the normal state.

The specific heat, C_V, is shown in Fig. 4. The specific heat increases in proportion to the temperature in the low-temperature region, while its increment decreases with temperature in the high-temperature region. The specific heat of the supernova matter with $Y_e = 0.5$ exceeds 1.5 at the high temperature due to the contribution of electrons. This figure shows that the specific heat of the pion condensed state is much smaller than that of the normal state and that the specific heat of the neutron matter is much smaller than that of the supernova matter. This is the first calculation of the specific heat of the pion-condensed state.

Since the nuclear interaction we adopt can be applied only up to the density of $\rho/\rho_0 \sim 6$, we connected our EOS to the one in which the sound velocity equals the light velocity. Then the EOS with π^0-condensed state for the cold neutron matter gives 1.47 $M_\odot$ for the critical mass of neutron stars barely satisfies the astrophysical constraint given by the binary pulsar PSR 1913+16, 1.44 $M_\odot$.[18]

Using our EOS, we simulate the collapse of iron cores using the model described in section 3. The evolutionary path of the central density and temperature during the collapse phase is plotted in Fig. 5. Although this is a preliminary result, this figure shows that the pion condensation is indeed realized in the collapsing core,

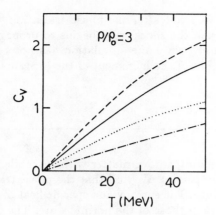

FIGURE 4 Specific heat per baryon C_V versus temperature T at the density of $\rho/\rho_0 = 3.0$ (ρ_0 denotes the nuclear density, $\rho_0 = 0.17\text{fm}^{-3}$). The specific heat of the neutron matter is shown by the dotted line for the normal state and by the dot-dashed line for the π^0-condensed state, and C_V for the supernova matter with $Y_e = 0.5$ is shown by the dashed line for the normal state and by the solid line for the π^0-condensed state.

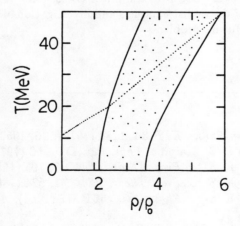

FIGURE 5 Evolutionary path of the central density on the phase diagram for supernova matter with $Y_e = 0.46$. The notation is the same as in Fig. 1 except for the path, shown by the dotted line.

and the core becomes very hot after the phase transition, as expected. Then the core bounces strongly, and the shock wave begins to propagate toward the surface of the core. Unfortunately, since the simulation has not been completed yet, we cannot discuss the amount of the increment of the explosion energy here.

5. SUMMARY

We calculate the EOS for supernova matter with π^0-condensation and investigate the thermodynamical properties. We find that the phase transition is considerably suppressed by temperature. The entropy and specific heat of the π^0-condensed state are considerably lower than those of the normal state. The nucleon pressure of the symmetric nuclear matter is lower than that of the neutron matter, while the total pressure of the supernova matter with $Y_e = 0.5$ is higher than that of the neutron matter due to the electron pressure. The specific heat of the supernova matter with $Y_e = 0.5$ is also higher than that of the neutron matter. This is the first calculation of the specific heat of the pion-condensed state.

We simulate the collapse of the iron core using our EOS for an idealized pre-supernova model. Although the phase transition to the π^0-condensed state is suppressed considerably by temperature, we find that this phase transition is realized in the collapsing core. Since the collapsing core evolves along the entropy-constant line, the core becomes very hot after the phase transition.

The calculation of the realistic models and investigation of the increment of the explosion energy are future problems.

REFERENCES

1. A. B. Migdal, *Sov. Phys.-JETP* **34**, 1184 (1972); **36**, 1052 (1973).
2. R. F. Sawyer and D. J. Scalpino, *Phys. Rev.* **D7**, 953 (1972).
3. M. Takahara and K. Sato, *Prog. Theor. Phys.* **68**, 795 (1982); **72**, 978 (1984).
4. M. Takahara and K. Sato, *Prog. Theor. Phys.* **71**, 524 (1984).
5. M. Takahara and K. Sato, *Phys. Lett.* **156B**, 17 (1985); *Astrophys. J.* **335**, 301 (1988).
6. E. Baron, J. Cooperstein, and S. Kahana, *Nucl. Phys.* **A440**, 744 (1985); *Phys. Rev. Lett.* **55**, 126 (1985); **59**, 736 (1987).
7. W. Hillebrandt, *High Energy Phenomena Around Collapsed Stars*, F. Pacini (ed.) (D. Reidel, Dordrecht, 1987).
8. H. Bethe and J. R. Wilson, *Astrophys. J.* **295**, 14 (1985).
9. T. Takatsuka and R. Tamagaki, *Prog. Theor. Phys.* **58**, 694 (1977).

10. T. Takatsuka, K. Tamiya, T. Tatsumi, and R. Tamagaki, *Prog. Theor. Phys.* **59**, 1933 (1978).
11. R. Tamagaki, *Nucl. Phys.* **A328**, 352 (1979).
12. D. W. L. Sprung and P. K. Banerjee, *Nucl. Phys.* **A168**, 273 (1971).
13. T. Takatsuka, *Prog. Theor. Phys.* **72**, 252 (1984).
14. T. Kunihiro, T. Takatsuka, and R. Tamagaki, *Prog. Theor. Phys.* **79**, 120 (1988).
15. T. Kunihiro and T. Tatsumi, *Prog. Theor. Phys.* **65**, 613 (1981).
16. T. A. Weaver, G. B. Zimmerman, and S. E. Woosley, *Astrophys. J.* **225**, 1021 (1978).
17. K. Nomoto, Y. Kamiya, K. Yokoi, and S. Miyaji, private communication.
18. J. H. Taylor and J. M. Weisberg, *Astrophys. J.* **345**, 434 (1989).

Hideyuki Suzuki†and Katsuhiko Sato‡
†KEK, National Laboratory for High Energy Physics, 1-1 Oho, Tsukuba, Ibaraki 305, Japan; and ‡Department of Physics, Faculty of Science, University of Tokyo, 7-3-1 Hongo, Bunkyo, Tokyo 113, Japan

The Cooling of Proto–Neutron Stars and Neutrino Bursts

We performed numerical simulations of the quasi-static evolution of a proto–neutron star with neutrino transfer using a multigroup flux-limited diffusion scheme. The second half of the supernova neutrinos are emitted in this stage. We found (1) the nucleon bremsstrahlung process is important as the source of ν_μ's, (2) numerical spectra of emitted neutrinos can be interpreted as the superposition of scattered spectra from thermalization spheres defined for each neutrino energy, and (3) the time-integrated energy spectra of ν_e's and $\bar{\nu}_e$'s can be roughly fitted by the Fermi-Dirac spectra with positive chemical potentials while those of the other neutrinos can be fitted by Maxwell-Boltzmann spectra.

1. INTRODUCTION

Detection of the neutrino burst from SN 1987A by KAMIOKANDE-II[1] and IMB[2] gave us an opportunity to test the current picture of the collapse-driven supernova explosion and the birth of a neutron star. The duration time of the burst, the average energy of neutrino events, the total number of events, and the angular distribution of events are all consistent with theoretical predictions.[3,4,5] The scenario that neutrinos once trapped in the core diffuse out in a time scale of 10 s[6,7] is confirmed. The limited number of data (only 19 events), however, prevents us from studying more details.

Theoretically, the time evolution of the supernova neutrinos is divided into two stages: a dynamical stage of core bounce and explosion and a quasi-static cooling stage of the proto–neutron star just born. So far many numerical simulations have been performed. However, most of them model the first stage because their main purpose was to make clear the mechanism of the collapse-driven supernova explosion. In order to make clear the whole structure of the neutrino burst—in particular, the duration of the burst—it is obviously necessary to investigate the second stage in detail. Burrows and Lattimer[8] investigated the cooling stage of the proto–neutron star in detail, but using simplified LTE neutrino transfer. They assumed that neutrino distribution in the core is a Fermi-Dirac one and calculated the neutrino flow in the energy-integrated form. They showed that in general the neutrino burst from SN 1987A agrees with their model within the small statistics.[5] Their method, however, is unsatisfactory if one wants to know the time profile of the neutrino burst more quantitatively, since the neutrino opacity depends strongly on the neutrino energy. It is hoped that properties of the neutrino burst, including neutrino spectra and their evolution, will be obtained by future experiments such as Super-KAMIOKANDE[9] if the next supernova appears in our Galaxy. It is, therefore, urgent to present a theoretical prediction. We can expect fruitful information on the explosion mechanism and on the recently born neutron star by comparing theoretical prediction with observations. In order to investigate the neutrino burst from a proto–neutron star more precisely and in detail, we performed numerical simulations of the proto–neutron star cooling with multigroup neutrino transfer. Here we present our numerical results on the second half of the supernova neutrinos.

2. NUMERICAL METHOD

We simulate the quasi-static evolution of the proto–neutron star with neutrino transfer using a multigroup flux-limited diffusion scheme (MGFLD). Hydrodynamical treatment is not necessary because the dynamical time scale of the proto–neutron star ($\sim$ ms) is much less than the cooling time scale ($\sim$ 10 s). The general relativistic hydrostatic structure of the proto–neutron star, neutrino transfer, deleptonization of the matter, and energy transfer between the matter and neutrinos are solved simultaneously.[10] While ν_e's and $\bar{\nu}_e$'s are treated separately, ν_μ's, $\bar{\nu}_\mu$'s, ν_τ's, and $\bar{\nu}_\tau$'s are treated together as an averaged species (hereafter ν_μ). Included neutrino interactions are as follows:

$$
\begin{array}{llllll}
e^- \, p & \longleftrightarrow \nu_e \, n & , & e^+ \, n & \longleftrightarrow \bar{\nu}_e \, p & , & e^- \, A & \longleftrightarrow \nu_e \, A' \\
e^+ \, e^- & \longleftrightarrow \nu \, \bar{\nu} & , & \text{plasmon} & \longleftrightarrow \nu \, \bar{\nu} & , & \nu \, e^\pm & \longleftrightarrow \nu \, e^\pm \\
\nu \, p & \longleftrightarrow \nu \, p & , & \nu \, n & \longleftrightarrow \nu \, n & , & \nu \, A & \longleftrightarrow \nu \, A \\
n \, n & \longrightarrow n \, p \, e^- \, \bar{\nu}_e & , & n \, p \, e^- & \longrightarrow n \, n \, \nu_e \\
n \, n & \longrightarrow n \, p \, \mu^- \, \bar{\nu}_\mu & , & n \, p \, \mu^- & \longrightarrow n \, n \, \nu_\mu \\
n \, n & \longleftrightarrow n \, n \, \nu \, \bar{\nu} & , & n \, p & \longleftrightarrow n \, p \, \nu \, \bar{\nu}
\end{array}
$$

where ν represents all species of neutrinos. Interaction rates are basically taken from Bruenn's paper[11] with a correction[12] except for the plasmon process, the modified URCA process, and the nucleon bremsstrahlung process. In particular neutrino-electron scattering and pair process are treated exactly without approximations such as the Fokker-Planck approximation.[12,13] Plasmon, modified URCA, and nucleon bremsstrahlung processes are treated in a rough manner only to reproduce the calculated energy loss rate.[14,15] Our multigroup neutrino transfer scheme includes general relativistic effects such as the redshift of neutrino energy and the time dilation. Although many types of flux limiters are proposed, we adopt representative ones in the present study: Mayle and Wilson's[16] and Levermore and Pomraning's.[17] The finite temperature equation of state (hereafter EOS) calculated by Wolff with the Hartree-Fock method[18] is used. This is a somewhat stiff EOS in the high-density region. The initial structure of the proto–neutron star and the initial neutrino distribution are prepared by referring to outputs of Mayle and Wilson's hydrodynamical simulation at 0.4 s after the bounce.[19]

The present model is constructed by 84 radial meshes and by 16 energy meshes for one species of neutrinos. The total energy and the total (electron type) lepton number are found to be conserved during the first 19 s, with small errors less than $10^{-1}\%$ and $10^{-3}\%$ respectively.

3. RESULTS AND DISCUSSION

We show the time evolution of a neutrino signal (Fig. 1) for the case in which Mayle and Wilson's flux limiter was used. Qualitative features of the thermal history of the proto–neutron star are similar to the results of the simple LTE neutrino transfer calculation.[8] Note that the gravitational mass of our initial model is $1.55 M_\odot$, and that of the neutron star at $t = 19$ s is $1.48 M_\odot$, whereas its baryon mass is $1.62 M_\odot$. This means that our initial model corresponds to a proto–neutron star which has already radiated out the energy of $0.07 M_\odot$ (1.3×10^{53} ergs) during the dynamical stage before it settled into hydrostatic equilibrium. We should keep in mind that our results concern only half of the total supernova neutrinos. This is because we use the results of Mayle and Wilson's hydrodynamical simulation as our initial model, in which the deleptonization has already proceeded at the outer core. On the other

The Cooling of Proto–Neutron Stars and Neutrino Bursts

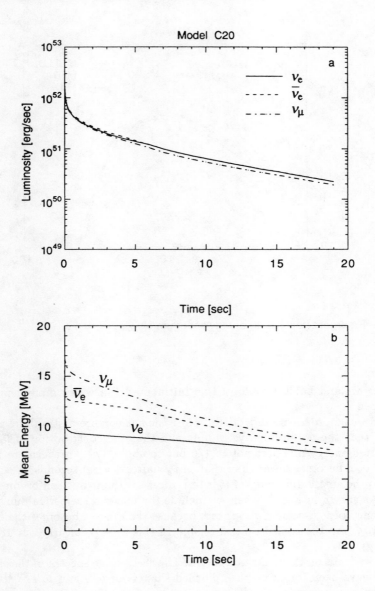

FIGURE 1 (a) Time evolution of the emergent neutrino luminosity. (b) Time evolution of the mean energy of the neutrino flux. All values are what the infinite-distance observers would observe.

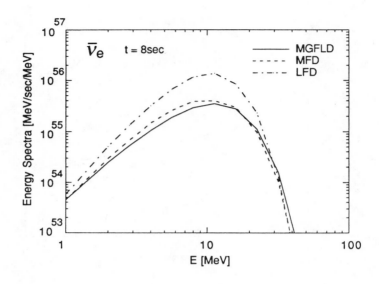

FIGURE 2 Emergent (redshifted) energy spectrum of $\bar{\nu}_e$'s at $t = 8$ s.

hand Burrows and Lattimer adopted an initial model in which many leptons were still trapped.

When we ignore the modified URCA and the nucleon bremsstrahlung processes, we find[10] that the ν_μ flux decays rapidly and that the contribution of $\bar{\nu}_e$'s to the total emitted energy amounts not to 1/6 but to about 1/4. Furthermore the mean energy of ν_μ's becomes lower than that of $\bar{\nu}_e$'s after several seconds. These results are due to the small emissivity of ν_μ's in the case without nucleon bremsstrahlung process. As shown in Fig. 1, when we include the nucleon bremsstrahlung process, the deficiency of ν_μ's almost disappears because the process becomes the dominant source for ν_μ's in the cooled central region. In this case, $\bar{\nu}_e$'s carry out 18% of the total energy.

Next we focus on the neutrino spectra. The early-stage spectra of the supernova neutrinos have been found to have pinched shapes for ν_e's and $\bar{\nu}_e$'s.[20,21,22] These spectra can be interpreted as the superposition of the Fermi-Dirac spectra (LFD: local Fermi-Dirac spectra) from various neutrino spheres defined for each neutrino energy. However, the ν_μ spectra and the late-stage spectra cannot be fitted by LFD spectra. We present $\bar{\nu}_e$ spectra at $t = 8$ s in Fig. 2. The solid line is our numerical result and the dot-dashed line is the LFD spectrum. In order to understand the spectra, we should take into account the scattering processes. By referring to the modified blackbody spectra from the scattering stellar atmosphere, we consider the modified Fermi-Dirac spectra (hereafter MFD) as follows:

The Cooling of Proto–Neutron Stars and Neutrino Bursts

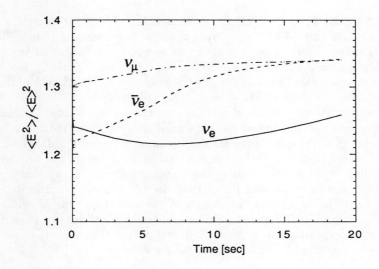

FIGURE 3 Time evolution of the peak width of the neutrino spectra

$$\left.\frac{dL_\nu}{dE}\right|_{MFD} = \frac{c}{4} \, 4\pi R_\nu^2(E) \, \frac{4\pi}{(hc)^3} \, f_{th}(\omega, R_\nu(E))\omega^2 E \cdot S(E) \quad , \tag{1}$$

where $R_\nu(E)$ is the radius of the thermalization sphere for neutrinos with redshifted energy of E, ω is the corresponding neutrino energy at the sphere, and $f_{th}(\omega, R_\nu(E))$ is the thermal neutrino distribution function at that sphere. Note that f_{th} is not necessarily the chemically equilibrated one. The thermalization sphere is defined as the sphere where the effective optical depth (τ_{th}) for thermalizing processes becomes nearly unity. $S(E)$ represents the effect of scattering above the thermalization sphere and is defined in terms of the total optical depth, τ_T, and τ_{th} as

$$S(E) \equiv \frac{2\tau_{th}(E, R_\nu)}{\tau_T(E, R_\nu) + \tau_{th}(E, R_\nu)} \quad . \tag{2}$$

In Fig. 2, the MFD spectrum is shown as the dashed line. It reproduces the numerical result well. Thus, we can understand the spectral shape in terms of the scattered radiation.

In Fig. 3, the time evolution of the peak width of the emitted neutrino spectra is shown. Using the above discussion on the MFD spectra, we can interpret the evolution in terms of the variation of the relative strength of various channels of neutrino interaction. Meanwhile, we can fit the time-integrated spectra for the 19 s with single-temperature Fermi-Dirac spectra. Fitting temperatures are 2.5 MeV for

ν_e's, 3.4 MeV for $\bar\nu_e$'s, 4.4 MeV for ν_μ's; and degeneracy parameters are 2.1 for ν_e's, 1.1 for $\bar\nu_e$'s, and -10 for ν_μ's. Note that the degeneracy parameter for ν_μ's is arbitrarily taken to have a large negative value and the fitting spectrum for ν_μ's is nearly identical to the Maxwell-Boltzmann spectrum with $T = \langle E_{\nu_\mu}\rangle/3$. However, we should not take this result as the fixed one because our treatment of the nucleon bremsstrahlung is insufficient to discuss details of the spectra.

To summarize, we performed detailed numerical simulations of proto–neutron star cooling with neutrino transfer using a multigroup flux-limited diffusion scheme. Although the qualitative features are similar to earlier results of Burrows and Lattimer, we found that the nucleon bremsstrahlung processes are important even in the cooling stage of the proto–neutron star. Furthermore, we investigated the emergent neutrino spectra and found that the spectra can be interpreted as the modified Fermi-Dirac spectra. Evolution of the spectral shape is related to the neutronization and cooling of the proto–neutron star. When we tried to fit the time-integrated spectra, the required chemical potentials are positive for ν_e's and $\bar\nu_e$'s and $-\infty$ (Maxwell-Boltzmann spectra) for ν_μ's.

ACKNOWLEDGMENTS

The authors thank R. Mayle, J. R. Wilson, M. Fukugita, and R. G. Wolff for offering them their data. Numerical studies were done on computers at KEK and CCUT. This work is supported in part by the Grant-in-Aid for Scientific Research Fund of the Japanese Ministry of Education, Science, and Culture (No.01629504, No. 01790167) and the Japan-U.S. Cooperative Science Program (MPCR-185).

REFERENCES

1. Hirata, K., et al., *Phys. Rev. Letters*, **58**, 1490 (1987).
2. Bionta, R. M., et al., *Phys. Rev. Letters*, **58**, 1494 (1987).
3. Sato, K., and Suzuki, H., *Phys. Rev. Letters*, **58**, 2722 (1987).
4. Bahcall, J. N., Piran, T., Press, W. H., and Spergel, D. N., *Nature*, **327**, 682
5. Burrows, A., and Lattimer, J. M., *Astrophys. J. (Letters)*, **318**, 63 (1987).
6. Sato, K., *Prog. Theor. Phys.*, **53**, 595 (1975).
7. Burrows, A., Mazurek, T. J., and Lattimer, J. M., *Astrophys. J.*, **251**, 325
8. Burrows, A., and Lattimer, J. M., *Astrophys. J.*, **307**, 178 (1986).
9. Totsuka, Y., *Proc. Last Workshop on Grand Unification*, P. H. Frampton (ed.) (World Scientific, Singapore, 1989), in press.
10. Suzuki, H., Ph.D. thesis, University of Tokyo (1990).
11. Bruenn, S. W., *Astrophys. J. Suppl.*, **58**, 771 (1985).

12. Myra, E. S., et al., *Astrophys. J.*, **318**, 744 (1987).
13. Bowers, R. L., and Wilson, J. R., *Astrophys. J. Suppl.*, **50**, 115 (1982).
14. Itoh, N., et al., *Astrophys. J.*, **339**, 354 (1989).
15. Maxwell, O. V., *Astrophys. J.*, **316**, 691 (1987). (In this paper, neutrino emissivity is presented only for the degenerate matter.)
16. Mayle, R., Wilson, J. R., and Schramm, D. N., *Astrophys. J.*, **318**, 288 (1987).
17. Levermore, C. D., and Pomraning, G. C., *Astrophys. J.*, **248**, 321 (1981).
18. Hillebrandt, W., and Wolff, R. G., *Nucleosynthesis: Challenges and New Developments*, W. D. Arnett, and J. W. Truran (eds.) (University of Chicago Press, Chicago, 1985), p. 131.
19. Mayle, R., and Wilson, J. R., private communication (1988).
20. Myra, E. S., Lattimer, J. M., and Yahil, A., *Proc. Supernova 1987A in the Large Magellanic Cloud*, M. Kafatos and A. Michalitsianos (eds.) (Cambridge University Press, Cambridge, 1988), p. 213.
21. Giovanoni, P. M., Ellison, D. C., and Bruenn, S. W., *Astrophys. J.*, **342**, 416 (1989).
22. Janka, H.-T., and Hillebrandt, W., *Astron. Astrophys.*, **224**, 49 (1989).

Humitaka Sato
Department of Physics, Kyoto University, Kyoto 606, Japan

Pulsar Cavity

Types of nonthermal energy from a pulsar can be studied through its interaction with the dense ejecta of a supernova. Two topics, the observational search for the high-energy particles in SN 1987A and the relation between the light curve and the pulsar power, are discussed.

1. PULSAR SURROUNDED BY SUPERNOVA EJECTA

SN 1987A has provided us with various phenomena to check the theories on supernova and neutron stars by observation. One of these phenomena is the interaction between the supernova ejecta and whatever is liberated from the pulsar in nonthermal energy modes, which have not been seen. It must be emphasized that such an interaction will be observable only in the early phase after the supernova explosion like in SN 1987A. There are three types of pulsar environment: single radio pulsar, binary X-ray pulsars, and the pulsars surrounded by thick ejecta of supernovae. In the former two cases, this interaction is not observable and a transient period in the last case becomes a very valuable chance to check this interaction.

It is widely believed that the rotational energy of a neutron star is liberated by nonthermal energy modes through a catalysis of the magnetic fields. However, the species of this nonthermal energy is still quite uncertain. It may be magnetic fields, strong-wave electron-positron plasma, high-energy particles, or a combination. Most of these modes of energy are not directly detectable by us because they are not radiations. The pulsed component of the flux from pulsars is much less than the total liberation rate of the period and its drift as $L_p = I\Omega\dot\Omega$. In the case of the Crab nebula, the total synchrotron flux from the extended nebula is known to be comparable to L_p. There is also evidence which suggests an acceleration of the nebula's expansion by the energy from the pulsar. If this is true, we may say that L_p has become visible by the interaction between the ejecta and L_p.

Liberation of the rotational energy will start as soon as the magnetic fields have been formed. If it is just after the birth of the neutron star and if most of L_p is not absorbed directly in the ejecta, the liberated energy will be gathered in a cavity surrounding the pulsar in the expanding ejecta. We call this the *pulsar cavity*.[1]

In this paper, two subjects related to the interaction between the supernova ejecta and the pulsar energy are discussed; an observational check of the high-energy particle acceleration power and the relation between L_p and the light curve. We report on our observations of SN 1987A.

2. SEARCH FOR HIGH-ENERGY γ-RAYS

The origin of the cosmic rays is a long-standing unsolved problem. Its close relation to supernovae and neutron stars has been suspected for a long time. In fact, the synchrotron flux from supernova remnants is clear evidence of particle acceleration. Relativistic electrons are observable through the synchrotron flux, but the nucleonic components of cosmic rays are not observable by it. However, the nucleonic components will be observable in the early stage after the supernova explosion through their interaction with the supernova ejecta; their energy is converted into high-energy neutrinos and γ-rays via multiple pion production. If the particles are radially injected from the pulsar into the expanding ejecta, the time scale of this stage is on the order of several years.[2,3] Thus, we might be able to see this interaction for SN 1987A.[4,5,6]

The energy of these secondary components will be above GeV. The γ-rays above 1 TeV are observable on the ground through extensive air shower but those in GeV–TeV are observed only by detectors loaded on a satellite such as the planned *GRO* or those loaded on airplanes. Under the ground, the high-energy neutrinos are detected by a water Cherenkov detector such as the KAMIOKANDE and the IMB. This detection is much easier than detection of a supernova neutrino burst because of the higher-energy neutrinos involved.

In order to search for the high-energy γ-rays from SN 1987A, a collaboration group was organized among cosmic-ray physicists and high-energy physicists in Japan, New Zealand, and Australia; the project was named as JANZOS. I was a spokesman for this project. JANZOS consists of two air shower detectors: a scintillator array for the energy region above 100 TeV and the Cherenkov light mirror system for that above 3 TeV. We installed these detection systems in the Black Birch Range in South Island of New Zealand.

The observation started as early as October 1987 for the scintillator array[7] and in December 1987 for the Cherenkov mirrors.[8] By 1990, we have checked only the upper limit of the high-energy γ-ray flux from SN 1987A.[9] The upper limit is about 10^{37} ergs in the energies above 3 TeV. This limit is much smaller than the upper limit obtained from the observation of up-going muons in the KAMIOKANDE, which is of the order of 10^{42} ergs in the energies above 2 GeV.[10] The γ-ray upper limit is already much smaller than the L_p of a young pulsar. But this result does not imply the weakness of the pulsar in SN 1987A. It only implies either a smallness of the generation power of the nucleonic high-energy particles or a smallness of the interaction due to magnetic shielding of the high-energy particles from the ejecta.

Besides the upper limit, we detected an indication of a burst signal from SN 1987A with the Cherenkov mirror system in the middle of January 1988.[8] Just at this time, *Ginga* observed an X-ray flare.[11] The X-ray flare can be explained by the collision of the supernova shock wave with the circumstellar clouds around SN 1987A.[12] It is well known that a particle acceleration at the shock wave front works actively. The energy released by the shock wave collision is estimated to be 10^{48} ergs during the flare. In order to get the observable level of the high-energy γ-rays, about 0.02 of this amount must be converted into the high-energy particles, which is comparable to the amount of energy emitted by the X-ray flare. This efficiency is not so absurd. But the estimation of the time scale gives us a severe constraint on the minimum strength of the magnetic fields of 10 mG at the shock front.[13] This value might be high if we estimate the strength by a reasonable law such as $B \simeq (r_0/r)B_0$, where $B_0 \simeq 1$G at the progenitor's surface ($r_0 \simeq 10^{12.5}$ cm) and $r \simeq 10^{16.5}$ cm is the radius of the site at the shock wave collision.

3. PULSAR CAVITY

We come back to the question, In which energy mode is the pulsar's rotational energy liberated? If this energy is dissipative at the interaction with the ejecta, it will contribute to the supernova light curve. Here we can consider two extreme cases: "total absorption" in the ejecta and "no absorption." If the mode is a magnetic field, it is not dissipative and will not contribute to the heating of the ejecta.

We are interested in the relation between L_p and the light curve L_l. If all of L_p is absorbed, it will contribute to the light curve as well as the radioactive source in the ejecta. In this case, we can estimate the upper limit of L_p for SN 1987A from the observed L_l for it. There have been done many such estimation but this will not be necessarily true. Now we consider the other extreme case, of "no absorption". How much does L_p contribute to L_l? The contribution is not zero even in this case.

This is because the pulsar energy in the cavity pushes the surrounding ejecta, and a fraction of the pulsar energy must be dissipated by the shock compression in front of the cavity's surface.

In the early phases of a pulsar study, the dynamics of pulsar cavity was studied assuming uniform sphere for supernova ejecta.[14,15,16] Furthermore it is assumed that L_p is a relativistic gas. This simple model admits an analytic similarity solution such as

$$R = V_0 t \left(\frac{125}{99}\right)^{0.2} \left(\frac{t}{T_0}\right)^{0.2} \quad \text{with} \quad T_0 = \frac{M_0 V_0}{L_p},$$

where R is a radius of the cavity, V_0 is a velocity at $M_r = M_0$, and L_p is a power of the pulsar. According to this solution, $(5/11)$ of L_p is stored in the cavity, $(35/66)$ of L_p is used to accelerate the ejecta's expanding motion, and the rest $(1/66) L_p$ ($\equiv L_d$) is dissipated at the shock front.[14] Then the heat source for L_l is not L_p but L_d.

Assuming a magnetic braking as a mechanism for L_p, the observed L_l is sometimes used to get the upper limit of $B\Omega^2$. This argument would not be correct if L_p is a nondissipative mode of energy. The upper limit could be increased by the factor $\sqrt{66}$. Therefore it is important to estimate this factor for a more realistic model.[17]

The basic equations of the expansion of the pulsar cavity are given as follows:[14]

$$M_s \frac{d^2 R}{dt^2} = 4\pi R^2 \left(P - \rho(v-V)^2\right), \quad \frac{d}{dt}\left(4\pi R^4 P\right) = L_p R,$$

$$\frac{dM_s}{dt} = 4\pi R^2 \rho(v - V),$$

where $v = dR/dt$, M_s is the mass of the shell at the surface of the cavity, V is the ejecta's expansion velocity, and $3P$ is the energy density in the cavity. We write the density profile of the ejecta by $f(x)$ as

$$\rho(t,x) = \frac{3M_0}{4\pi (V_0 t)^3} f(x) \quad \text{and} \quad \int_0^1 f(x) x^2 dx = \frac{1}{3},$$

where M_0 is the mass of the ejecta within $x = 1$, and $x = R/V_0 t$ is the comoving coordinate in the ejecta. Introducing the quantities such as[17]

$$\tau = \frac{t}{T_0}, \quad m(x) = \frac{M_s}{M_0}, \quad p = \frac{4\pi P (V_0 T_0)^3}{M_0 V_0^2},$$

the above equations are rewritten as

$$m\left(x'' + 2\frac{x'}{\tau}\right) = x^2 \tau p - m' x', \quad m' = 3f(x) x^2 x', \quad p' + 4p\left(\frac{1}{\tau} + \frac{x'}{\tau}\right) = \frac{1}{(\tau x)^3},$$

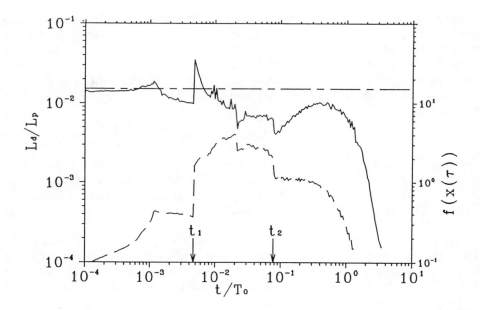

FIGURE 1 The time evolution of the ratio L_d/L_p is shown by the solid line; the dot-dashed line denotes the level of (1/66). The dashed line denotes $f(x(t))$, and the right ordinate gives the scale.

where $m' = dm/d\tau$ and so on. T_0 is given as

$$T_0 = \frac{M_0 V_0^2}{L_p} = 15 \text{ yr } \left(\frac{M_0}{M_\odot}\right)\left(\frac{V_0}{500 \text{ km s}^{-1}}\right)^2 \left(\frac{10^{40} \text{ erg s}^{-1}}{L_p}\right).$$

Taking a realistic $f(x)$ obtained for SN 1978A,[18] we computed the dissipation rate, L_d/L_p, at the shock wave in front of the cavity.[17] The ejecta has a hollow structure at the center, and it is surrounded by a dense shell formed by the compression of the C+O and He shell of the progenitor. The result is shown in Figure 1.

The cavity's surface arrives at the inner edge of this shell at $t_1 = 19 L_{40}^{-1}$yr and leaves the shell at $t_2 = 320 L_{40}^{-1}$yr, ($L_{40} \equiv 10^{40}$ ergs). During the passage through this shell, the dissipation rate is much smaller than 10^{-2}, which implies an increase the estimation of $B\Omega^2$ by a factor more than 10.

REFERENCES

1. H. Sato, *Prog. Theor. Phys.* **80**, 96 (1988).
2. H. Sato, *Prog. Theor. Phys.* **62**, 549 (1977).
3. V. S. Berezinsky and O. P. Prilutsky, *Astron. Astrophys.* **66**, 325 (1978).
4. H. Sato, *Modern Phys. Lett.* **A2**, 801 (1987).
5. T. K. Gaissar, A. Hardig and T. Stanev, *Nature* **329**, 314 (1987).
6. V. S. Berezinsky and V. L. Ginzburg, *Nature* **329**, 809 (1987).
7. I. A. Bond et al. (JANZOS), *Phys. Rev. Letters* **60**, 1110 (1988).
8. I. A. Bond et al. (JANZOS), *Phys. Rev. Letters* **61**, 2292 (1988).
9. I. A. Bond et al. (JANZOS), *Astrophys. J.* **344**, L17(1989).
10. Y. Ohyama et al. (KAMIOKANDE), *Phys. Rev. Letters* **59**, 2604 (1987).
11. Y. Tanaka, *Big Bang, Active Galactic Nuclei and Supernovae*, S. Hayakawa and K. Sato (eds.) (Universal Academy Press, 1988), p. 481.
12. K. Masai, S. Hayakawa, H. Inoue, H. Ito K. Nomoto, *Nature* **335**, 805 (1988).
13. M. Honda, H. Sato, and T. Terasawa, *Prog. Theor. Phys.* **82** 315 (1989)
14. J. P. Ostriker and J.E.Gunn, *Astrophy.J.* **164**, L5 (1971).
15. F. Pacini and F. Salvati, *Asrophys. J.* **186**, 249 (1973).
16. M. J. Rees and J. E. Gunn, *Month. Not. R. Astron. Soc.* **167**, 1 (1974).
17. H. Sato and Y. Yamada, *Prog.Theor.Phys.* **85**, No. 3 (1991).
18. K. Nomoto and T. Shigeyama, *Astron.Astrophys.* **196**, 141 (1988).

T. Shigeyama,*‡ T. Kozasa,† and K. Nomoto‡
*Department of Astronomy, University of Tokyo, Tokyo 113, Japan; †Max-Planck Institut für Kernphysik, D6900 Heidelberg 1, F.R.G.; and ‡Max-Planck Institut für Astrophysik, Garching, F.R.G.

How Will the Pulsar in SN 1987A Emerge?

1. INTRODUCTION

Observations of SN 1987A have shown that its radiation has been emitted mostly in the infrared bands after ~ day 600 (Suntzeff et al. 1991). The observed spectrum on day 1030 is well represented by a blackbody function with the color temperature ~ 160 K at $\lambda \leq 20$ μm. The total luminosity of this blackbody radiation is ~ 2×10^{38} ergs s^{-1}, which exceeds that due to any radioactive decays of nuclei in the ejecta. Suntzeff et al. showed that the contribution of *the echo* is at most 20% of the total flux.

From the infrared luminosity L and color temperature T_c obtained as above, the photosphere is estimated to be located at the radius $R_{\rm ph} = (L/4\pi\sigma T_c^4)^{1/2} \sim 2.1 \times 10^{16}$ cm, where σ is the Stefan-Boltzmann constant. Since the ejecta is homologously expanding, the photospheric velocity $v_{\rm ph}$ is evaluated as

$$v_{\rm ph} = R_{\rm ph}/t_{\rm obs} \sim 2.3 \times 10^3 \text{ km s}^{-1} \left(\frac{L}{2 \times 10^{38} \text{ergs s}^{-1}}\right)^{1/2} \left(\frac{T_c}{160 \text{ K}}\right)^{-2} \left(\frac{t_{\rm obs}}{1030 \text{d}}\right)^{-1}.$$

How Will the Pulsar in SN 1987A Emerge?

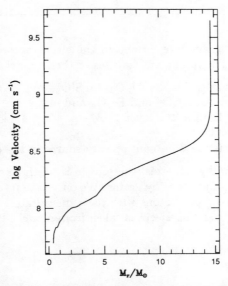

FIGURE 1 The velocity distribution versus mass of a homologously expanding ejecta model, 14E1 (Shigeyama and Nomoto 1990)

According to Shigeyama and Nomoto (1990), the layer with this velocity lies in hydrogen-rich envelope of the ejecta (Fig. 1).

The presence of such an optically thick region in the ejecta may be due to the formation of dust grains, as has been indicated from the infrared light curve and the deformation of redshifted components of atomic emission line features (Lucy et al. 1990). To examine when and at which wavelength the ejecta will be optically thin, we calculate the thermal structure of the ejecta in the radiative equilibrium and obtained the spectra of radiation based on a dust formation model by Kozasa et al. (1990). From the comparison of the calculated spectra including dust features with the observations, we estimate the rate of energy deposition which is necessary to account for the observed fluxes.

Recent ESO observations have shown that the luminosity is decreasing by a factor of ~ 2 in ~ 50 days (Bouchet et al. 1991). Assuming a rapidly changing central energy source, we calculate infrared light curves and discuss the relation between the evolution of the energy source and the emergent luminosities.

We briefly describe our models in the next section. Calculated spectra are compared with the observations in section 3, and we discuss constraints on the dust-forming region of the ejecta. In section 4, the evolution of infrared emission from SN1987A and how the pulsar will emerge are discussed.

2. MODELS

We assume that dust grains are formed in the ejecta according to the model by Kozasa et al. (1990). The important features of their model are

1. Main species of dust grains are Al_2O_3, $MgSiO_3$, and Fe_3O_4. Their typical sizes are $\sim 10^{-3}\mu$m for Al_2O_3 and Fe_3O_4 and $\sim 7 \times 10^{-3}\mu$m for $MgSiO_3$.
2. These dust grains form in the layer $1.0 M_\odot \leq M_r \leq 4.4 M_\odot$, that is, inside the He core of the ejecta.
3. Silicate ($MgSiO_3$) is the dominant opacity source in the ejecta.

Because of the small grain sizes, the albedo is much smaller than unity. Thus we neglect scatterings of photons by grains. We only take into account interactions (absorption and emission) of photons with these dust grains. The density structure of the homologously expanding ejecta is taken from model 14E1 by Shigeyama and Nomoto (1990).

3. SPECTRA

Spectra at day 1000 are calculated by assuming that the ejecta is in radiative equilibrium and that there exists some clumpiness in the ejecta. We solve the radiative equilibrium by Λ iteration method and use the Feautrier scheme with variable Eddington factors to get fluxes (Mihalas 1978). Some clumpiness in the heavy-element layer has been inferred from the observed X-ray and γ-ray light curves (e.g., Kumagai et al. 1989). We describe clumpiness simply by multiplying all the opacities by a factor of $F_{CL} < 1$ (0.1 in this calculation).

As shown in Fig. 2, the calculated spectra are approximately blackbody ones, though they have some silicate features. The temperature distributions are shown in Fig. 3. When we assume that the Co decays are the only energy sources, the calculated luminosities in the photometric bands N0, N1, N2, and Q0 (the dashed curve in Fig. 2) are much lower than the observations. The total infrared luminosity is $\sim 4 \times 10^{37}$ ergs s^{-1}. Clearly, *some additional energy sources are required* to account for these observed fluxes.

If an energy source exists in the center of the ejecta to make the total infrared luminosity 2×10^{38} ergs s^{-1}, the calculated emergent spectrum is harder than the observations (solid curve in Fig. 2). To obtain a softer spectrum, the photosphere should be larger, that is, located at the hydrogen-rich layer. This implies that *we need a mixing of heavy elements in the core into the hydrogen-rich envelope*. The optical depth of the ejecta at each frequency is shown in Fig. 4. The ejecta will be optically thin at frequencies less than 10^{13} Hz on $\sim$ day 2200.

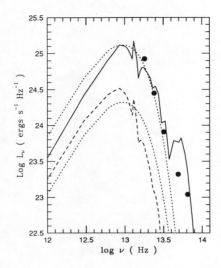

FIGURE 2 The spectra from the dust layer at 1000 d after explosion. The solid curve assumes the central energy source of 2×10^{38} ergs s^{-1}, and the dashed curve assumes that the energy source is due only to Co decays in the ejecta. Filled circles are the observations of ESO at 1030 d (Bouchet et al. 1991). The dotted curves are the blackbody with the temperature of 160 K and the same luminosities as those of the calculated spectra.

4. LIGHT CURVES

Here we assume the existence of a central energy source from which energy is deposited in the ejecta at a constant rate $L_{\rm dep} = 10^{38}$ ergs s^{-1}. The same degree of clumpiness as in section 3 ($F_{\rm CL} = 0.1$) is assumed.

Figure 5 shows the changes in the photospheric temperature ($T_{\rm ph}$; solid curve) and the Rosseland mean optical depth (dashed curve). The ejecta will be optically thin on ~ day 3400, when it will be almost isothermal with a temperature of ~110 K. If the pulsar had a spectrum similar to that of the Crab pulsar, hard X-rays (>~10 keV) would be detected after day ~ 2000.

Recent observations at ESO show that the total luminosity has been decreasing since day 1030 with the time scale as short as ~ 50 d. We calculate the evolution of the emergent luminosities for the clumpiness factors $F_{\rm CL} = 1$, 0.1, and 0.01 (solid curves in Fig. 6) assuming that $L_{\rm dep}$ decreases with the time scale of 50 d (dotted curve in Fig. 6). Fig. 6 clearly shows that the emergent luminosity hardly changes even when $L_{\rm dep}$ changes rapidly, if there is little clumpiness. This is because the diffusion time scale in the ejecta is rather long owing to the large optical depth. *We need significant clumpiness ($F_{\rm CL} \sim 0.01$) to account for the observed rapid change*

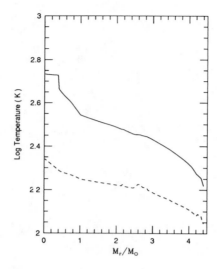

FIGURE 3 The temperature distributions in the ejecta at 1000 d.

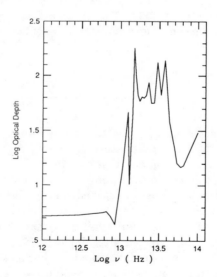

FIGURE 4 The optical depth of the ejecta at each frequency on day 1000 with the clumpiness factor $F_{CL} = 0.1$

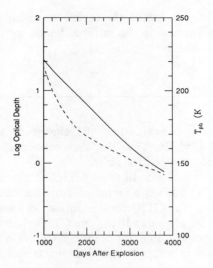

FIGURE 5 The time evolutions of the photospheric temperature (*solid curve*) and Rosseland mean optical depth of the ejecta (*dashed curve*)

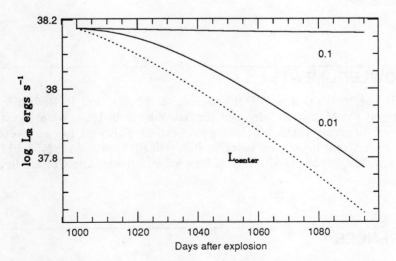

FIGURE 6 The time change in the emergent luminosities (*solid curves*). The numbers indicate the clumpiness factors (see section 4). The dotted curve is the time change in the rate of energy deposition from the central source.

in the total luminosity. We can expect the earlier emergence of the pulsar not only in the hard X-ray bands but also in the infrared bands with such clumpiness.

5. DISCUSSION

If the clumpiness factor $F_{\rm CL}$ is as small as 0.01, the ejecta is already optically thin for the radiation at $\nu \leq 10^{13}$ Hz. Thus the observations at $\lambda \geq 30$ μm could give us some information about the central pulsar.

We should consider the observations of CTIO at the same epoch. The total luminosity of CTIO on day 1055 is a factor of 3 smaller than that of ESO on day 1030 (Suntzeff et al. 1991). The CTIO total luminosities have not shown such rapid changes as observed by ESO after day 1030. The CTIO observations can be ascribed to the Co decays (the necessary abundance ratio ^{57}Co/^{56}Co is about twice as much as the solar abundance) and a significant clumpiness such as $F_{\rm CL} \sim 0.01$ is not necessary.

As shown in Fig. 4, the optical depth of the dust layer strongly depends on frequencies. To discuss the infrared emission from SN 1987A after 1000 d using the photometric data, which cover a part of the energy range of emissions, it is necessary to calculate infrared spectra from the ejecta based on a reliable dust formation model.

ACKNOWLEDGMENTS

We would like to thank Drs. W. Hillebrandt, E. Müller, and P. Höflich for their hospitality at the Max-Planck Institut für Astrophysik in 1990. We are grateful to Dr. Suntzeff for giving us the data before publication. This work has been supported in part by the Grant-in-Aid for Scientific Research (01540216, 01652503, 01790169, 02234202, 02302024) of the Ministry of Education, Science, and Culture in Japan.

REFERENCES

1. Bouchet, P., et al., 1991 in *SN 1978A and Other Supernovae*, ed. I. J. Danziger (Garching: ESO), *in press*.
2. Kozasa, T., Hasegawa, H., and Nomoto, K., 1990, *submitted to Astron. and Astrophys*.
3. Kumagai, S., Shigeyama, T., Nomoto, K., Itoh, M., Nishimura, J., and Tsuruta, S., 1989, *Ap. J.*, **345**, 412.

4. Lucy, L. B., Danziger, I. J., Gouiffes, C., and Bouchet, P., 1990, in *Structure and Dynamics of Interstellar Medium, IAU Colloq. No. 120*, eds. G. Tenorio-Tagle, M. Mole, and J. Melnick, p. 164.
5. Mihalas, D., 1978, in *Stellar Atmospheres* (San Francisco: W. H. Freeman), p. 250.
6. Shigeyama, T., and Nomoto, K., 1990, *Ap. J.*, **360**, 242.
7. Suntzeff, N. B., et al., 1991, preprint.

Friedrich-Karl Thielemann*, Ken'ichi Nomoto[†] and Masa-aki Hashimoto[‡]
*Harvard-Smithsonian Center for Astrophysics, 60 Garden St., Cambridge, MA 02138, USA; [†]Department of Astronomy, Faculty of Science, University of Tokyo, Bunkyo-ku, Tokyo 113, Japan; and [‡]Department of Physics, College of General Education, Kyushu University, Rapponmatsu, Fukuoka 810, Japan

Neutron Star Masses from Supernova Explosions

We utilize models for 13, 15, 20, and $25 M_\odot$ stars by Nomoto and Hashimoto (1988) to perform explosive nucleosynthesis calculations with explosion energies leading to kinetic energies of the remnant of 10^{51} ergs. Such values lie within the uncertainty range inferred from bolometric light curves. Considering constraints on the amount of ejected radioactive ^{56}Ni from the light curve of SN 1987A (a $20 M_\odot$ star), from Type Ib and Ic light curves (for 13–$15 M_\odot$ progenitor stars), and from the average nucleosynthesis by Type II supernovae in Galactic chemical evolution, we predict the position of the mass cut between the neutron star and the ejecta of these models. This is an indirect way, necessary until consistent collapse and explosion calculations with realistic explosion energies exist. The main error sources are uncertainties in explosion energies and the time interval between collapse and the emergence of a shock wave in delayed explosions. The resulting neutron star masses populate the whole range $1.1 < M/M_\odot < M_{max}$, where M_{max} is the maximum mass in units of $M_\odot$, allowed by the nuclear equation of state.

1. INTRODUCTION

Type II supernova explosions, the endpoint of the evolution of massive stars, lead to the formation of neutron stars or black holes. The long predicted Fe-core collapse to nuclear densities (see, e.g., Bethe 1990) was finally proved by the neutrino emission of supernova 1987A, detected in the Kamiokande and IMB experiments (Hirata et al. 1987; Bionta et al. 1987; see Burrows 1990 for an overview). The total energy released in neutrinos of $2-3 \times 10^{53}$ ergs equals the gravitational binding energy of a neutron star, because neutrinos are the particles with the longest mean free path, being able to carry away that energy in the fastest fashion. Whether a neutron star or black hole is formed depends on the size of the collapsing Fe core and the maximum neutron star mass, which is somewhat uncertain and related to our still limited understanding of the nuclear equation of state.

A major uncertainty is the lack of a complete understanding of the Type II supernova mechanism. Apparently, present calculations do not lead to successful explosions with the prompt mechanism, in which the shock wave, created at the edge of the collapsed core at bounce, can cause the ejection of the outer envelope (see, e.g., Bruenn 1989a,b; Cooperstein and Baron 1989; Baron and Cooperstein 1990; Myra and Bludman 1989). The remaining promising mechanism is a delayed explosion, caused by neutrino heating on a time scale of seconds (diffusion time scale of neutrinos from the core). Less than 1% of the total neutrino energy would suffice to explain the kinetic ejection energies, observed in supernova remnants (Wilson and Mayle 1988; Mayle and Wilson 1988, 1990; Bethe 1990). However, present calculations still have difficulties to explain such energies (Bruenn and Haxton 1991). Even though it is not possible to obtain such direct information about the mass cut between the forming neutron star and the ejected envelope, we can nevertheless make use of some indirect information.

Three sets of arguments are helpful with respect to neutron star masses:
1. Core Si burning leads typically to values of the electron fraction $Y_e = <Z/A>$ that produce an abundance composition that cannot be a major component of solar abundances (see Thielemann and Arnett 1985), nor does the Fe-group composition of low-metallicity stars (reflecting the early evolution of the Galaxy, during which only SN II events occurred) allow for such matter to be a main component of SN II ejecta (Gratton 1990). Only minor amounts, of the order of $10^{-5}-10^{-6} M_\odot$ of neutron-rich matter in the form of r-process elements or Fe-group nuclei, are allowed to be ejected per SN II event. This is negligible and does not affect any consideration of neutron star masses. Therefore, the absolute lower limit for neutron star masses from SN II explosions is given by the Fe-core masses of progenitor stars. This information is available from a grid of stellar models at the onset of Fe-core collapse (e.g., Nomoto and Hashimoto 1988).
2. Explosive nucleosynthesis calculations predict the products of explosive C, Ne, O, and Si burning. Si burning with complete Si exhaustion leads to the dominant formation of ^{56}Ni for $Y_e > 0.49$ (Woosley, Arnett, and Clayton 1973). This

is the ^{56}Ni which powers the light curve of all supernovae with its decay to ^{56}Co and ^{56}Fe. With known distances to a supernova one can deduce the required amount of ^{56}Ni. In the case of SN 1987A it was $0.075 \pm 0.01 M_\odot$ and in that case we also knew the mass of the progenitor star. As ^{56}Ni is produced in the innermost ejected zones, which experience the highest temperatures, an integration of the explosive nucleosynthesis products for a specific stellar model will imply the location of the mass cut between ejected envelope and the remaining neutron star (see, e.g., Thielemann, Hashimoto, and Nomoto 1990). While we also observe light curves and know the distances for other SN events, it is harder to make the connection between the supernova and its progenitor mass. However, recent understanding of the deviation of light curves from a pure ^{56}Co exponential decay gives an indication about the envelope masses involved and the distribution of ^{56}Co in the ejecta, mixed outward by Rayleigh-Taylor instabilities behind the shock wave (Shigeyama, Nomoto, and Hashimoto 1988; Woosley 1988; Arnett and Fu 1989; Arnett, Fryxell, and Müller 1989; Fryxell, Arnett, and Müller 1991; Hachisu et al. 1990; Herant and Benz 1991; Ensman and Woosley 1988; Shigeyama et al. 1990; Nomoto, Filippenko, and Shigeyama 1990).

3. SNe Ib and SNe Ic seem to be events related to the core collapse of massive stars but without (Ib) or only with a minute (Ic) H envelope, therefore classified as SNe I. The steepness of their light curves, deviating early from the pure exponential decline with the ^{56}Co half-life, indicates small He cores between 3 and $4 M_\odot$, which qualify them as originating from $13\text{--}15 M_\odot$ progenitor stars. Such stars do not lose their H envelope as single stars by a strong stellar wind (Wolf-Rayet stars) but only in binary systems (for details see the discussion in Nomoto, Filippenko, and Shigeyama 1990; Shigeyama et al. 1990; Wheeler and Harkness 1990). The light curve information requires ejected ^{56}Ni masses of $\approx 0.15 M_\odot$ for both events, larger than the $0.075 M_\odot$ for SN 1987A, a $20 M_\odot$ star. This gives us information for some grid points in the mass spectrum of SNe II progenitors. The available information for intermediate-mass elements like O and Ca, which are not affected by the choice of a mass cut, and the known ratios of [O/Fe] through [Ca/Fe] from abundance observations in low-metallicity stars, [Fe/H] ≤ -1, give an additional constraint on the permitted or required Ni(Fe) integrated over all SN II progenitor masses and thus set limits for the Fe ejecta of a $25 M_\odot$ star.

2. EXPLOSIVE NUCLEOSYNTHESIS AND Ni(Fe) EJECTA

Hashimoto, Nomoto, and Shigeyama (1988) and Thielemann, Hashimoto, and Nomoto (1990) performed explosive nucleosynthesis calculations for a $20 M_\odot$ star ($6 M_\odot$ He core) and obtained detailed nucleosynthesis products (for a comparison see also Woosley, Pinto, and Weaver 1988). The outer boundary of explosive Si burning with

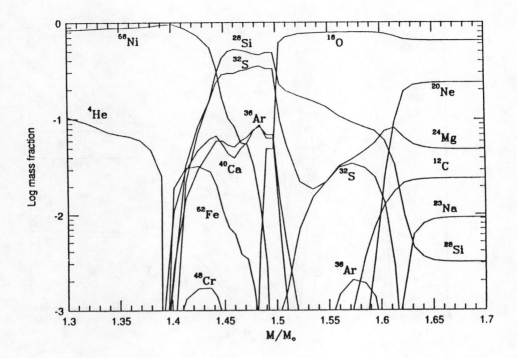

FIGURE 1 Mass fractions of a few major nuclei after passage of the supernova shock front through a $13 M_\odot$ star. Matter outside $1.65 M_\odot$ is essentially unaltered. Mass zones further in experience explosive Si, O, Ne, and C burning. In order to eject $0.15 M_\odot$ of ^{56}Ni the mass cut between neutron star and ejecta must be located at $1.29 M_\odot$.

complete Si exhaustion is located at $1.7 M_\odot$, where temperatures of 5×10^9 K are attained. A pure ^{56}Ni composition inside this boundary would require the mass cut to be at $1.63 M_\odot$, with the ejection of $0.07 M_\odot$ of ^{56}Ni. Making use of the slightly more neutron-rich composition within the vicinity of the Si-burning shell (smaller Y_e) reduces somewhat the mass fraction of ^{56}Ni and about 30% in mass are found in more neutron-rich Fe-group nuclei. In order to obtain the same ^{56}Ni ejection, a slightly deeper mass cut is required. Accounting also for observational uncertainties leads to a value of 1.59–$1.63 M_\odot$.

Recently Thielemann, Nomoto, and Hashimoto (1991) also performed similar calculations for 13, 15, and $25 M_\odot$ stars (3.3, 4, and $8 M_\odot$ He cores, respectively). The results for the whole sequence are shown in Figs. 1 through 4. Only the dominant abundances are plotted. The outer boundary of explosive Si burning with complete Si exhaustion is located in these cases at 1.43, 1.47, and $1.82 M_\odot$, respectively. When ejecting $0.15 M_\odot$ of ^{56}Ni from the 13 and $15 M_\odot$ stars, this would lead to mass cuts

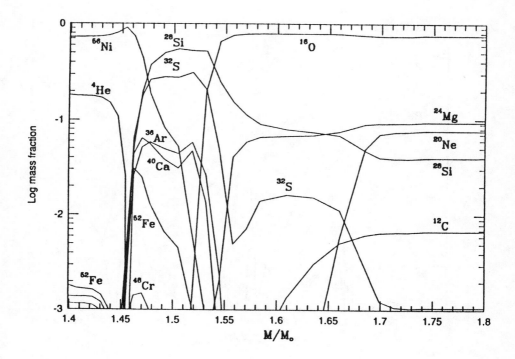

FIGURE 2 Mass fractions of a few major nuclei from explosive processing after the passage of the supernova shock front through a $15M_\odot$ star

at 1.28 and $1.32 M_\odot$ in the case of explosive Si burning with small neutron excess $\eta = 1 - 2Y_e$ or large $Y_e (> 0.49)$. Thielemann, Nomoto, and Hashimoto (1991) found that amounts of 0.05–$0.10 M_\odot$ of ^{56}Ni for the $25 M_\odot$ star are consistent with chemical evolution models and the [O/Fe] through [Ca/Fe] ratios observable in low-metallicity stars.

All of these calculations were performed by depositing energy at a radius inside the Fe core and letting the shock wave propagate outward, which causes explosive burning. In reality the stellar models *at the time t, when the successful shock wave is initiated,* have to be utilized. Instead they were taken at the onset of core collapse. Aufderheide, Baron, and Thielemann (1991) performed a calculation with a model at 0.29 s after core collapse for a $20 M_\odot$ star, when the prompt shock had failed, and found an increase of the mass cut by roughly 0.03–$0.05 M_\odot$. A delayed explosion would set in after a delay of about 1 s, with the exact time being somewhat uncertain.

As mentioned before, the outer boundary of explosive Si burning with Si exhaustion is the outer boundary of ^{56}Ni production. It was shown that this corresponds

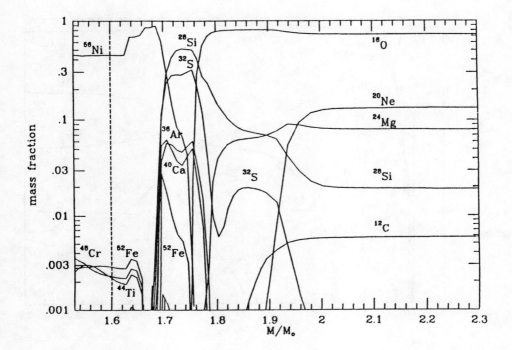

FIGURE 3 The same as Fig. 1 for a $20 M_\odot$ star

approximately to a radius R (Weaver and Woosley 1980; Thielemann, Hashimoto, and Nomoto 1990) with

$$E_{SN} = \frac{4\pi}{3} R^3 a T^4. \tag{1}$$

For $T = 5 \times 10^9$ K and $E_{SN} \approx 10^{51}$, this leads to $R = 3700$ km. The mass cut would then be at

$$M_{cut} = M(R) - M_{ej}(^{56}\text{Ni}). \tag{2}$$

In case of a delayed explosion, we have to ask the question from which radius $R_0(0)$ matter fell in, which is located at radius $R(t) = 3700$ km when the shock wave emerges at time t. Then the mass cut is not related to $M[R(0) = 3700 \text{ km}]$ but to $M[R(t) = 3700 \text{ km}] = M[R_0(0)]$ with

$$M_{cut} = M[R_0(0)] - M_{ej}(^{56}\text{Ni}) = M[R(0) = 3700 \text{ km}] + \Delta M_{acc} - M_{ej}(^{56}\text{Ni}). \tag{3}$$

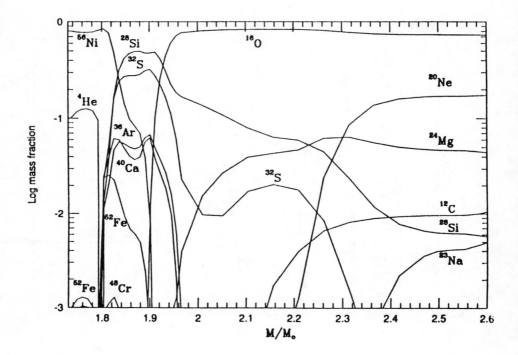

FIGURE 4 The same as Fig.1 for a 25M$_\odot$ star

When we assume free-fall velocity and an attracting mass M inside the radius R, the time for infall is related to $R(t)$ and $R_0(0)$ by

$$t_f = (2GM)^{-1/2} \int_R^{R_0} (\frac{1}{R'} - \frac{1}{R_0})^{-1/2} dR' \qquad (4)$$

(see Bethe 1990 and his Eq. 6.37). The solution given by Bethe is only valid for $R/R_0 \ll 1$; here we want to present the general result,

$$t_f = (2GM)^{-1/2} R_0^{3/2} [\arctan(\sqrt{R_0/R - 1}) - \sqrt{R/R_0(1 - R/R_0)}]. \qquad (5)$$

For typical core masses M, $(2GM)^{1/2}$ is of the order 2×10^{13} cm^3 s^{-2}, which we will take as a constant for all cores, because the square root has a weak mass dependence. Taking a typical time scale of 1 s for delayed explosions, this would lead to $R_0(0) = 11900$ km.

This, however, neglects the sound travel time to R_0 before infall can set in. If R_s marks the location of the outmoving rarefaction wave at core bounce, the delay time consists of two parts, $t = t_s(R_s, R_0) + t_f(R_0, R)$, and has to be solved for $t \approx 1$ s. With a typical sound velocity $c_s = (\Gamma P/\rho)^{1/2}$ of 5000 km s^{-1} close to the Fe core, a sound wave would only travel from $R = 3700$ km to 8700 km within 1 s and no infall would have started at 11900 km. Therefore, we find $R_0 \approx 7500$ km with $t_s \approx 0.7$ s and $t_f \approx 0.3$ s, approximating R_s by R. We can test the validity of this reasoning with a calculation we performed with a model taken at 0.29 s after bounce (Aufderheide, Baron, and Thielemann 1991). In that case $R_0 \approx 5000$ km for $t_s \approx 0.23$ s and $t_f \approx 0.06$ s, corresponding to $\Delta M_{acc} \approx 0.05 M_\odot$, as also found in the actual hydro-dynamic calculation.

3. NEUTRON STAR MASSES
3.1 BARYONIC MASSES

When taking into account the results of the previous section, we obtain baryonic neutron star masses M_b for the sequence of 13, 15, 20, and 25$M_\odot$ SN II progenitors as noted in Table 1. Column 1 indicates the progenitor mass. Column 2 gives the original Fe core mass, the absolute lower limit for M_b. The third column lists the outer boundary of explosive Si burning with Si exhaustion M_{Si-ex}, which represents the outer edge of ^{56}Ni production. The fourth column indicates the location of the mass cut under the assumption that the amount of ^{56}Ni ejected is that required from light curve and/or chemical evolution arguments, using Eq. (2) (i.e., assuming a prompt explosion). In addition one has to take into account the following error sources:

1. Y_e in the zones of explosive Si burning: A $Y_e < 0.49$ leads to a reduction of the fraction of ^{56}Ni produced in this zone and could require a deeper mass cut and a smaller neutron star mass in order to fulfill the same constraint on the ejected amount of ^{56}Ni.
2. Uncertainties in the explosion energy: The hydrodynamic calculations were performed under the assumption that the typical kinetic energy observed in supernova remnants is of the order 10^{51} ergs. This requires the deposition of a somewhat larger amount of energy inside the mass cut because some potential energy has to be lifted as well. The observed light curves for SN 1987A and SNe Ib/c underline this finding. The radius R for a given temperature T is related to the explosion energy (expanding radiation bubble, see Eq. [1]); that is, the radius at the outer boundary of complete Si burning with $T = 5 \times 10^9$ K (where only Fe-group nuclei are produced) is proportional to $E^{1/3}$. We assume here that the deposited energy is uncertain by 50%. The neutron star mass cut moves with the mass boundary M_{Si-ex}. The resulting uncertainties are listed in column 5.
3. A final and probably the largest source of uncertainty is our lack of a complete understanding of the SN II explosion mechanism. The present analysis, utilizing precollapse models and depositing the correct explosion energy, takes into

account the philosophy of a prompt explosion (see, e.g., Aufderheide, Baron, and Thielemann 1991). In the case of a delayed explosion, accretion onto the proto–neutron star will occur for a duration of seconds before a shock wave is formed, leading to the ejection of the outer layers. As Mayle and Wilson (1988) have shown, masses of the order of 0.1–$0.3 M_\odot$ can be accreted and the neutron star boundary would have to be moved outward, accordingly. We estimate ΔM_{acc} in column 6 by the difference in mass from Eqs. (2) and (3) and using $R_0 = 7500$ km.

3.2 GRAVITATIONAL MASSES

All of the previous dicussion was related to the baryonic mass of the neutron star. The proto–neutron star with a baryonic mass M_b will release a binding energy of about E_{bin} in the form of blackbody radiation in neutrinos during its contraction to neutron star densities. This has been observed with the Kamiokande II and IMB detectors for SN 1987A (Hirata et al. 1987; Bionta et al. 1987; for a general discussion see also Burrows 1990). The gravitational mass is then given by

$$M_g = M_b - E_{bin}/c^2. \tag{6}$$

For reasonable uncertainties in the equation of state Lattimer and Yahil (1989) obtained a relation between gravitational mass and binding energy

$$E_{bin} = (1.5 \pm 0.15) \left(\frac{M_g}{M_\odot}\right)^2 \times 10^{53} \text{ ergs}, \tag{7}$$

$$M_{bin} = \frac{E_{bin}}{c^2} = M_b - M_g = (0.0839 \pm 0.0084) \left(\frac{M_g}{M_\odot}\right)^2 M_\odot c^2. \tag{8}$$

Applying Eq. (8) results in a gravitational mass of the formed neutron star M_g as listed in column 3 of Table 2. This number does include the uncertainties discussed in section 3.1 and Table 1. ΔM_{acc} dominates the uncertainties and thus the lower bound is close to the result for a prompt explosion, while the upper bound represents a delayed explosion.

TABLE 1 Mass Cut in SN II Events

$M/M_\odot$	M_{core}	M_{Si-ex}	M_{cut}	ΔM_E	ΔM_{acc}
13	1.18	1.43	1.28	0.03	0.14
15	1.28	1.47	1.32	0.03	0.15
20	1.40	1.70	1.60	0.03	0.17
25	1.61	1.82	1.72	0.03	0.21

TABLE 2 Neutron Star Masses

$M/M_\odot$	M_b	M_g
13	1.25-1.45	1.13-1.32
15	1.29-1.50	1.16-1.36
20	1.57-1.80	1.39-1.60
25	1.69-1.96	1.49-1.73

4. DISCUSSION

It should first be noted that the discussion in this paper does not present straightforward results from detailed collapse and explosion calculations with completely understood physics of the explosion mechanism. It is based on the precollapse models from stellar evolution and constraints on the observed or required amount of ejected ^{56}Ni. The numbers discussed for SN 1987A agree remarkably well with the ones obtained from the observed neutrino emission (Burrows 1990). A number of error sources have been taken into account, the largest being introduced by accretion of matter during a delayed explosion. Nevertheless, the results indicate a clear spread of neutron star masses, and it is unlikely that a conspiracy of the different error sources would lead to a unique neutron star mass, for instance, due to an increasing amount of ^{56}Ni ejection with progenitor mass. It is also now apparent that such a spread in neutron star masses is found in observations (e.g., Nagase 1989; Page and Baron 1990) and not only because of large observational errors. Whether this observed spread already includes the uncertain upper mass limit of neutron stars due to the nuclear equation of state cannot be answered by the analysis. We also did not make an attempt to quantify uncertainties entering the progenitor models from the stellar evolution calculations.

ACKNOWLEDGMENTS

We want to thank our collaborators M. Aufderheide, E. Baron, T. Shigeyama, and T. Tsujimoto who contributed to the material presented here. We also want to thank M. Herant and W. Benz for stimulating discussions and comments on the manuscript. This research was supported in part by NSF grant AST 89-13799,

NASA grant NGR 22-007-272, the Grants-in-Aid for Scientific Research of the Ministry of Education, Science, and Culture in Japan (02234202, 02302024, 03218202), and the U.S.-Japan Cooperative Science Program (INT 88-15999/ EPAR-071) operated by the NSF and JSPS. The computations were performed at the National Center for Supercomputer Applications at the University of Illinois (AST 890009N).

REFERENCES

Arnett, W.D., Fryxell, B., Müller, E. 1989, *Ap. J. Lett.* **341**, L63
Arnett, W.D., Fu, A. 1989, *Ap. J.* **340**, 396
Aufderheide, M., Baron, E., Thielemann, F.-K. 1991, *Ap. J.* **370**, 630
Baron, E., Cooperstein, J. 1990, in *Supernovae*, ed S.E. Woosley, (Springer-Verlag, New York), p.342
Bethe, H.A. 1990, *Rev. Mod. Phys.* **62**, 801
Bionta, R.M. et al. 1987, *Phys. Rev. Lett.* **58**, 1494
Bruenn, S.W. 1989, *Ap. J.* **340**, 955
Bruenn, S.W. 1989, *Ap. J.* **341**, 385
Bruenn, S.W., Haxton, W.C. 1991, *Ap. J.* **376**, 678
Burrows, A. 1990, *Ann. Rev. Nucl. Part. Sci.* **40**, 181
Cooperstein, J., Baron, E. 1989, in *Supernovae*, ed A. Petschek, (Springer-Verlag, New York), p.213
Ensman, L., Woosley, S.E. 1988, *Ap. J.* **333**, 754
Fryxell, B.A., Arnett, W.D., Müller, E. 1991, *Ap. J.* **367**, 619
Gratton, R. 1991, *Evolution of Stars*, IAU Symp. 145, ed. G. Michaud, (Reidel, Dordrecht), in press
Hachisu, I., Matsuda, T., Shigeyama, T., Nomoto, K., 1990, *Ap. J. Lett.* **358**, L57
Hashimoto, M., Nomoto, K., Shigeyama, T. 1989, *Astron. Astrophys.* **210**, L5
Herant, M., Benz, W. 1991 *Ap. J. Lett.* **370**, L81
Hirata, K. et al. 1987, *Phys. Rev. Lett.* **58**, 1490
Lattimer, J.M., Yahil, A. 1989, *Ap. J.* **340**, 426
Mayle, R.W., Wilson, J.R. 1988, *Ap. J.* **334**, 909
Mayle, R.W., Wilson, J.R. 1990, in *Supernovae*, ed S.E. Woosley, (Springer-Verlag, New York), p.333
Myra, E.S., Bludman, S. 1989, *Ap. J.* **340**, 384
Nagase, F. 1989, *Publ. Astron. Soc. Japan* **41**, 1
Nomoto, K., Hashimoto, M. 1988, *Phys. Rep.* **163**, 13
Nomoto, K., Filippenko, A.V., Shigeyama, T. 1990, *Astron. Astrophys.* **240**, L1
Page, D., Baron, E. 1990, *Ap. J. Lett.*, **354**, L17
Shigeyama, T., Nomoto, K., Hashimoto, M. 1988, *Astron. Astrophys.* **196**, 141
Shigeyama, T., Nomoto, K., Tsujimoto. T.,Hashimoto, M. 1990, *Ap. J. Lett.* **361**, L23
Thielemann, F.-K., Arnett, W.D. 1985, *Ap. J.* **295**, 604

Thielemann, F.-K., Hashimoto, M., Nomoto, K. 1990, *Ap. J.* **349**, 222
Thielemann, F.-K., Nomoto, K., Hashimoto, M. 1991, in *Supernovae, Les Houches, Session LIV*, eds. J. Audouze, S. Bludman, R. Mochkovitch, J. Zinn-Justin (Elsevier, Amsterdam), in press
Weaver, T.A., Woosley, S.E. 1980, *Ann. N.Y. Acad. Sci.* **366**, 335
Wheeler, J.C., Harkness, R.P. 1990, *Rep. Prog. Phys.* **53**, 1467
Wilson, J.R., Mayle, R.W. 1988, *Phys. Rep.* **163**, 63
Woosley, S.E. 1988, *Ap. J.* **330**, 218
Woosley, S.E., Arnett, W.D., Clayton, D.D. 1973, *Ap. J. Suppl.* **26**, 231
Woosley, S.E., Pinto, P.A., Weaver, T.A. 1988, *Proc. Astron. Soc. Australia* **7**, 355

K. Nomoto, H. Yamaoka, and T. Shigeyama
Department of Astronomy, Faculty of Science, University of Tokyo, Tokyo 113, Japan

Evolutionary Origin of Binary Pulsars

Possible evolutionary scenarios for the formation of binary pulsars are presented. (1) In relation to the origin of low-mass X-ray binaries and binary millisecond pulsars, conditions for the occurrence of accretion-induced collapse of white dwarfs are examined. The outcome of the evolution of accreting white dwarfs is summarized as functions of accretion rate and the initial mass of the white dwarf. (2) We discuss neutron star formation in Type Ib/Ic supernovae (SNe Ib/Ic), which are likely to be due to core collapse of helium stars in binary systems. (3) The evolutionary origin of double neutron stars is discussed in relation to SNe Ib/Ic.

1. LOW-MASS BINARY PULSARS AND DOUBLE NEUTRON STARS

We present possible evolutionary scenarios for the formation of binary pulsars, in particular, low-mass binary pulsars and double neutron star systems. Theoretically, the final forms of the stars in interacting binary systems are predicted to be either white dwarfs or helium stars of mass $\gtrsim 2.5 M_\odot$ after the loss of their hydrogen-rich envelope by Roche lobe overflow. The helium stars evolve through the Fe core collapse and undergo supernova explosions, which would be observed as Type Ib/Ic supernovae. White dwarfs, if accreting matter from companion stars, would either explode or collapse depending on the conditions of binary systems. These scenarios of neutron star formation in binary stars have been suggested to be related to the origin of some binary pulsars as follows:

(1) An unexpectedly large number of low mass binary pulsars (LMBPs) have recently been discovered. The birth rate of LMBPs is now estimated to be about 100 times higher than that of low mass X-ray binaries (LMXBs) in both the Galactic disk (Kulkarni and Narayan 1988; Narayan et al. 1990) and the globular clusters (Kulkarni et al. 1990; Romani 1990). Since LMXBs have been thought to be the progenitors of LMBPs, this birthrate discrepancy has raised a serious question about the evolutionary origin of LBMPs. Two scenarios have been proposed to resolve this problem: (a) accretion-induced collapse (AIC) of white dwarfs in close binaries (Michel 1987; Chanmugam and Brecher 1987; Bailyn and Grindlay 1990; Romani 1990; Ray and Kluzniak 1990), and (b) shortening of the LMXB phase due to the evaporation of the companion star (e.g., Tavani 1991 and references therein). Further, combinations of AIC and the tidal capture of neutron stars have been suggested as an explanation for the very high incidence of LMBPs in globular clusters (Romani 1990; Ray and Kluzniak 1990).

(2) Recently the masses of the component stars in the binary pulsar system 1532+12 have been determined strongly suggest a neutron star companion (Wolszczan 1991). The observed binary parameters raise some interesting questions about the evolutionary origin of such a double neutron star system, which might be related to the questions regarding the progenitors of Type Ib/Ic supernovae.

2. CONDITIONS FOR ACCRETION-INDUCED COLLAPSE OF WHITE DWARFS

The scenario that possibly brings a close binary system to a Type Ia supernova (SN Ia) or AIC is as follows (although the exact binary system has not been identified): Initially the close binary system consists of two intermediate mass stars ($M \lesssim 8 M_\odot$). As a result of Roche lobe overflow, the primary star of this system becomes a white dwarf composed of C+O or O+Ne+Mg. When the secondary star evolves, it begins to transfer hydrogen-rich matter over to the white dwarf.

The mass accretion onto the white dwarf releases gravitational energy at the white dwarf surface. Most of the released energy is radiated away from the shocked region as UV and does not contribute much to heating the white dwarf interior. The continuing accretion compresses the previously accreted matter and releases gravitational energy in the interior. A part of this energy is transported to the surface and radiated away from the surface (radiative cooling) but the rest goes into thermal energy of the interior matter (compressional heating). Thus the interior temperature of the white dwarf is determined by the competition between compressional heating and radiative cooling; that is, the white dwarf is hotter if the mass accretion rate $\dot{M}$ is larger, and vice versa (e.g., Nomoto 1982a).

When a certain amount of hydrogen ΔM_H is accumulated on the white dwarf surface, hydrogen shell burning is ignited (Nariai and Nomoto 1979; Nomoto 1982a). Its outcome depends on $\dot{M}$ as follows:

(1) For slow accretion ($\dot{M} \lesssim 1 \times 10^{-9} M_\odot \text{yr}^{-1}$), hydrogen shell burning is unstable to *flash*, which leads to the ejection of most of the accreted matter from the white dwarf; the strongest flash grows into a *nova* explosion (e.g, Nariai et al. 1980 and references therein). For these cases, the white dwarf does not become a supernova since its mass hardly grows. In other words, it seems rather unlikely that novae are the precursors of supernovae or AIC.

(2) If the accreting material is helium for $\dot{M} \lesssim 1 \times 10^{-9} M_\odot \text{yr}^{-1}$, the material is too cold to ignite helium burning, thereby increasing the white dwarf mass. An exception is the case with $M_{CO} \lesssim 1.1 M_\odot$ where pycnonuclear helium burning is ignited (Nomoto 1982a,b).

(3) For $\dot{M}_{det} \gtrsim \dot{M} \gtrsim 10^{-9} M_\odot \text{yr}^{-1}$, off-center helium detonation prevents the white dwarf mass from growing (Nomoto 1982b; Woosley et al. 1986). Here we adopt $\dot{M}_{det} \sim 1 \times 10^{-8} M_\odot \text{yr}^{-1}$, since the $^{14}N(e^-, \nu)^{14}C(\alpha, \gamma)^{18}O$ (NCO) reaction ignites weak helium flashes (Hashimoto et al. 1986) if the mass fraction of NCO elements in the accreting material exceeds 0.005. For smaller NCO abundances, the NCO reaction is not effective and thus $\dot{M}_{det} \sim 4 \times 10^{-8} M_\odot \text{yr}^{-1}$ (Nomoto 1982a).

(4) For intermediate accretion rates ($3 \times 10^{-6} M_\odot \text{yr}^{-1} \gtrsim \dot{M} \gtrsim 1 \times 10^{-8} M_\odot \text{yr}^{-1}$), hydrogen flashes and the subsequent helium flashes are of moderate strength, thereby recurring many times to increase the white dwarf mass. When the white dwarf mass becomes close to the Chandrasekhar mass, either thermonuclear explosion or collapse occurs (§3).

(5) If the accretion rate is higher than $\sim 2 \times 10^{-6} M_\odot \text{yr}^{-1}$, the accreted matter is too hot to be swallowed by the white dwarf (Nomoto et al. 1979b). The matter forms a common envelope, which is eventually lost from the system. As a result of mass and angular momentum losses from the system, some binaries form a pair of white dwarfs. Further evolution of such a double white dwarf system is driven by gravitational wave radiation and leads to a Roche lobe overflow of the smaller mass white dwarf (Iben and Tutukov 1984; Webbink 1984).

(6) Though the fate of these merging white dwarfs is not clear yet (i.e., whether SN Ia or AIC), we adopt the following scenario for $\dot{M} \gtrsim 2 \times 10^{-6} M_\odot \text{yr}^{-1}$: First, merging of double C+O white dwarfs forms a thick disk around the more massive component (Benz et al. 1990). Subsequent heat generation at the boundary layer ignites off-center carbon burning (Mochkovitch and Livio 1990), which burns the entire C+O white dwarf into O+Ne+Mg quietly (Nomoto and Iben 1985; Saio and Nomoto 1985). Eventually the O+Ne+Mg white dwarf collapses.

(7) For C+O white dwarfs, the outcome of accretion depends not only on $\dot{M}$ but also on the initial mass M_{CO}. For $M_{CO} < 1.2 M_\odot$, substantial heat inflow from the surface layer into the central region ignites carbon at relatively low central density

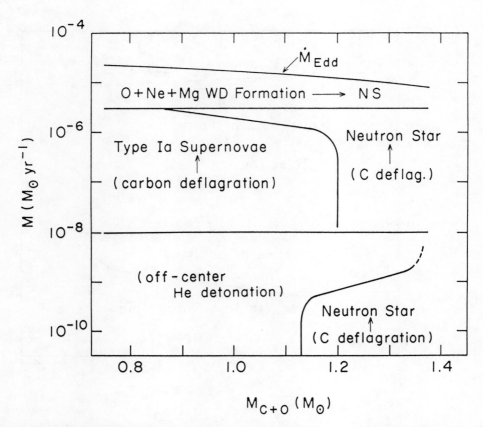

FIGURE 1 The final fate of accreting C+O white dwarfs expected for their initial mass and accretion rate $\dot{M}$ (Nomoto and Kondo 1991).

($\rho_c \sim 3 \times 10^9$ g cm^{-3}), which make SNe Ia (Nomoto et al. 1984). On the other hand, if the white dwarf is sufficiently massive and cold at the onset of accretion, the central region is compressed only adiabatically, thereby being cold (and solid) when carbon is ignited in the center of density as high as 10^{10} g cm^{-3}. This case of high ignition density is expected to lead to AIC (§3.1).

The fate of the white dwarf after ignition at high densities will be discussed in §3. With these results, we draw boundaries for AIC in a diagram of mass accretion rate ($\dot{M}$) versus mass of the white dwarf at the onset of accretion (M_{CO} and M_{ONeMg}) in Fig. 1 and Fig. 2 (Nomoto 1986; Nomoto and Kondo 1991). We note that the boundaries must be regarded as relatively optimistic ones for the growth of white dwarfs since wind-type mass loss associated with shell flashes of hydrogen and helium is not fully taken into account (e.g., Kato and Hachisu 1989).

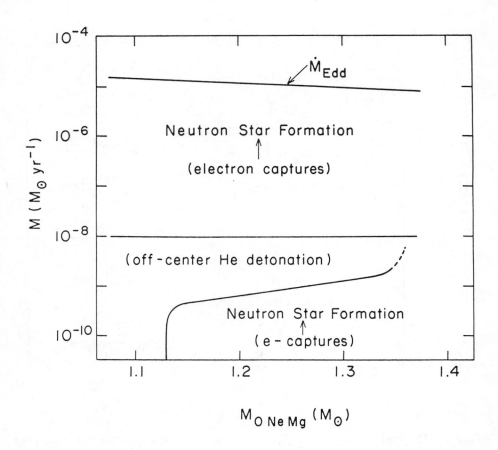

FIGURE 2 Same as Figure 1 but for O+Ne+Mg white dwarfs. Collapse is triggered by electron capture on ^{24}Mg and ^{20}Ne (Nomoto and Kondo 1991).

Figure 2 shows that for a relatively wide parameter range the O+Ne+Mg white dwarf can increase its mass. Since M_{ONM} can be very close to the Chandrasekhar mass, only a small increase in mass is enough to trigger a collapse. Such very massive O+Ne+Mg white dwarfs would give rise to recurrent novae (Nariai and Nomoto 1979).

Figures 1 and 2 clearly show that close binaries with relatively high $\dot{M}$ and high initial white dwarf mass are favored for AIC. This leads to the possibility that LMBPs with relatively long orbital periods may originate from AIC (Nomoto 1987b; Romani 1990; Ray and Kluzniak 1991) since the mass transfer rate from giant stars may be relatively high. On the other hand, Wheeler (1990) suggested that many of the white dwarfs in cataclysmic variables could possibly increase their masses to the Chandrasekhar mass, ending up with $\dot{M} \sim 10^{-9} M_\odot \mathrm{yr}^{-1}$ at the lower right-hand corners of Figures 1 and 2; resultant systems could be short-orbital period LMBPs. Also if we consider various types of helium star companions (helium main sequence, helium subgiants, helium white dwarfs, etc.) as well as companions surrounded by a common envelope (Hachisu et al. 1989), LMBPs with relatively short orbital periods could be formed from AIC.

3. COLLAPSE OF WHITE DWARFS INDUCED BY DEFLAGRATION

Possible models for AIC previously advanced include solid C+O white dwarfs (Canal et al. 1980; Isern et al. 1983) and O+Ne+Mg white dwarfs (Nomoto et al. 1979a), whose masses could grow to the Chandrasekhar mass limit for a white dwarf. In the AIC models, collapse of the white dwarf is induced by electron capture that effectively reduces the Chandrasekhar mass limit.

However, since the white dwarf contains nuclear fuel, whether the white dwarf undergoes collapse or explosion depends on which is faster behind the deflagration wave, nuclear energy release or electron capture. The energy generation rate is determined mainly by the propagation velocity of the deflagration wave, v_{def}, while the electron capture rate depends on the density. If v_{def} is lower than a certain critical speed, electron capture induces collapse. If on the other hand v_{def} is sufficiently high, complete disruption results. It is important to determine the critical velocity that divides collapse and explosion (Nomoto 1986).

3.1. SOLID C+O WHITE DWARFS

It is possible that the accreting C+O white dwarfs could collapse rather than explode, depending on the conditions of the white dwarfs. As described in §2 (7), compression of the white dwarf by the accreted matter first heats up a surface layer because of the small pressure scale height there. Later, heat diffuses inward (Nomoto et al. 1984). The diffusion time scale depends on $\dot{M}$ and is short for larger $\dot{M}$ because of the large heat flux and steep temperature gradient generated by rapid accretion. For example, the time it takes for the heat wave to reach the central region is $\sim 2 \times 10^5$ yr for $\dot{M} \sim 10^{-6} M_\odot$ yr^{-1} (Nomoto and Iben 1985) and 5×10^6 yr for $\dot{M} \sim 4 \times 10^{-8} M_\odot$ yr^{-1} (Nomoto et al. 1984).

If the initial mass of the white dwarf, M_{CO}, is smaller than 1.2 $M_\odot$, the entropy in the center increases substantially due to the heat inflow and thus carbon ignites at relatively low central density ($\rho_c \sim 3 \times 10^9$ g cm^{-3}). On the other hand, if the white dwarf is initially more massive than 1.2 $M_\odot$ and cold at the onset of accretion, the central region is compressed only adiabatically and thus is cold when carbon is ignited in the center. In the latter case, the ignition density is as high as 10^{10} g cm^{-3} (e.g., Isern et al. 1983) and the white dwarf may well have a solid core. For such a case, it is necessary to determine the critical condition for which a carbon deflagration induces collapse rather than explosion.

For solid C+O white dwarfs, recent work finds that no significant separation between carbon and oxygen occurs during solidification (Barrata et al. 1988; Ichimaru et al. 1988). This in turn leads to the ignition of explosive carbon burning at much lower densities than in models which postulate chemical separation (Isern et al. 1983).

In solid cores, carbon ignition takes place in the pycnonuclear reaction regime (Ogata et al. 1991) and develops into explosive burning at $\rho_c \sim 1 \times 10^{10}$g cm^{-3}. After thermal runaway of carbon burning, it is likely that a conductive deflagration wave propagates in the solid core. A detonation wave would not form because of steep temperature gradient in the central solid region. Convection would not be effective unless the solid core is melted by the heating from nuclear burning or neutrinos (Canal et al. 1990a).

The conductive deflagration in the solid C+O white dwarf is calculated assuming a constant ratio of v_{def}/v_s for conductive deflagration wave (Nomoto 1986, 1987b; Nomoto and Kondo 1991; Canal et al. 1990b).

Figure 3 shows the change in the central density associated with the propagation of the deflagration wave. Three cases with $v_{\text{def}}/v_s = 0.05$, 0.03, and 0.01 are calculated and the latter two slow cases undergo collapse. This implies that the critical velocity, v_{crit}, that divides collapse and explosion is $v_{\text{crit}} \sim 0.04 v_s$ for $\rho_c \sim 10^{10}$g cm^{-3}. Since a realistic value of conductive deflagration speed is $v_{\text{def}} \sim 0.01 v_s$ (Woosley and Weaver 1986), collapse is the most likely outcome for the solid white dwarf.

3.2. O+Ne+Mg WHITE DWARFS

The O+Ne+Mg white dwarfs are formed from stars of main-sequence masses of 8–12$M_\odot$ in close binaries with initial masses as large as 1.1–1.37$M_\odot$ (Nomoto et al. 1979a; Nomoto 1984a). After mass accretion from the companion star, the mass of the white dwarf increases toward the Chandrasekhar mass for a certain range of accretion rate (Fig. 2). When ρ_c exceeds 4×10^9g cm^{-3}, the O+Ne+Mg white dwarf undergoes electron captures ^{24}Mg(e^-,ν)^{24}Na(e^-,ν)^{20}Ne and ^{20}Ne(e^-,ν)^{20}F(e^-,ν)^{20}O (Fig. 5). Electron capture not only reduces the effective Chandrasekhar mass but also releases heat due to γ-ray emission, which eventually ignites oxygen deflagration at a certain central density.

In the previous AIC models (Nomoto et al. 1979a), oxygen is ignited at $\rho_{ig} \sim 2.5 \times 10^{10}$g cm^{-3} after the initiation of collapse (Miyaji et al. 1980; Nomoto 1987a). At such central densities, electron capture is much faster than oxygen burning, thus promoting further collapse. However, ρ_{ig} has been found to depend on the time scale of semiconvective mixing in the electron capture region (Nomoto 1984b; Mochkovitch 1984; Miyaji and Nomoto 1987). If semiconvective mixing is negligible and the heating due to γ-ray emission is confined to the very central region,

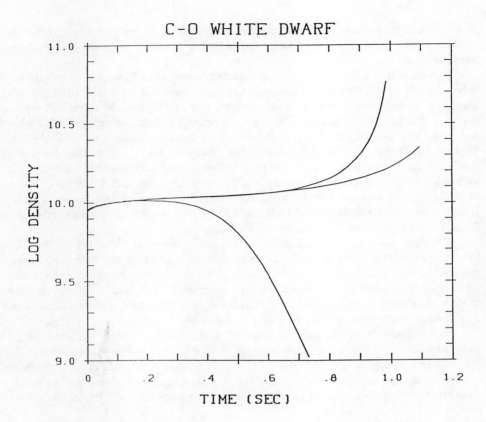

FIGURE 3 Change in the central density of the C+O white dwarfs following the propagation of the conductive carbon deflagration wave in the initially solid core. Three cases with v_{def}/v_s = 0.05, 0.03, and 0.01 are shown and the latter two undergo collapse (Nomoto and Kondo 1991).

oxygen burning is ignited at $\rho_{ig} \sim 9.95 \times 10^9$g cm^{-3} before collapse (Miyaji and Nomoto 1987). Hydrodynamical calculation is carried out to see whether this model leads to collapse or explosion (Nomoto and Kondo 1991).

The heat released by electron capture on ^{24}Mg results in the formation of a liquid core even if the white dwarf initially had a solid core (Mochkovitch 1984; Miyaji and Nomoto 1987; Canal et al. 1990a). When γ-rays resulting from electron capture on ^{20}Ne ignite explosive oxygen burning, several modes of the subsequent propagation of the explosive burning front are possible. Among them the formation of a detonation wave is very unlikely in the present case because negligible semi-convective mixing forms a very steep temperature gradient when explosive oxygen burning starts.

Therefore it is likely that an oxygen deflagration wave forms to propagate at a subsonic velocity. The propagation velocity, v_{def}, depends on which mode of heat transport is faster, conductive or convective. For conductive deflagration, we apply $v_{def} = 0.01 v_s \sim 100 \text{km s}^{-1}$, where v_s denotes the local sound velocity, which is a good approximation of v_{def} obtained by numerical calculations (Woosley and Weaver 1986). For the convective deflagration wave, we apply a time-dependent mixing length prescription using the ratio between the mixing length and the pressure scale height (or radial distance) $\alpha = \ell/\min(H_p, r) = 0.7, 1.4$, and 2 (Nomoto et al. 1984). For small α, the deflagration speed in the very central region is slower than the conductive deflagration because of small buoyancy force across the burning front; then the minimum v_{def} is set to be $0.01 v_s$.

Whether this will lead to collapse or explosion depends on which is faster behind the deflagration wave, nuclear energy release or electron capture. The energy generation rate is determined mainly by the propagation velocity of the deflagration wave, v_{def}, while the electron capture rate depends on the density. If v_{def} is lower than a certain critical speed, electron capture induces collapse. If on the other hand v_{def} is sufficiently high, complete disruption results.

Figure 4 shows the changes in the central density associated with the propagating deflagration front for three cases. It is seen that the slowest case of $\alpha = 0.7$ leads to increasing ρ_c (i.e., collapse), while the propagation with $\alpha = 2$ results in explosion; the intermediate case with $\alpha = 1.4$ is marginal. The fate of the convective deflagration wave depends mainly on whether v_{def} exceeds $\sim 0.03 v_s$ in the central region at $M_r <\sim 0.1-0.2 M_\odot$.

Though the determination of v_{def} may require multi-dimensional calculations, the carbon deflagration model for Type Ia supernovae favors $\alpha = 0.7$; the model with $\alpha = 0.7$ can nicely account for many of the observed features of Type Ia supernovae, while the propagation with $\alpha = 0.8$ is a little too fast to be consistent with their spectral features (Nomoto et al. 1984). For oxygen deflagration, $\alpha < 1$ may also be the case for the same prescription of deflagration; then the collapse of O+Ne+Mg white dwarfs would be the most likely outcome since $\alpha \sim 1.4$ is marginal between collapse and explosion. (If total disruption results from the white dwarf with central density of $\sim 10^{10} \text{g cm}^{-3}$, such an explosion should be extremely rare since the ejection of too much neutron-rich iron-peak elements would not be compatible with solar isotopic ratios. In addition, the explosion with such low energy as $\sim$ a few times 10^{50} ergs due to large neutrino losses does not match any subclass of SN I frequently observed.)

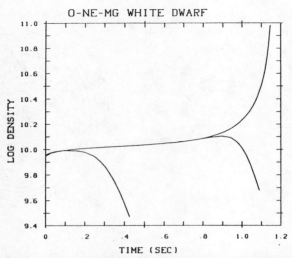

FIGURE 4 Same as Figure 3 but for the O+Ne+Mg white dwarfs following the propagation of the oxygen deflagration wave for three cases with $\ell/\min(H_p, r) = 2.0$, 1.4, and 0.7. For the slowest case of $\ell/H_p = 0.7$, the central density increases, i.e., the white dwarf undergoes collapse. Faster propagation induces an explosion of the white dwarf (Nomoto and Kondo 1991).

4. TYPE Ib/Ic SUPERNOVAE

Wolf-Rayet stars with a wide range of masses have been proposed for the progenitors of SNe Ib and Ic, since most SNe Ib/Ic are associated with star-forming regions (Wheeler and Harkness 1990 for a review). Recently Shigeyama et al. (1990), Hachisu et al. (1991), Nomoto et al. (1990), and Yamaoka and Nomoto (1991) have calculated the evolution, nucleosynthesis, Rayleigh-Taylor instabilities, and optical light curves of exploding helium stars. They have suggested that the helium stars of 3–5$M_\odot$ (which form from stars with initial masses $M_i \sim 12$–$18 M_\odot$ in binary systems) are the most likely progenitors of typical SNe Ib/Ic and that SNe Ic progenitors may be slightly less massive than those of SNe Ib. Such low mass helium star models can account for the observations that (1) the light curves of SNe Ic decline faster than SNe Ib, and (2) the early time spectra of SNe Ic show the presence of hydrogen (Jeffery et al. 1991), while hydrogen is absent in SNe Ib. It remains an open question how the presence of hydrogen causes the difference between SNe Ib and Ic in their early time spectra.

FIGURE 5 (*Left*) The expanding supernova matter is excited by the decay of ^{56}Co into ^{56}Fe. (*right*) Collapse of an O+Ne+Mg core is induced by electron capture (H. Nomoto 1989).

4.1. EVOLUTION OF INTERACTING BINARIES

The difference in the spectra of SNe Ic and Ib may be due to the presence of a thin envelope of hydrogen in SNe Ic immediately prior to the explosion. By evolving massive stars in close binary systems, we examine whether hydrogen can be left on the helium stars after mass exchange and wind-type mass loss. Following are some preliminary results for two cases 13A and 18A, where the initial masses of the primary stars are $M_i = 13 M_\odot$ (13A) and $18 M_\odot$ (18A), and their Roche lobe radii are $50 R_\odot$ (Yamaoka and Nomoto 1991).

After hydrogen exhaustion, the star undergoes Roche lobe overflow, forming a helium star of $\sim 3.4 M_\odot$ (13A) or $\sim 5 M_\odot$ (18A). Yamaoka and Nomoto (1991) found that a significant amount of hydrogen remains in a relatively thick layer below the surface ($0.5 M_\odot$ for 13A and $1 M_\odot$ for 18A).

Whether this hydrogen layer will be lost from the helium-rich star depends on the Roche lobe radius and wind mass loss rate. If the helium star is detached from the Roche lobe during helium burning, the star loses mass in a wind. If the mass loss rate depends on mass as $\dot M \propto M^{2.5}$ (Langer 1989), it may lead to different surface abundances in 13A and 18A. For case 13A hydrogen may still remain in the layer down to $\sim 0.2 M_\odot$ below the surface, whereas for case 18A all hydrogen may be lost in a wind (Yamaoka and Nomoto 1991). If a common envelope forms or the Roche lobe radius has become small enough for the mass and angular momentum to be lost from the system, then all hydrogen will be lost from the star.

Although more parameter study is needed, the present results suggest that more hydrogen may remain on the stellar surface if the helium star had initially smaller main-sequence mass. Also more compact binary systems may lose more hydrogen through common envelope evolution.

4.2. LIGHT CURVES

Figure 6 shows the observed bolometric light curves of SNe Ia 1972E and 1981B (Graham 1987) and SN Ib 1983N (Panagia 1987), and the approximate bolometric light curve of SN Ic 1987M constructed from flux-calibrated spectra (Filippenko et al. 1990; Nomoto et al. 1990). In each case, the observed light curve has been shifted along the abscissa to match the corresponding theoretical curve. The peak bolometric luminosities assume $H_0 = 60 \mathrm{km\ s^{-1} Mpc^{-1}}$.

The previous Wolf-Rayet star models have some difficulties (1) in reproducing the light curves of typical SNe Ib, which decline as fast as those of SNe Ia (Panagia 1987), and (2) in producing enough ^{56}Ni to attain the maximum luminosities of SNe Ib in relatively low mass helium star models (Ensman and Woosley 1988). In particular, Figure 6 demonstrates an important feature of SN Ic 1987M; that is, its brightness fell somewhat more rapidly than those of SNe Ia and SN Ib 1983N (also SN Ic 1983I reported by Tsvetkov 1985). Maximum brightness of 1987M is not significantly different from those of SNe Ib if we take the extinction estimated by Jeffery et al. (1991).

Figure 6 also shows the calculated bolometric light curves of the exploding helium star models with $M_\alpha = 3.3 M_\odot$ for SNe Ic and $4 M_\odot$ for SNe Ib as well as the white dwarf model W7 for SNe Ia (Nomoto et al. 1984). The amount of ^{56}Ni is $0.58 M_\odot$ (W7), $0.15 M_\odot$ (SN Ic), and $0.15 M_\odot$ (SN Ib). The helium star models assume a uniformly mixed distribution of elements from the center through the layer at $0.2 M_\odot$ beneath the surface for both cases. Such a mixing may be due to the Rayleigh-Taylor instability during the explosion (Hachisu et al. 1991).

The calculated bolometric light curves of helium stars are powered by the radioactive decays of ^{56}Ni and ^{56}Co (Fig. 4). Maximum brightness is higher if the ^{56}Ni mass is larger and the date of maximum earlier. After the peak, the optical light curve declines at a rate that depends on how fast γ-rays from the radioactive decays escape from the star without being thermalized, thereby declining faster if the ejected mass is smaller and if ^{56}Ni is mixed closer to the surface.

The resulting bolometric light curves of $M_\alpha = 3.3 M_\odot$ and $4 M_\odot$ are in good agreement with SN Ic 1987M and SN Ib 1983N, respectively. Compared with the $4 M_\odot$ model, the light curve of the 3.3 $M_\odot$ model declines faster due to the smaller ejected mass, just as observed in SN Ic 1987M. (See Thielemann et al. for the neutron star masses inferred from SNe Ib/Ic.)

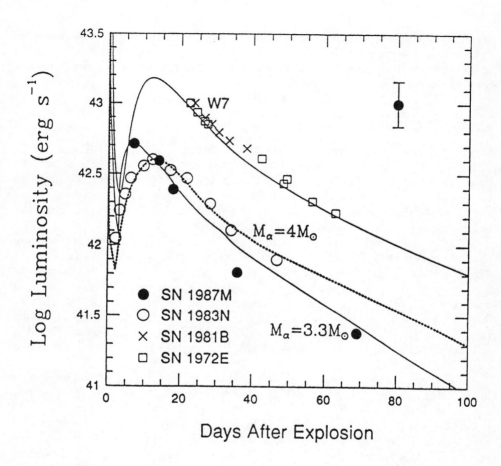

FIGURE 6 Approximate bolometric light curve of SN Ic 1987M, and the bolometric light curves of SNe Ia 1972E and 1981B and of SN Ib 1983N. The predicted curves of the 3.3 $M_\odot$ model for SN 1987M, the $4M_\odot$ model for SN Ib, and the W7 model for SNe Ia are indicated by solid and dotted lines. The error bar illustrates the 2σ photometric uncertainty in the SN 1987M points (Nomoto et al. 1990).

5. FORMATION OF DOUBLE NEUTRON STARS AND SNe Ib/Ic

The low-mass helium stars considered for the progenitors of SNe Ib/Ic would mostly occur in close binaries, because the 12–$18M_\odot$ stars would not lose their entire hydrogen-rich envelope by wind mass loss. Meurs and van den Heuvel (1989) predicted that more than 70% of massive star explosions would occur in close binaries.

This estimate predicts that the occurrence frequencies of SNe Ib/Ic are higher than SNe II, which might be consistent with the increasing number of SNe Ic recently discovered.

The binary scenario suggests that SNe Ib/Ic might be closely related to the formation of binary pulsars and X-ray binaries. If the binary system is not disrupted by the supernova mass ejection, a neutron star is left to orbit a companion star (a main-sequence star or a helium star). Many of them would become Be X-ray binaries.

Recently the masses of the component stars in the binary pulsar system 1532+12 have been determined and strongly suggest a neutron star companion (Wolszczan 1991). Their masses, eccentricity, semimajor axis, and orbital period are summarized in Table 1 together with those of the first binary pulsar 1913+16 (Taylor 1991). A possible evolutionary scenario for such systems is as follows (e.g., van den Heuvel 1991): (1) Two main-sequence stars 1 and 2; (2) Roche lobe overflow of star 1, which becomes a helium star 1; (3) the first supernova explosion of helium star 1 to form a neutron star 1; (4) Roche lobe overflow of star 2, which leads to a spiraling in of neutron star 1 into star 2 and thus to a considerable shrinkage of the system due to the losses of angular momentum and mass from the system; the system now consists of the recycled neutron star 1 and a helium star 2; (5) the second supernova explosion of the helium star 2; this forms a two-neutron star system in an eccentric orbit.

Given the observed orbital parameters in Table 1 and the assumption of a circular orbit for the pre-explosion helium star 2–neutron star 1 system, the mass of the helium star 2 M_α and the possible kick velocity v_{kick} at the explosion can be calculated. If the explosion is spherical (i.e., $v_{\text{kick}} = 0$), $M_\alpha \sim 2.1 M_\odot$ (Wolszczan 1991). This is smaller than the minimum mass of the helium star that can form a neutron star ($\sim 2.5 M_\odot$ [Nomoto 1984a]–$\sim 2.2 M_\odot$ [Habets 1986] depending on the treatment of overshooting and semiconvection). This suggests that either the explosion is not spherical or the exploding star had lost even its helium layer before the explosion.

If we introduce a finite v_{kick} for the explosion of the helium star 2, the kick velocity and its direction can be calculated as functions of assumed M_α and the initial orbital radius a_0 before the explosion, where $a_f(1-e) < a_0 < a_f(1+e)$ (Fig. 7; Yamaoka et al. 1991). It should be noted that smaller helium stars have larger radii at the collapse, which is $\sim 3 R_\odot$ for $M_\alpha \sim 3.3 M_\odot$ (Nomoto and Hashimoto 1988). For the solid lines in Figures 7 and 8, the radius of the helium star is equal to its Roche lobe radius. As far as the star 2 is a helium star, therefore, M_α should be larger than $5 M_\odot$ to underfill the Roche lobe. If this is the case, a kick velocity of $v_{\text{kick}} \sim 180$–240 km s^{-1} is necessary to avoid the disruption of the binary system. The same relation is obtained for 1913+16 in Figure 8, where $v_{\text{kick}} \sim 340$–420 km s^{-1} is necessary (see also Burrows and Woosley 1986).

Two possible extreme cases are (1) the star 2 is a helium star more massive than $\sim 5 M_\odot$, for which the explosion produces a large kick velocity, and (2) star 2, being initially a helium star of smaller than $5 M_\odot$, loses its helium envelope to become an almost bare C+O star. The masses of C+O stars are 6.0, 3.8, 2.1, and $1.8 M_\odot$ for

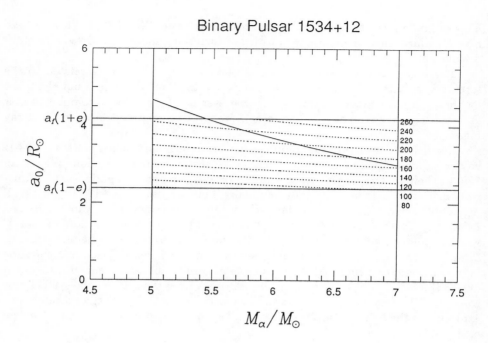

FIGURE 7 Kick velocity imparted to the neutron star at the explosion of the helium star 2 of mass M_α as a function of M_α and the initial orbital radius a_0 before the explosion (Yamaoka et al. 1991). Here $a_f(1-e) < a_0 < a_f(1+e)$. For the solid line, the radius of the helium star is equal to its Roche lobe radius, so that only the upper-right part of the parameter space is allowed. For PSR 1534+12, $v_{kick} \sim 180 - 240$ km s^{-1} is necessary to avoid the disruption of the binary system.

$M_\alpha = 8, 6, 4, 3.3 M_\odot$, respectively (Nomoto and Hashimoto 1988). For $M_\alpha \lesssim 4 M_\odot$, therefore, the explosion of star 2 could be spherical with $v_{kick} = 0$.

If the main difference between SNe Ib and Ic is the existence of a hydrogen-rich envelope, the first explosion could be SN Ic because of the possible presence of hydrogen, while the second explosion might be SN Ib if star 2 loses its helium envelope. This binary scenario also suggests that SNe Ic occur more frequently than SNe Ib.

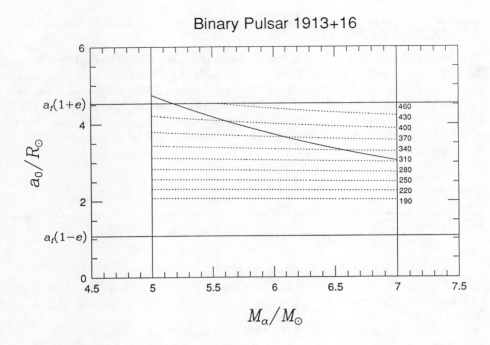

FIGURE 8 Same as Figure 7 but for PSR 1913 + 16, where $v_{kick} \sim 340 - 420$ km s^{-1} is necessary (Yamaoka et al. 1991).

TABLE 1 Binary Pulsars

PSR	M_p ($M_\odot$)	M_c ($M_\odot$)	e	a_f ($R_\odot$)	P_b (hours)	$P_b/\dot{P}_b$ (years)
1913+16	1.4421±0.0012	1.3875±0.0012	0.617127	2.8	7.752	3×10^8
1532+12	1.32±0.03	1.36±0.03	0.274	3.28	10.1	$\sim 1\times10^9$

ACKNOWLEDGMENTS

K N would like to thank Ramon Canal, Jordi Isern, Shri Kulkarni, Marco Tavani, Ed van den Heuvel, and Stan Woosley for stimulating discussion during his visit to Barcelona (December 1990) and Santa Barbara (January 1991). This work has been supported in part by the Grant-in-Aid for Scientific Research (01540216, 02234202, 02302024, 03218202) of the Ministry of Education, Science, and Culture in Japan, and by the Japan-U.S. Cooperative Science Program (EPAR-071/88-15999) operated by the JSPS and the NSF.

REFERENCES

1. Bailyn, C. D., and Grindlay, J. E. 1990, *Ap. J.*, **353**, 159.
2. Barrat, J. L., Hansen, J. P., and Mochkovitch, R. 1988, *Astr. Ap.*, **199**, L15.
3. Benz, W., Bowers, R. L., Cameron, A. G. W., and Press, W. 1990, *Ap. J.*, **348**, 647.
4. Burrows, A., and Woosley, S. E. 1986, *Ap. J.*, **308**, 680.
5. Canal, R., Garcia, D., Isern, J., and Labay, J. 1990b, *Ap. J. (Letters)*, **356**, L51.
6. Canal, R., Isern, J., and Labay, J. 1980, *Ap. J. (Letters)*, **241**, L33.
7. Canal, R., Isern, J., and Labay, J. 1990a, *Ann. Rev. Astr. Ap.*, **28**, 183.
8. Chanmugam, G., and Brecher, K. 1987, *Nature*, **329**, 696.
9. Ensman, L., and Woosley, S. E. 1988, *Ap. J.*, **333**, 754.
10. Filippenko, A. V., Porter, A. C., and Sargent, W. L. W. 1990, *A. J.*, **100**, 1575.
11. Habets, G. M. H. J. 1986, *Astr. Ap.*, **167**, 61.
12. Hachisu, I., Kato, M., and Saio, H. 1989, *Ap. J. (Letters)*, **342**, L19.
13. Hachisu, I., Matsuda, T., Nomoto, K., and Shigeyama, T. 1991, *Ap. J. (Letters)*, **368**, L27.
14. Hashimoto, M., Nomoto, K., Arai, K., and Kaminisi, K. 1986. *Ap. J.*, **307**, 687.
15. Iben, I., Jr., and Tutukov, A. 1984, *Ap. J. Suppl.*, **55**, 335.
16. Ichimaru, S., Iyetomi, H., and Ogata, S. 1988, *Ap. J. (Letters)*, **334**, L17.
17. Isern, J., Labay, J., Hernanz, M., and Canal, R. 1983, *Ap. J.*, **273**, 320.
18. Jeffery, D., Branch, D., Filippenko, A. V., and Nomoto, K. 1991, *Ap. J. (Letters)*, **377**, L89.
19. Kato, M., and Hachisu, I. 1989, *Ap. J.*, **346**, 424.
20. Kulkarni, S. R., and Narayan, R. 1988, *Ap. J.*, **335**, 755.
21. Kulkarni, S. R., Narayan, R., and Romani, R. W. 1990, *Ap. J.*, **356**, 174.
22. Langer, N. 1989, *Astr. Ap.*, **220**, 135.

23. Meurs, E. J. A., and van den Heuvel, E. P. J. 1989, *Astr. Ap.*, **226**, 88.
24. Michel, F. C. 1987, *Nature*, **329**, 310.
25. Miyaji, S., and Nomoto, K. 1987, *Ap. J.*, **318**, 307.
26. Miyaji, S., Nomoto, K., Yokoi, K., and Sugimoto, D. 1980, *Pub. Astr. Soc. Japan*, **32**, 303.
27. Mochkovitch, R. 1984, in *Problems of Collapse and Numerical Relativity*, ed. D. Bancel and M. Signore (Dordrecht: Reidel), p. 125.
28. Mochkovitch, R., and Livio, M. 1990, *Astr. Ap.*, **236**, 378.
29. Narayan, R., Fruchter, A. S., Kulkarni, S. R., and Romani, R. W. 1990, in *Accretion-Powered Compact Binaries*, ed. C. Mauche (Cambridge University Press), p. 451.
30. Nariai, K., and Nomoto, K. 1979, in *IAU Colloquium 53, White Dwarfs and Variable Degenerate Stars*, ed. H. M. Van Horn and V. Weidemann (Rochester: Univ. of Rochester), p. 525.
31. Nariai, K., Nomoto, K., and Sugimoto, D. 1980, *Pub. Astr. Soc. Japan*, **32**, 473.
32. Nomoto, H. 1989, *Exploring SN 1987A* (in Japanese), (Tokyo: Kodansha).
33. Nomoto, K. 1982a, *Ap. J.*, **253**, 798.
34. Nomoto, K. 1982b, *Ap. J.*, **257**, 780.
35. Nomoto, K. 1984a, *Ap. J.*, **277**, 791.
36. Nomoto, K. 1984b, *Problems of Collapse and Numerical Relativity*, ed. D. Bancel and M. Signore (Dordrecht: Reidel), p. 89.
37. Nomoto, K. 1986, *Prog. Part. Nucl. Phys.*, **17**, 249.
38. Nomoto, K. 1987a, *Ap. J.*, **322**, 206.
39. Nomoto, K. 1987b, in *IAU Symposium 125, The Origin and Evolution of Neutron Stars*, ed. D. J. Helfand and J.-H. Huang (Dordrecht: Reidel), p. 281.
40. Nomoto, K., Filippenko, A. V., and Shigeyama, T. 1990, *Astr. Ap.*, **240**, L1.
41. Nomoto, K., and Hashimoto, M. 1987, *Ap. Space Sci.*, **131**, 395.
42. Nomoto, K., and Hashimoto, M. 1988, *Physics Reports*, **163**, 13.
43. Nomoto, K., and Iben, I., Jr. 1985, *Ap. J.*, **297**, 531.
44. Nomoto, K., and Kondo, Y. 1991, *Ap. J. (Letters)*, **367**, L19.
45. Nomoto, K., Miyaji, S., Sugimoto, D., and Yokoi, K. 1979a, in *IAU Colloquium 53, White Dwarfs and Variable Degenerate Stars*, ed. H. M. Van Horn and V. Weidemann (Rochester: Univ. of Rochester), p. 56.
46. Nomoto, K., Nariai, K., and Sugimoto, D. 1979b, *Pub. Astr. Soc. Japan*, **31**, 287.
47. Nomoto, K., Thielemann, F.-K., and Yokoi, K. 1984, *Ap. J.*, **286**, 644.
48. Ogata, S., Iyetomi, H., and Ichimaru, S. 1991, *Ap. J.*, **372**, 259.
49. Panagia, N. 1987, in *High Energy Phenomena Around Collapsed Stars*, ed. F. Pacini (Dordrecht: Reidel), p. 33.
50. Ray, A., and Kluzniak, W. 1990, *Nature*, **344**, 415.
51. Romani, R. W. 1990, *Ap. J.*, **357**, 493.
52. Saio, H., and Nomoto, K. 1985, *Astr. Ap.*, **150**, L21.

53. Shigeyama, T., Nomoto, K., Tsujimoto, T., and Hashimoto, M. 1990, *Ap. J. (Letters)*, **361**, L23.
54. Tavani, M. 1991, *Ap. J.*, in press.
55. Taylor, J. 1991, in *NATO ARW, X-Ray Binaries and the Formation of Binary and Millisecond Radio Pulsars*, ed. E. P. J. van den Heuvel (Kluwer), in press.
56. Thielemann, F.-K., Nomoto, K., and Hashimoto, M. 1991, this volume.
57. Tsvetkov, D. Yu. 1985, *Sov. Astr.*, **29**, 211.
58. Van den Heuvel, E. P. J. 1991, in *NATO ASI, Neutron Stars: Theory and Observation*, ed. J. Ventura and D. Pines (Kluwer), p. 171.
59. Webbink, R. 1984, *Ap. J.*, **277**, 355.
60. Wheeler, J. C. 1991, in *Frontiers of Stellar Evolution*, ed. D. L. Lambert (San Francisco: Astron. Soc. of Pacific), in press.
61. Wheeler, J. C., and Harkness, R. 1990, *Rep. Prog. Phys.*, **53**, 1467.
62. Wolszczan, A. 1991, *Nature*, **350**, 688.
63. Woosley, S. E., Taam, R. E., and Weaver, T. A. 1986, *Ap. J.*, **301**, 601.
64. Woosley, S. E., and Weaver, T. A. 1986, *Lecture Notes in Physics*, **255**, 91.
65. Yamaoka, H., and Nomoto, K. 1991, in *SN 1987A and Other Supernovae*, ed. I.J. Danziger and K. Kjär (Garching: ESO), p. 193..
66. Yamaoka, H., Shigeyama, T., and Nomoto, K. 1991, in preparation.

Stuart L. Shapiro and Saul A. Teukolsky
Center for Radiophysics and Space Research, Cornell University, and Departments of Astronomy and Physics, Cornell University, Ithaca NY 14853

Compact Stars on the Supercomputer

During the past several years we have made use of the availability of supercomputers to tackle a number of problems involving the physics of compact stars. Here we will summarize the results of four studies in this area.

1. EXPLODING NEUTRON STARS NEAR THE MINIMUM MASS

Extensive studies on the gravitational equilibrium states of *cold catalyzed* matter have laid the basis for current models of neutron stars and white dwarfs.[1] To determine the structure of these compact objects two distinct categories of input physics are involved. The global response of matter to gravity is described macroscopically by the Oppenheimer-Volkoff (OV) equation of hydrostatic equilibrium, from which the mass M of a spherical star can be determined as a function of one parameter, such as the central density ρ_c. The local microphysical properties are then fully incorporated in the equation of state, which is usually derived under the assumption that matter is in a state of local thermodynamic equilibrium at zero temperature. Consequently, all reactions occurring within the star are assumed to be completely catalyzed.

Though the properties and the structure of the configurations of hydrostatic equilibrium depend upon the details of the adopted equation of state, one feature is universal: the existence of a minimum-mass neutron star configuration. On an M versus ρ_c equilibrium curve the minimum mass marks the boundary between the unstable white dwarf branch at low densities and the stable neutron star branch at higher densities. Along the unstable branch, equilibrium matter is controlled by inverse beta-decay reactions. Electron capture leads to a drop in the mean adiabatic index below 4/3 and the configurations are unstable to radial oscillations. These electron captures effect a transition from white dwarf matter dominated by degenerate electrons to neutron star matter dominated by degenerate neutrons. At high enough density, the transition is sufficiently advanced that radial stability is restored along the equilibrium sequence. This point is precisely located at the minimum neutron star mass.

The identification of the minimum mass with the transition from instability to stability along the M versus ρ_c curve depends crucially through the equation of state on the hypothesis that matter when perturbed remains cold, in equilibrium, and completely catalyzed. However, this assumption fails unless the time scales associated with the reactions maintaining chemical equilibrium are shorter than the dynamical time scale on which the perturbations grow. Indeed, the weak beta and inverse beta-decay reactions might not fulfill this constraint since neutron star matter is stabilized against beta-decay by the filled Fermi sea of electrons. If these processes are responsible for the onset of the dynamical instability it thus seems appropriate to examine their rates carefully to assess the stability of an actual configuration at the minimum neutron star mass. Remarkably, the dynamical instability might disappear if beta-decays are prohibited from occurring. It is therefore of interest to investigate and to isolate the microphysical mechanisms triggering the instability and to explore the subsequent evolution in terms of a full dynamical computation. It is of further interest to determine the final state of the star once the instability has driven matter out of gravitational equilibrium. Does the star explode and disperse to infinity, or does it ultimately settle into a new, bound, hydrostatic equilibrium configuration?

In collaboration with Colpi, we have investigated these issues in the framework of a simplified, but concrete, *dynamical* model.[2,3] The motivation for tackling this problem is twofold: Theoretically, the dynamical evolution of an unstable low-mass neutron star has never been tracked in detail. Page[4] has suggested that beta-decay reactions can trigger the instability, but that matter needs to be somewhat out of beta-equilibrium for the instability to grow. Furthermore, he noted that the total binding energy per baryon of the star at the minimum-mass configuration is larger than the binding energy of a hydrogen cloud dispersed at rest to infinity, but smaller than the binding energy of a corresponding cloud of ^{56}Fe. Therefore, the final configuration of the initially expanding star depends in a crucial way on the composition reached at late stages of its evolution. The star can settle into a bound, low-density configuration if it is composed of sufficiently light elements. Conversely, it may disperse to infinity with a nonzero kinetic energy if the matter transforms into ^{56}Fe, the nucleus with the highest binding energy per baryon at low densities.

A further complication is the energy loss due to antineutrinos generated during beta-decay. The extent of this energy loss can be evaluated only by a dynamical calculation. It could, in principle, lower the total energy of the star and favor the formation of a bound state. In particular, if matter remains catalyzed during dynamical evolution and enough energy is dissipated one can speculate that a white dwarf forms at the endpoint of the evolution.

The second motivation for studying this problem is astrophysical. The star can become a potentially significant source of antineutrinos. The emergent luminosity could in principle attain values comparable to those predicted in a supernova event. A plausible site for the occurrence of this process is a close binary system consisting of two neutron stars whose orbits decay via gravitational radiation. Clark and Eardley[5] and Blinnikov et al.[6] have studied the final evolution of this double compact binary system. They envision that the lightest neutron star, in tidal interaction with the companion, suffers slow-mass stripping. In this process the star loses mass as it evolves through states of quasi-hydrostatic equilibrium. Eventually, the light star may reach the minimum-mass configuration, thereby becoming dynamically unstable to expansion. It is plausible that the explosive event that may be triggered by such an instability could have observable astronomical consequences.

To explore these issues, we follow the growth of the instability in a neutron star at the minimum-mass configuration. We determine the star's subsequent evolution and ultimate fate. We first examine the radial stability of the equilibrium at the minimum mass, exploring two simple equations of state that model the interior of a compact object. We verify Page's conjecture that beta-decay reactions trigger the dynamical instability and demonstrate that in cold matter these weak interaction processes cannot proceed in equilibrium as matter expands. Under these conditions the modeling of the star microphysics is complex since matter is not catalyzed. To simplify the hydrodynamics, we employ the Newtonian equations of hydrodynamics for a *homogeneous* sphere that remains homogeneous and spherical during subsequent evolution (*Homogeneous Sphere Approximation*, hereafter HSA). The equation of state incorporates entropy evolution due to neutrino dissipation.

We consider two distinct equations of state: (1) an ideal gas of neutrons, electrons, and protons (the *n-e-p model*) and (2) a two-phase system consisting of nuclei, free electrons, and dripped neutrons based on the Harrison-Wheeler nuclear model (the *HW model*[7]). We thus analyze dynamical evolution governed by two very different mechanisms for beta-decay. In the n-e-p model, the weak interaction processes are controlled by the modified URCA reactions and are characterized by a time scale exceedingly large compared to the dynamical time scale. By contrast, in the HW model beta-decays occur in nuclei in a time that can approach the dynamical time. We therefore expect the whole dynamical evolution of the star to be affected by this large difference between the beta-decay time scales. The two models are further distinguished by their relative tractability. The n-e-p model presents no uncertainties and is simple. However, the HW model is more realistic and provides an analytical expression for the bulk energy of the nucleus. For our nonequilibrium HW model, our starting point is the approach of Lattimer et al.[8] As these authors note, in cold neutron star matter the processes that would lead to a nuclear size corresponding to lowest energy cannot proceed rapidly and nuclear statistical equilibrium is not attained by the expanding matter. Instead, the number of nuclei is conserved during dynamical evolution. This condition gives rise to a new equation of state for noncatalyzed matter and we incorporate this effect in our treatment.

Our analysis of the problem proceeds through four different stages: (1) Given the microphysical model for cold catalyzed matter we determine within the HSA the M versus ρ relation to locate the minimum-mass neutron star configuration. (2) We examine the stability of the equilibria showing that the star is stable against radial perturbations if beta-decays do not occur. We check that other noncatalyzed processes do not affect the stability analysis. (3) We estimate the actual magnitude of the time for beta- and inverse beta-decay and we make a comparison with the dynamical time. (4) We derive the evolution equations that model the local properties of matter during dynamical motion, and we evolve the system of equations from the appropriate initial condition. We then estimate the neutrino luminosity and compare the efficiencies of the two models.

The emphasis throughout this calculation is on determining the ultimate fate of the star. We show that the expansion cannot be reversed if the velocity ever exceeds the escape velocity from the star. This result is a consequence of the simple form assumed by the dynamical equations in the HSA and holds independently of the equation of state employed.

The key findings of our idealized calculations are the following: In the n-e-p model, a star initially driven toward a state of beta-disequilibrium undergoes a *secular expansion*, evolving through a sequence of states of hydrostatic equilibrium. The duration of this epoch can be extremely long, $\sim 10^9$ yr. Because of the steady growth of kinetic energy induced by beta-decay reactions, the star eventually enters the regime of dynamical instability. Matter then suffers abrupt acceleration while releasing a copious amount of antineutrinos with a luminosity $L_{\bar{\nu}} \approx 10^{45}$ ergs s^{-1}. The associated luminosity decays following the decay of neutrons. As the star expands to very low densities free neutron decay dominates. The star continues to expand and disperse toward a diluted state of hydrogen, since the matter attains escape velocity almost instantly during the phase of dynamical expansion. In the more realistic HW model, the star explodes catastrophically on a time scale between a few milliseconds and a few seconds, depending on the initial conditions. The event is accompanied by a burst of antineutrinos and terminates in our computations when nuclei become unstable to spontaneous fission. The antineutrino burst has a peak luminosity $L_{\bar{\nu}} \approx 10^{51}$ ergs s^{-1} and the total kinetic energy of the explosion exceeds 10^{45} ergs. The process thus resembles a minisupernova event. This result holds over a wide range of initial conditions.

We have also analyzed[2] the explosion of a *rotating* neutron star just below the minimum mass. We adopt the HW equation of state and follow the dynamical evolution of the star by solving the Newtonian equations of motion for a homogeneous, uniformly rotating spheroid with internal pressure and gravity. While expanding, the oblateness of the spheroid decreases as the centrifugal force weakens with increasing radius. We find antineutrino luminosities of $\sim 10^{50-52}$ ergs s^{-1} and bulk kinetic energies $\sim 10^{49}$ ergs. Gravitational radiation and gravitational wave amplitudes are computed as perturbations in the weak-field, slow-motion limit of general relativity. For a wide set of initial conditions, the emission in gravitational waves is found to be small, with GR energy efficiencies $E_{GR}/M_B \sim 10^{-14}$ and wave amplitudes $h_+ \sim 5 \times 10^{-23}$ for a source distance of 10 kpc.

We also estimate the photon luminosity from the explosion via a simple two-zone core-atmosphere model. The photon luminosity calculated in the diffusion approximation varies between 10^{36} and 10^{38} ergs s^{-1}, with the photons emerging in the hard UV and soft X-ray bands.

2. SPINUP OF A RAPIDLY ROTATING STAR

The spin-down of a rapidly rotating star is qualitatively different from that of a slowly rotating one. A slowly rotating star is essentially spherical and loss of angular momentum J results in a decrease in the angular rotation rate Ω but no substantial change in the moment of inertia I. A rapidly rotating star, however, may be highly oblate and the loss of angular momentum can result in a significant decrease in I. Like a skater who pulls in her arms while spinning, the star can actually spin up while losing angular momentum. Since $\Omega = J/I$, the sign of the rate of change of Ω clearly depends on the relative rates of decrease of J and I.

The energy and angular momentum of a rotating equilibrium star can slowly decrease by mechanisms such as the emission of electromagnetic or gravitational radiation, or of neutrinos. For example, the emission of magnetic dipole radiation is generally accepted as the dominant spin-down mechanism for radio pulsars. For a uniformly rotating equilibrium star that slowly loses energy E and angular momentum J while its baryon rest mass M, entropy S, and chemical composition remain constant, the changes in E and J are simply related by

$$dE = \Omega dJ. \tag{1}$$

This is a general result, proved by Ostriker and Gunn[9] for Newtonian configurations and extended by Hartle[10] to relativistic stars. Rewriting equation (1) in the form

$$\frac{dE}{dt} = \frac{1}{(\partial \Omega/\partial J)_{M,S}} \Omega \frac{d\Omega}{dt}, \tag{2}$$

we see that the sign of $d\Omega/dt$ is determined by the sign of $(\partial \Omega/\partial J)_{M,S}$, since the rate of energy loss dE/dt is negative. The derivative $(\partial \Omega/\partial J)_{M,S}$ depends only on the properties of the equilibrium sequence along which the star evolves, and not on the actual energy loss mechanism.

For the sequence of Maclaurin spheroids, it is straightforward to show that $(\partial \Omega/\partial J)_{M,S}$ changes sign from positive to negative for sufficiently rapidly rotating configurations. But Maclaurin spheroids are homogeneous, incompressible, and uniformly rotating. Moreover, the sign change occurs at $T/|W| = 0.2379$, well beyond the onset of secular instability at $T/|W| = 0.1375$. Here T is the rotational kinetic energy and W is the gravitational potential energy of the star. Does a sign change occur for more realistic inhomogeneous configurations governed by a compressible equation of state? And if it does occur, are the corresponding configurations stable?

Finn and Shapiro[11] modeled rotating stars as homogeneous, uniformly rotating, compressible spheroids whose central pressure is related to the density by a polytropic equation of state. They found that $(\partial\Omega/\partial J)_{M,S}$ becomes negative for polytropic indices sufficiently close to $n = 3$ (adiabatic index sufficiently close to 4/3). Thus, these stars spin up as they lose angular momentum. Does the same result apply to real (inhomogeneous) uniformly rotating polytropes with n close to 3? We know the result holds for $n = 0$, the Maclaurin spheroids. Does it apply to values of n intermediate between 0 and 3?

In Figure 1 we plot Ω versus J in nondimensional units for equilibrium sequences of uniformly rotating stars of constant M and S for various values of polytropic index n. The $n = 0$ curve is given by the Maclaurin equations. The other curves are determined from the tabulations of James[12] and Hachisu.[13]

Each polytropic sequence is plotted up to the termination point at which the centrifugal force balances the gravitational attraction at the equator. (The termination point for $n = 0$ is off the scale of the plot.) There exist[13,14] additional sequences of ring and highly flattened, concave-hamburger structures beyond this point, but they do not smoothly join onto the slowly rotating sequence. In addition, they have high values of $T/|W|$ and hence are unstable.

The figure shows the turning point in Ω versus J for the $n = 0$ Maclaurin sequence. It also reveals the existence of a turnover for the nearby sequences $n = 0.1$, 0.5, *and* for the sequences with n close to 3. Interestingly, there are no turnovers for values of n well inside the interval 0–3.

Are any of the models beyond the turning points on the equilibrium curves stable? Several authors have analyzed the stability of uniformly rotating polytropes. Usually only the $m = 2$ nonaxisymmetric mode is important,[15] but it is only reached when $n \lesssim 0.808$. For rapidly rotating stars like millisecond pulsars, the modes with $m \lesssim 5$ are relevant, because they are the modes driven unstable by gravitational radiation.[16] We have used the critical values of $T/|W|$ tabulated by Managan,[17] Imamura, Friedman, and Durisen,[18] and Ipser and Lindblom[19] to plot the curves in Figure 1 at which the configurations become unstable for various modes m. For n close to 0 all models beyond the turning points are clearly unstable. However, for n close to 3 there are models beyond the turning points that are stable. Thus we have shown[20] that *if an isolated uniformly rotating star is observed to be spinning up, then it must have an equation of state with n very close to 3, that is, adiabatic index Γ very close to 4/3.*

The above conclusion could be altered if the star is undergoing *differential* rotation. After all, the angular velocity of a satellite in Keplerian orbit about a central mass increases if the satellite loses angular momentum. We thus expect that spin-up of a differentially rotating star losing J might occur under more general

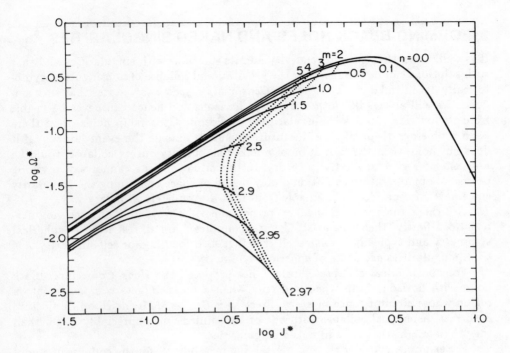

FIGURE 1 Variation of Ω versus J along uniformly rotating equilibrium sequences of constant M and S. The solid lines show the sequences for various values of polytropic index n. The dotted lines mark the onset of nonaxisymmetric instabilities of various mode number m.

circumstances than for a uniformly rotating star. In fact, there exist numerical models of differentially rotating stars which exhibit spin-up. While some are known to be stable to the $m = 2$ mode, their stability to higher m modes is not known. It nevertheless seems plausible that *there may exist stable, differentially rotating stars which can undergo spin-up for n quite different from 3*.

Nonaccreting millisecond pulsars are the most promising candidates for observing this behavior. These objects are rapidly rotating neutron stars that can lose angular momentum by electromagnetic and gravitational radiation, keeping their baryon mass constant and their entropy essentially zero. When it is detected, the newborn neutron star in SN 1987A may also be a candidate.

3. FORMING BLACK HOLES AND NAKED SINGULARITIES

It is wellknown that general relativity admits solutions with singularities and that such solutions can be produced by the gravitational collapse of nonsingular, asymptotically flat initial data. The *Cosmic Censorship Hypothesis*[21] states that such singularities will always be clothed by event horizons and hence can never be visible from the outside (no naked singularities). If Cosmic Censorship holds, then there is no problem with predicting the future evolution outside the event horizon. If it does not hold, then the formation of a naked singularity during collapse would be a disaster for general relativity theory. In this situation, one cannot say anything precise about the future evolution of any region of space containing the singularity since new information could emerge from it in a completely arbitrary way.

Are there guarantees that an event horizon will always hide a naked singularity? No definitive theorems exist. Counterexamples[22] are all restricted to spherical symmetry and typically involve shell crossing, shell focusing, or self-similarity. Are these singularities accidents of spherical symmetry?

For nonspherical collapse Thorne[23] has proposed the *Hoop Conjecture:* Black holes with horizons form when and only when a mass M gets compacted into a region whose circumference in *every* direction is $C \lesssim 4\pi M$. If the Hoop Conjecture is correct, aspherical collapse with one or two dimensions appreciably larger than the others might then lead to naked singularities.

For example, consider the Lin-Mestel-Shu instability[24] for the collapse of a nonrotating, homogeneous spheroid of collisionless matter in Newtonian gravity. Such a configuration remains homogeneous and spheroidal during collapse. If the spheroid is slightly oblate, the configuration collapses to a pancake, while if the spheroid is slightly prolate, it collapses to a spindle. Although in both cases the density becomes infinite, the formation of a spindle during prolate collapse is particularly worrisome. The gravitational potential, gravitational force, tidal force, kinetic and potential energies all blow up. This behavior is far more serious than mere shell crossing, where the density alone becomes momentarily infinite. For collisionless matter, prolate evolution is forced to terminate at the singular spindle state. For oblate evolution the matter simply passes through the pancake state, but then becomes prolate and also evolves to a spindle singularity.

Does this Newtonian example have any relevance to general relativity? We already know that *infinite* cylinders do collapse to singularities in general relativity, and, in accord with the Hoop Conjecture, are not hidden by event horizons.[23,25] But what about *finite* configurations in asymptotically flat space-times?

Previously, we constructed[26] an analytic sequence of momentarily static, prolate and oblate collisionless spheroids in full general relativity. We found that in the limit of large eccentricity the solutions all become singular. In agreement with the Hoop Conjecture, extended spheroids have no apparent horizons. Can these singularities arise from the collapse of nonsingular initial data? To answer this, we have performed[27] fully relativistic dynamical calculations of the collapse of these spheroids, starting from nonsingular initial configurations.

We find that the collapse of a prolate spheroid with sufficiently large semi-major axis leads to a spindle singularity without an apparent horizon. *Our numerical computations suggest that the Hoop Conjecture is valid, but that Cosmic Censorship does not hold, because a naked singularity may form in nonspherical relativistic collapse.*

Figure 2 shows our candidate for the formation of a naked singularity. It describes the collapse of a prolate spheroid whose initial semimajor axis is $10M$ and initial eccentricity is 0.9. The configuration collapses to a spindle singularity at the pole without the appearance of an apparent horizon.

To measure the growth of any singularity that might arise, we compute the Riemann invariant

$$I \equiv R_{\alpha\beta\gamma\delta} R^{\alpha\beta\gamma\delta} \tag{3}$$

at every spatial grid point. Here $R_{\alpha\beta\gamma\delta}$ is the Riemann curvature tensor, which is a relativistic generalization of the Newtonian tidal force. *If I blows up at any point, the space-time has a singularity.*

Figure 3 shows the profile of the Riemann invariant I in a meridional plane for the collapse shown in Figure 2. The key point is that its peak value is $\gg 1$ and that the peak occurs *outside* the matter on the pole. This is a much more serious type of singularity than those arising from simple shell crossing in spherical symmetry. *The absence of an apparent horizon suggests that the spindle is a naked singularity.*

4. RADIATION FROM STELLAR COLLAPSE TO A BLACK HOLE

Calculating the radiation spectrum from a collapsing star is a highly nontrivial exercise in general relativity. The computation calls for the simultaneous solution of the fully relativistic Boltzmann equation for the radiation field, the equations of relativistic hydrodynamics for the matter, and Einstein's equations for the gravitational field. Integrating any one of these sets of partial-differential equations is quite challenging computationally. The coupled system of equations is technically difficult to solve self-consistently even when the collapse is restricted to spherical symmetry and the integrations are performed on a supercomputer.

The recent detection[28] of neutrinos from SN 1987A has made calculations of the radiation spectrum from stellar core collapse much more significant. In the case of supernovae the dominant radiation mechanism is neutrino emission. Core collapse is intrinsically relativistic, so the treatment of the neutrino transport, hydrodynamics, and gravitational field must be fully general relativistic. Although the detailed neutrino spectrum could not be measured from SN 1987A, the prospect of another, possibly closer, supernova and the construction of more sensitive detectors have stimulated renewed activity in solving the problem of radiation transport during relativistic collapse. A number of authors[29] have recently constructed new numerical codes designed to treat this problem carefully in spherical symmetry. Because of

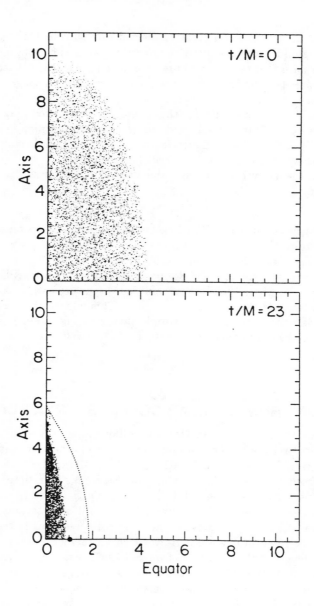

FIGURE 2 Snapshots of the particle positions at the initial and final times for prolate collapse. The minimum exterior polar circumference is shown by a dotted line. The minimum equatorial circumference, which is a circle, is indicated by a solid dot. The minimum polar circumference is $\mathcal{C}_{\text{pole}}^{\min}/4\pi M = 2.8$. There is no apparent horizon, in agreement with the Hoop Conjecture. This is a good candidate for a naked singularity, which would violate the Cosmic Censorship Hypothesis.

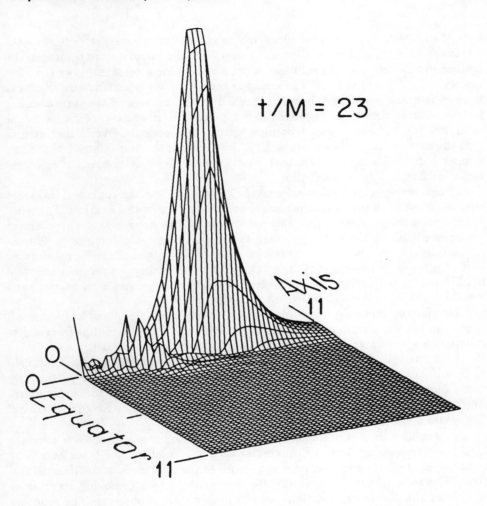

FIGURE 3 Profile of I in a meridional plane for the collapse shown in Figure 2. The peak value of I is $24/M^4$ and occurs on the axis just outside the matter.

the complexity of the coupled system of equations, as well as uncertainties in the adopted microphysics, most of these codes are still being refined and tested.

Here we discuss a simple example of thermal radiation transport during relativistic stellar collapse. By design, our analysis[30] is analytic, although our example is highly idealized and our treatment very approximate. Specifically, we consider the spherical collapse of a homogeneous star from rest to a Schwarzschild black hole. We neglect the dynamical effects of gas and radiation pressure and follow the implosion as if the star were a spherical dust ball. We calculate the transport and emission of radiation in the diffusion approximation for a constant, gray (i.e., frequency independent) opacity. The emission can be photons, neutrinos or any other form of thermal radiation. Thus, the thermal energy content of the star, assumed to be radiation dominated, is considered a perturbation which contributes to the radiation flux but not to any dynamically significant pressure. We further assume that the matter and radiation are in LTE. We focus on the case in which the initial temperature profile in the gas sphere is isothermal and we determine the emergent radiation flux as a function of time.

In our example the dynamical equations decouple from the radiation transport equations, which leads to a considerable simplification in the analysis. For a radiation pressure-dominated gas, this decoupling will be appropriate whenever the emergent stellar luminosity is less than the critical Eddington luminosity. For homogeneous collapse from rest, the resulting implosion is then closely approximated by the exact Oppenheimer-Snyder solution[31] for dust collapse to a Schwarzschild black hole. This solution then provides the background, dynamical matter, and gravitational fields through which the radiation flows.

Two factors motivate our study. First, it yields a simple, albeit crude, model which can provide a quick estimate of the radiation as a function of time from stellar collapse in a strong gravitational field. Second, it provides a test-bed calculation for the numerical codes now under construction to solve this important problem in detail. As our collapse leads to a black hole, our solution may be particularly useful for testing those fully relativistic transport codes designed to handle spacetimes with fluid flows approaching the speed of light, arbitrarily strong gravitational fields, and black holes.

As seen by a distant observer, the intensity at late times appears as a constant blackbody originating from a shrinking annular region about the black hole. The corresponding effective temperature is a simple function of the initial stellar parameters. The total luminosity typically appears constant to a comoving observer at late times but decays exponentially according to a distant observer. The resulting exponential decay rate of the total luminosity as measured by a distant observer as $t_\infty \to \infty$ is a distinguishing feature of collapse to black holes. If observed, it could help discriminate between stellar collapse to a black hole and collapse to a neutron star.

We have made a video movie using color graphics of the collapse of a star to a black hole. The movie illustrates the spectrum as seen both by a comoving observer as well as by a distant observer. This is a powerful tool for visualizing the results of calculations of the emitted spectrum.

ACKNOWLEDGMENTS

This work has been supported in part by National Science Foundation grants AST 87-14475, PHY 90-07834, and INT 88-15793, and by a NASA grant in Theoretical Astrophysics at Cornell University. Computations were performed on the Cornell National Supercomputer Facility.

REFERENCES

1. See, e.g., S. L. Shapiro and S. A. Teukolsky, *Black Holes, White Dwarfs, and Neutron Stars: The Physics of Compact Objects* (John Wiley, New York, 1983).
2. M. Colpi, S. L. Shapiro, and S. A. Teukolsky, *Astrophys. J.* **339**, 318 (1989).
3. M. Colpi, S. L. Shapiro, and S. A. Teukolsky, *Astrophys. J.*, **369**, 422 (1991).
4. D. N. Page, *Phys. Lett.* **A91**, 201 (1982).
5. J. P. A. Clark and D. M. Eardley, *Astrophys. J.* **215**, 311 (1977).
6. S. I. Blinnikov, I. D. Novikov, T. V. Perevodchikova, and A. G. Polnarev, *Sov. Astron. Lett.* **10**, 177 (1984).
7. B. K. Harrison, K. S. Thorne, M. Wakano, and J. A. Wheeler, *Gravitation Theory and Gravitational Collapse* (University of Chicago Press, Chicago, 1965).
8. J. M. Lattimer, F. Mackie, D. G. Ravenhall, and D. N. Schramm, *Astrophys. J.* **213**, 225 (1977).
9. J. P. Ostriker and J. E. Gunn, *Astrophys. J.* **157**, 1395 (1969).
10. J. B. Hartle, *Astrophys. J.* **161**, 111 (1970).
11. L. S. Finn and S. L. Shapiro, *Astrophys. J.* (1990).
12. R. A. James, *Astrophys. J.* **140**, 552 (1964).
13. I. Hachisu, *Astrophys. J. Suppl.* **61**, 479 (1986).
14. T. Fukushima, Y. Eriguchi, D. Sugimoto, and G. S. Bisnovatyi-Kogan, *Progr. Theoret. Phys.* **63**, 1957 (1980).
15. See J.-L. Tassoul, *Theory of Rotating Stars* (Princeton University Press, Princeton, 1978), for discussion and references.
16. J. R. Ipser and L. Lindblom, *Phys. Rev. Lett.* **62**, 2777 (1989), and references therein.
17. R. A. Managan, *Astrophys. J.* **294**, 463 (1985).
18. J. N. Imamura, J. L. Friedman, and R. H. Durisen, *Astrophys. J.* **294**, 474 (1985).
19. J. R. Ipser and L. Lindblom, *Astrophys. J.* **355**, 226 (1990).
20. S. L. Shapiro, S. A. Teukolsky, and T. Nakamura, *Astrophys J. (Letters)* **357**, L17 (1990).
21. R. Penrose, *Rivista del Nuovo Cimento* **1** (Numero Special), 252 (1969).

22. See, e.g., D. S. Goldwirth, A. Ori, and T. Piran, in *Frontiers in Numerical Relativity*, C. R. Evans, L. S. Finn, and D. W. Hobill (eds.) (Cambridge University Press, Cambridge, 1989), p. 414, for discussion and references.
23. K. S. Thorne, in *Magic Without Magic: John Archibald Wheeler*, J. Klauder (ed.) (Freeman, San Francisco, 1972), p. 1.
24. C. C. Lin, L. Mestel, and F. H. Shu, Astrophys. J. **142**, 1431 (1965).
25. C. W. Misner, K. S. Thorne, and J. A. Wheeler, *Gravitation* (Freeman, San Francisco, 1973), p. 867.
26. T. Nakamura, S. L. Shapiro, and S. A. Teukolsky, *Phys. Rev.* **D38**, 2972 (1988).
27. S. L. Shapiro and S. A. Teukolsky, *Phys. Rev. Letters* **66**, 994 (1991).
28. R. Bionata et. al., *Phys. Rev. Letters* **58**, 1494 (1987) (IMB collaboration); K. Hirata et. al., *Phys. Rev. Letters* **58**, 1490 (1987) (Kamiokande II collaboration).
29. J. R. Wilson, *Astrophys. J.* **163**, 209 (1971); P. Schinder, *Phys. Rev.* **D38**, 1673 (1988); E. Baron, E. S. Myra, J. Cooperstein, and L. J. van den Horn, *Astrophys. J.* **339**, 978 (1989); A. Mezzacappa and R. A. Matzner, *Astrophys. J.* **343**, 853 (1989).
30. S. L. Shapiro, *Phys. Rev.* **D40**, 1858 (1989).
31. J. R. Oppenheimer and H. Snyder, *Phys. Rev.* **56**, 455 (1939).